Fossil Treasures of the Anza-Borrego Desert

The past few million years of evolutionary history is revealed through the spectacular fossil archive preserved in the badlands of the Anza-Borrego Desert State Park. This comprehensive review documents a tumultuous period in Earth history, during which the biosphere was faced with the most daunting environmental challenges since the great extinction of the giant dinosaurs 65 million years ago.

Humanity itself arose during this period of mass extinction, and we have been no less profoundly shaped by the cold, sere winds of the Pleistocene. So let us reflect carefully on the lessons to be learned from this fascinating time and place, as the fate of our own species may very well depend upon how deeply we understand them.

Jacques Gauthier, Yale University,
Yale Peabody Museum of Natural History

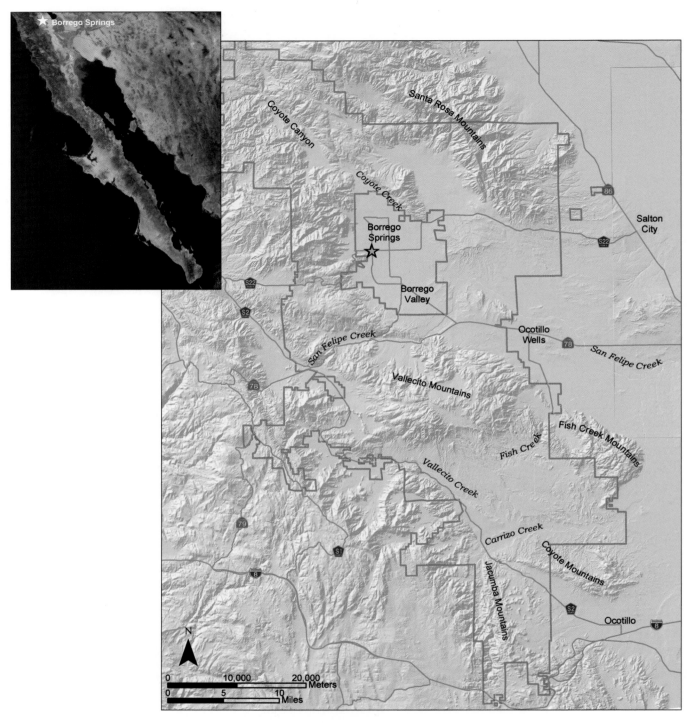

Anza-Borrego Desert State Park

The Park is outlined in red. This convention is used throughout the book on all maps. Also displayed are the location and place names of major cultural and geographic features found throughout the text. Inset map portrays the Gulf of California with the Salton Trough and Anza-Borrego Desert region in upper left corner.

Fossil Treasures
of the
Anza-Borrego Desert

Edited by

GEORGE T. JEFFERSON AND

LOWELL LINDSAY

California State Parks®
Colorado Desert District
Anza-Borrego Desert State Park®

Sunbelt Publications
San Diego, California

The Anza-Borrego Foundation and Institute

Fossil Treasures of the Anza-Borrego Desert
Copyright © 2006 by California State Parks

Sunbelt Publications, Inc.
All rights reserved. First edition 2006
Edited by George T. Jefferson and Lowell Lindsay
Production editing by Jennifer Redmond
Book packaged and designed by The Marino Group
Book composition by Ultratype & Graphics, Inc.
Printed in the United States of America

Sunbelt Publications, Inc.
P.O. Box 191126
San Diego, CA 92159-1126
(619) 258-4911, fax: (619) 258-4916
www.sunbeltbooks.com

09 08 07 06 05   5 4 3 2 1

"Adventures in the Natural History and Cultural Heritage of the Californias"
– A series edited by Lowell Lindsay

Library of Congress Cataloging-in-Publication Data

Fossil Treasures of the Anza-Borrego Desert / edited by George T. Jefferson and Lowell Lindsay.
    p. cm.
    Includes index.
    ISBN-13: 978-0-932653-50-5
    ISBN-10: 0-932653-50-2
    1. Fossils – California – Anza-Borrego Desert State Park®. 2. Paleontology – California –
Anza-Borrego Desert State Park. 3. Paleontology – Stratigraphy. 4. Anza-Borrego Desert State
Park® (Calif.) – Guidebooks. I. Jefferson, George T., II. Lindsay, Lowell.
    QE747.C2F67 2005
    560'.9794'98 – dc22

                                2005019609

Cover Photograph "Badland Vista" by Paul Remeika
Cover Painting "Bad Day for Prey" by John Francis

Contents

Major Maps and Illustrations

Authors and Contributors

Michael Cassiliano	Collections Manager, Department of Geology, University of Wyoming, Laramie
Shelley M. Cox	Laboratory Supervisor, Page Museum at the La Brea Tar Pits, Los Angeles, California
Thomas A. Deméré	Curator of Paleontology, San Diego Natural History Museum, California
Rebecca Dorsey	Professor, Department of Geological Sciences, University of Oregon, Eugene
Philip Gensler	Paleontologist, U.S. National Park Service, Hagerman Fossil Beds National Monument, Idaho
Lindsey T. Groves	Collection Manager, Malacology Section, Natural History Museum of Los Angeles County, California
Barbara Marrs	Researcher and Photographer, Creative Imaging Photography, Phelan, California
George E. McDaniel	Researcher and Laboratory Supervisor, Colorado Desert District Stout Research Center, California State Parks
H. Gregory McDonald	Senior Curator of Natural History, U.S. National Park Service, Fort Collins, Colorado
Lyndon K. Murray	Doctoral Candidate, Department of Geological Sciences, University of Texas, Austin
Kesler Randall	Collections Manager, Vertebrate Paleontology, San Diego Natural History Museum
Paul Remeika	Park Ranger, Anza-Borrego Desert State Park, California State Parks
Mark A. Roeder	Senior Paleontologist, Paleo Environmental Associates, San Diego, California
N. Scott Rugh	Collections Manager, Invertebrate Paleontology, San Diego Natural History Museum
Eric Scott	Curator of Paleontology, San Bernardino County Museum, Redlands, California
Christopher A. Shaw	Collections Manager, Page Museum at the La Brea Tar Pits, Los Angeles, California
Howard Spero	Professor, Department of Geology, University of California, Davis
Sharron Sussman	Researcher and Science Editor, Colorado Desert District Paleontology Program, California State Parks
S. David Webb	Distinguished Research Curator and Professor, Museum of Natural History, University of Florida, Gainesville
John A. White	Late Professor and Curator Emeritus, Idaho Museum of Natural History, Pocatello
Hugh Wagner	Research Associate, Vertebrate Paleontology, San Diego Natural History Museum
George T. Jefferson	District Paleontologist, Colorado Desert District, California State Parks
Lowell Lindsay	CEO and Publisher, Sunbelt Publications, Inc., San Diego, California
Pat Ortega	Anatomical Artist, Pat Ortega Studio, Los Angeles, California
John Francis	Landscape Artist, John Francis Studio, London, England
Jerry Marino	Managing Director, The Marino Group, San Diego, California

Sponsors

Without the support of the following, this book would not have been possible:

Elizabeth "Betty" Stout

California State Parks Foundation

Takahashi Family Fund of the San Diego Foundation

Sempra Energy (San Diego Gas and Electric Company)

Introduction

*Taken as a whole it [the Colorado Desert] is by far
the largest deposit of fossils yet reported
from the Pacific Slope.*

Stephen Bowers (1901)
Stephen Bowers, working for the California State Mining Bureau
(presently the California Geological Survey),
was the first geologist to describe in relative detail
the geological and paleontological resources of the
then Colorado Desert Mining District.

George T. Jefferson
Lowell Lindsay

A Summary of Our Story

Encounter ancient landscapes where today spreads a vast desert, and discover a prehistoric world teeming with wildlife. When you think of Anza-Borrego Desert State Park, you may envision wildflowers, bighorn sheep, or sandy, windblown lowlands framed by rugged mountains. Few realize that the expanses of Anza-Borrego's eroded badlands provide a very different view; one that opens windows into the region's long-vanished past. In fact, it contains the most continuous history of life for the last seven million years in North America. It is indeed one of the richest, most varied fossil records of its time in the western hemisphere.

The Colorado Desert of southeastern California was not always a seemingly barren wilderness. This was once a verdant landscape – an environment of rivers and streams, lakes, forest, and savanna. Before that it held an inland ocean. The key to understanding and engaging this prehistoric world is paleontology, the study of the fossilized remains of ancient life. Anza-Borrego has an exceptional fossil record, which preserves leaf impressions, shells, bones, and teeth, and in some places, even animal tracks.

The Anza-Borrego Desert badlands contain a record of changing environments and habitats that includes more than 550 types of fossil plants and animals, ranging from preserved microscopic plant pollen and algal spores to baleen whale bones and mammoth skeletons. Most of the species are extinct and some are only known from fossil remains recovered from the Park. Combined with a long and complete sedimentary depositional sequence, these diverse fossil assemblages are an unparalleled paleontologic resource of international importance. Both the Pliocene-Pleistocene Epoch geological time boundary (~1.8 million years ago) and the Blancan-Irvingtonian North American Land Mammal Age boundary (~1.5 million years ago) fall within the long geological record from the Anza-Borrego Desert. Environmental changes associated with these time divisions are probably better tracked by fossils from Anza-Borrego than in any other North American continental deposits. These changes herald the beginning of the Ice Ages, and the sedimentary strata undoubtedly contain fossil clues to the origin and development of our modern southwestern deserts.

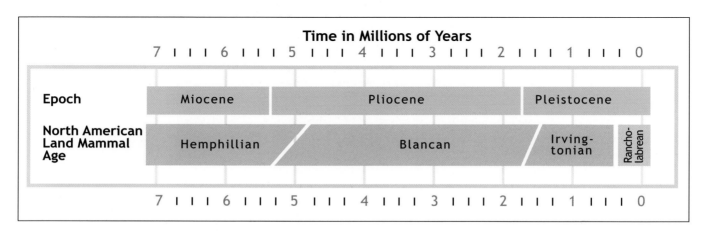

Figure 2
Anza-Borrego Desert Region Timeline.
This record of life extends from the Miocene into the Pleistocene Epoch and from the Hemphillian through the Rancholabrean of North American Land Mammal Ages (NALMA). It is the longest continuous record of life for that period in North America.

This extensive record of ancient life exists because Anza-Borrego lies in a unique geologic setting along the western margin of the Salton Trough. This is a major topographic depression, with elevations greater than 70 meters (230 ft) below sea level at the surface of the Salton Sea. This trough forms the northernmost end of an active rift valley and marks a portion of the boundary between the North American and Pacific tectonic plates. The Salton Trough extends north from the Gulf of California to San Gorgonio Pass and from the eastern edge of San Diego County's Peninsular Ranges east to the San Andreas fault zone. Over the past 7 million years, a relatively complete geologic record of more than 6,000 meters (19,000 ft) of fossil-bearing sediment has been deposited within the Park along the rift valley's western margin. Here, paleontologic remains are widespread and exceedingly diverse, and are found scattered over hundreds of square kilometers of eroded badlands terrain extending south from the Santa Rosa Mountains into northern Baja California, Mexico. This long and rich fossil record tells a story of both marine and terrestrial environments and their changes through time.

The oldest reported fossils from the region occur in the metamorphic rocks of the Santa Rosa, San Ysidro, and Coyote Mountains. These remains of microscopic marine animals are more than 450 million years old. However, there is a 430 million-year-long gap in the record of life between them and the next youngest. Either no sediments were deposited during this time, which includes the age of dinosaurs, or the deposits were since eroded away. The oldest terrestrial vertebrate fossils from the Colorado Desert are over 9 million years old, and predate the marine invasion of the proto-Gulf of California, that occurred about 6 million years ago. These very rare fossils, which include an elephant-like mammal, and a small camel, were collected from nearshore lake deposits. Most Anza-Borrego fossils date from between 6 to less than a half million years, comprising a continuous 5½ million-year record.

Six million years ago the ancestral Gulf of California filled the Salton Trough, extending northward past what would eventually become the city of Palm Springs. These tropical waters supported a profusion of marine organisms. Fossil assemblages from this Imperial Group of marine sediments include calcareous nannoplankton and dinoflagellates, foraminifera, corals, polychaetes, clams, gastropods, urchins and sand dollars, and crabs and shrimp. The deposits also yield the remains of marine vertebrates such as sharks and rays, bony fish, sea turtles, baleen whale, walrus, and dugong (sea cow). Carbonate platform, outer and inner shelf, and nearshore marine environments are all represented. As the sea became shallow, estuarine and brackish marine conditions prevailed, typified by thick channel deposits of oyster and pecten shell coquina. Many of these marine fossils are closely related to forms from the Caribbean Sea. They document a time before the Isthmus of Panama formed, when the warm Gulf Stream of the western Atlantic connected to eastern Pacific Ocean waters. Anza-Borrego was gradually changing from a predominately marine environment to a system of interrelated terrestrial habitats.

Through time, an immense volume of sediment, eroding during the formation of the Grand Canyon, spilled into the Salton Trough. Bit by bit, the ancestral Colorado River built a massive delta across the seaway, forming a natural dam that blocked Gulf marine water from filling the now-below-sea-level Salton Trough to the north. Fossil hardwoods from the deltaic deposits suggest that the region received three times as much rainfall as today.

North of the delta, and intermittently fed by the Colorado River, a sequence of fresh water lakes has persisted for over 3 million years. At the same time, sediments eroding from the growing Peninsular Ranges and Santa Rosa Mountains spread east into the Salton Trough. It is these sediments, first deposited more than 3.5 million years ago, that provide an almost unbroken terrestrial fossil record, ending only a half million years ago. Here, the deposits of ancient streams and rivers trapped the remains of wildlife that inhabited a vast brushland savanna laced with riparian woodlands.

North and South America were connected when the Isthmus of Panama formed about 3 million years ago, fostering terrestrial faunal migrations on a continental scale – the "Great American Biotic Interchange." Evidence of this is revealed by fossils found in the southwestern U.S.: animals like giant ground

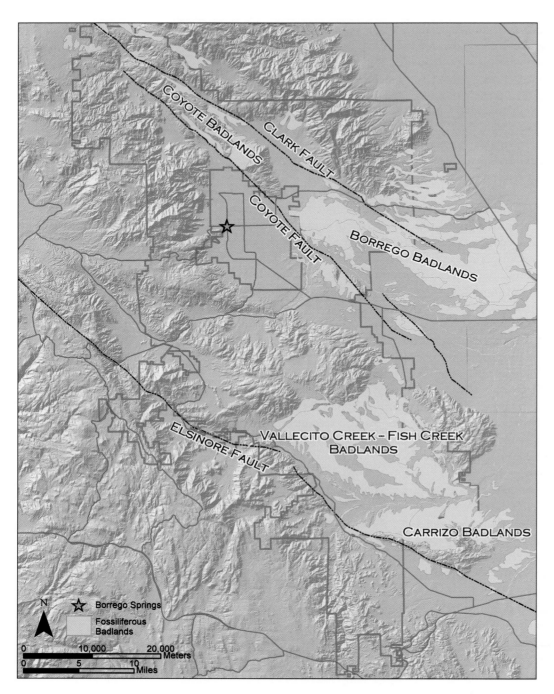

Figure 3
Fossiliferous Badlands.
Major fossiliferous badlands in the Park are colored green (also see Paleolandscape index maps). The map is based on the outcrop patterns of sedimentary formations, which are often bounded by major faults. These badlands are frequently mentioned by our authors.

sloths and porcupines make their first appearance in North America while llamas spread into South America. North American predators moving south tended to be very successful in preying on South American herbivores, leading to their extinction. Similar intercontinental migrations occurred throughout the Pleistocene Ice Ages as lowered sea levels opened the Bering Land Bridge between Siberia and Alaska. The rich Anza-Borrego fossil record provides evidence that ancestral horses and camels crossed westward into Asia while mammoths and oxen crossed east and south into North America. Anza-Borrego discoveries help to delimit the timing of incoming migrations including those of mammoths and ground sloths. Interestingly, some of the descendants of North American emigrants are found only in their adopted lands, such as llamas and tapirs in South America and camels and horses in Eurasia.

The most significant and abundant vertebrate fossils in Anza-Borrego have been recovered from the riverine and floodplain deposits exposed through erosion in the Vallecito Creek-Fish Creek and Borrego Badlands (Figure 3). These fossil assemblages have been dated using layers of volcanic ash as well as paleomagnetic methods.

The bestiary for this savanna landscape reads like a *Who's Who* for some of the most unique creatures to inhabit North America. Animals like: *Hesperotestudo,* a bathtub-sized tortoise; *Aiolornis incredibilis,* the largest flying bird of the northern hemisphere, with a 5 meter (16 ft) wing span; giant ground sloths, some with bony armor within their skin like *Paramylodon; Pewelagus,* a very small rabbit (paleontologists do have a sense of humor); *Borophagus,* a hyena-like dog; *Arctodus,* a giant short-faced bear; *Smilodon,* a sabertooth cat; *Miracinonyx,* the North American cheetah-like cat; *Mammuthus columbi,* the largest known mammoth; *Tapirus,* an extinct tapir; *Gigantocamelus*, a giant camel; and *Capromeryx,* the dwarf pronghorn.

The history of paleontologic discovery in the Salton Trough dates from the mid-nineteenth century. The first fossils – marine shells from the ancient Gulf of California and freshwater shells from prehistoric Lake Cahuilla, precursor of the Salton Sea – were collected and described by William Blake in 1853. Blake was the geologist and mineralogist for the *U.S. Pacific Railroad Survey*, commissioned by Congress and President Pierce, to find a railroad route to the Pacific. In fact, Blake named the Colorado Desert for the arid portions traversed by its namesake river.

Stephan Bowers' 1901 *Reconnaissance of the Colorado Desert Mining District* discusses the paleontological resources of the region. His narrative contains the first account of vertebrate fossils from Anza-Borrego. Shark teeth and whale fossils are reported from the Coyote Mountains, and the remains of an extinct zebra-like horse from the eastern Borrego Badlands. In the late 1930s, Guy Hazen and his field party, from the Frick Laboratory of the American Museum of Natural History, were the first to systematically survey portions of the Borrego and Vallecito Creek Badlands for vertebrate fossils. After World

War II, in the mid-1950s, Ted Downs, Harley Garbani, John White, and others of the Natural History Museum of Los Angeles County initiated the most productive phase of paleontologic research in the region to that time, lasting through the mid-1980s. In the early 1970s, George Miller, who had previously worked with Downs, established a paleontology program based at Imperial Valley College Museum in El Centro, California. Active work by Miller and his students continued through the late 1980s.

The extensive paleontological collections amassed by the Imperial Valley College Museum and the Natural History Museum of Los Angeles County were recently united. The collection now resides in the Anza-Borrego Desert State Park's Colorado Desert District Stout Research Center in Borrego Springs. Here, the fossils have been the focus of ongoing research, study, and interpretation since the mid-1990s. Although paleontological exploration of Anza-Borrego has stepped firmly into the 21st century with the application of Geographic Information Systems and computer-assisted analyses that aid field surveys and resource management, many questions still remain. Expanding the detail and clarity of our paleontological view of the region's vanished past and improving our understanding of its significance is an ongoing challenge.

The Chapters and Topics

As the drama unfolds in the following chapters, contributed by researchers from across the U.S., we will encounter two basic types of presentations. One type, clustered toward the beginning and end of the book, explores background themes and concepts such as paleontologic research, the geologic setting, stratigraphic dating, and intercontinental connections. The other type of chapter explores the creatures themselves, in their natural groups.

Our first thematic presentation is by Barbara Marrs on *The History of Anza-Borrego Desert State Park's Fossil Collections.* This story ranges from William Blake's pioneering work before the Civil War to today's state-of-the-art paleontology Stout Research Center, and the extensive human resources including professional staff, certified volunteers, and visiting academics that have made this story possible. They, along with readers and students of paleontology, are time travelers of a sort, engaged in an exciting detective story of the ages.

Next, Thomas Deméré focuses on the 6 million-year-old ancestral Gulf of California in *The Imperial Sea*, and with N. Scott Rugh, on the *Invertebrates of the Imperial Sea*, a rich shellfish fauna. Predating the formation of the Isthmus of Panama, these warm tropical waters hosted animals that were more closely related to Caribbean forms than to those in the eastern Pacific. These chapters provide an excellent overview of the interplay of geological and paleontological factors that have shaped the changing landscapes of the Anza-Borrego Desert.

Paul Remeika shifts the story from the marine record to a terrestrial setting with *Ancestral Woodlands of the Colorado River Delta Plain.* As the Colorado River built a massive delta across the marine embayment of the proto-Gulf of California, trees, similar to those that grow along California's central coast today, were being buried. Their fossils tell of a wetter, cooler climatic regime than that of today's desert. Plant pollens and freshwater protozoans that lived during the time of the dinosaurs on the Colorado Plateau, are found in the stratified badlands of Anza-Borrego. These eroded and re-deposited fossils are evidence that the ancestral Colorado River commenced its work of carving out the Grand Canyon some 4 million years ago.

Rebecca Dorsey next discusses the physical setting and the last 25 million years of geological development of the region in *Stratigraphy, Tectonics, and Basin Evolution in the Anza-Borrego Desert Region.* Emphasis is on three geological stages:

1) 10-million-year and older pre-marine continental sedimentation, well studied and spectacularly exposed in Split Mountain Gorge, and volcanism, evident in the southern part of the Park;

2) 10-million through 1-million-year rifting and "unzipping" of the Gulf of California which created a marine environment in the northern Salton Trough, the westward building of the Colorado River Delta "dam," and eastward filling of Anza-Borrego's major fossil-rich sedimentary basin;

3) 2-million-year to modern faulting and folding on the huge San Andreas fault system causing the rugged topography of mile-high mountains, numerous appearances of freshwater Lake Cahuilla (currently the Salton Sea) fed by the capricious Colorado River, and uplifted and eroded basins and badlands.

Turning to questions like, "How old are the bones?" and asking about deep time, geologic ages, and the fossil record, Paul Remeika presents *Dating, Ashes, and Magnetics.* While relative dating of fossil assemblages in the 19th century was a great leap forward in the solving of prehistoric mysteries, 20th century technologies provide the tools for precision dating of prehistoric and historic events. The absolute dating methods discussed include:

1) Radiometric isotope analyses, which measure the decay rate of radioactive elements to establish an age;

2) Chemical and trace element analyses of volcanic ash beds or "tephra," providing distinct marker beds of certain ages;

3) Paleomagnetic studies that correlate the initial magnetic fields of ancient sediments with changes in the earth's magnetic field of known ages, thereby dating the sediments.

Michael Cassiliano applies the relative dating methods of *Mammalian Biostratigraphy in the Vallecito Creek and Fish Creek Basins* to the rich fossil record of the southern Park badlands. This adds another important tool of correlation, North American Land Mammal Ages, to the classic geologic time scales of epochs and ages. Biostratigraphy provides a framework for understanding the distribution and timing of individual fossil animals in the Anza-Borrego record, and for correlating Anza-Borrego with other fossil sites in the southwest.

Having developed an understanding of the background and geological setting for Anza-Borrego's paleontologic drama, the central chapters of this book introduce the real stars of the story – the plants and animals themselves.

In the chapter *The Fossil Lower Vertebrates*, Philip Gensler and others examine the fossil record of the fish, amphibians, and reptiles that lived in the region following the retreat of the Imperial Sea some 4.5 million years ago. The fish, all from the ancestral Colorado River drainage system, corroborate the role of this great river in contributing to the Anza-Borrego landscape. The diversity of fossil reptiles from Anza-Borrego is one of the richest of its time in North America, attesting to once-tropical conditions in the region.

George Jefferson continues developing the theme that fossils provide windows into the past with *The Fossil Birds of Anza-Borrego.* Flamingos and other water birds comprise more than half of the Anza-Borrego fossil avifauna, evidence not only of local freshwater environments but also of a flyway linkage to other now-extinct riverine and lake systems throughout the western U.S.

With *The Ground Sloths,* H. Gregory McDonald introduces a most unique large mammal group. These animals were very successful and very visible on Anza-Borrego landscapes for millions of years, and went extinct only about ten thousand years ago. The three successive families of this group, overlapping in time and space, indicate a complex of different habitats and environments. The browser, *Megalonyx,* suggests woody forests and riparian habitats. Another browser, *Nothrotheriops,* probably lived in the drier, scrubbier local habitats. The third and largest of the three ground sloths, *Paramylodon,* grazed on grasses in the drier open savanna.

Mammalian predators are presented in two separate chapters: *The Large Carnivorans: Wolves, Bears, and Big Cats,* contributed by Chris Shaw and Shelley Cox, followed by *The Small Carnivorans: Canids, Felids, Procyonids, Mustelids* by Lyndon Murray. Interestingly, today's most feared and respected local predator, the puma, cougar or mountain lion, shared a common ancestor with the North American cheetah-like cat as recently as 3.2 million years ago. The cougars are actually more closely related to the smaller cats than to the sabertooth or Jaguar from Anza-Borrego's past.

The themes of adaptation, evolution, and extinction, and the influences of changing environments on the animals and landscapes, are intertwined in following chapters that discuss Anza-Borrego's famed megafauna. These chapters include *Mammoths and Their Relatives* by George McDaniel, *Extinct Horses and Their Relatives* by Eric Scott, *Extinct Camels and Llamas of Anza-Borrego* by S. David Webb and others, and their smaller cloven-hoofed relatives in *The Smaller Artiodactyls* by Lyndon Murray. Of the latter, the Park's namesake "Borrego" or mountain sheep arrived too late from Siberia (less than 300 thousand years ago) to be entombed in the region's sediments. Of the three former groups (mammoths, horses, and camels), all became extinct in North America coincidental with the end of the Ice Ages and with the arrival of the bipedal predator, *Homo sapiens.*

Dramatically distinct from the megafauna discussed in the previous chapters, in *The Small Fossil Mammals,* the remains of smaller animals address the question "Is bigger better?" The late John White and his colleagues argue "not necessarily." Specimens of small mammals are far more numerous than their larger counterparts because their maturation and gestation period is very short with consequent greater production of offspring. The home range of small creatures is limited and therefore their fossils are more indicative of past local ecological settings. These attributes allow in-depth studies of the great and small questions of evolution and ecological change and permit more accurate descriptions of past local habitat conditions.

In the chapter *Fossil Footprints,* Paul Remeika offers yet another kind of fossil evidence. The footprints of large mammalian herbivores and carnivorans demonstrate that these animals were indeed active on the ancient landscape. Bones or teeth may wash in from afar, but footprints and trackways were made on-site. The study of footprints or "ichnology" complements the methodologies of biostratigraphy, systematics, and taxonomy.

Included within each of the above chapters is an icon-like chart that displays the stratigraphic and geologic time ranges for the animals discussed. Thus the fossil record, from chapter to chapter in this long geological story, can be compared and related. It will be seen that some animals are associated in paleofaunas that appear early and disappear before the end of the Anza-Borrego story. Others, appearing later on the changing landscapes, become part of today's desert fauna.

In *The Great American Biotic Interchange,* H. Gregory McDonald treats the topic of continental connections, which allowed migrations of terrestrial animals across the Isthmus of Panama land bridge starting some 3.5 million years ago, as well as across the Bering Land Bridge during the Pleistocene Ice Ages. In both cases, it was a two-way street into and out of North America.

In *Paleoclimates and Environmental Change in the Anza-Borrego Desert Region,* Sharron Sussman and others discuss changes in climate through time. Driving forces are seen in the Earth's own orbit. Furthermore, chemical analyses of the carbon and oxygen isotopes extracted from fossil horse teeth can be used to track major environmental changes. These isotopes also indicate the types of plants that the horses ate and, therefore what the overall vegetation was like.

A Note on Using and Enjoying This Book

Because the topics and concepts treated by our authors are probably new to some readers, the technical language may at first appear intimidating. We don't want this to detract from our story. So, we have tried to explain these terms in the text where possible, and definitions of the technical jargon have been compiled in a *Glossary* at the back of the book.

Also at the end of the text, in the *Appendix* are *Tables* that provide the names of the geologic formations used by our authors, and *Tables* that list the fossil plants and animals from the different geological periods and deposits. Although placed with relevant chapters, *Paleolandscapes 1- 5*, depicting the major episodes and habitats of our story, are also available separately from the Publisher.

Our authors commonly refer to publications and references, for both technical and popular articles, for the information that they use in their chapters. These citations, in parentheses, not only credit the source of the data presented, a standard practice in scientific papers, but also provide a useful contact for those readers who wish to further explore literature in fields of paleontology and geology. The *Literature Cited* at the back of the book lists these references and includes others with a focus on the fossil record of the Salton Trough region.

Refer to these aids frequently to enhance your understanding and appreciation of Anza-Borrego's fossil creatures and extinct landscapes.

Acknowledgements

The editors would like to acknowledge the following persons, without whose invaluable aid and input this book would not have been as accurately or as clearly presented. Our deepest appreciation goes to:

Larry D. Agenbroad, Director, Mammoth Site of Hot Springs, South Dakota; Professor Emeritus, Northern Arizona University; **Gary Axen**, Professor, New Mexico Tech University; **Brett Cox**, Research Scientist, U.S. Geological Survey; **Harry Daniel**, Chairman Emeritus of the Anza-Borrego Desert Natural History Association; **Jacques Gauthier**, Professor, Yale University and Curator, Division of Vertebrate Paleontology, Yale Peabody Museum of Natural History; **Gary Girty**, Professor, San Diego State University; **Dr. Kenneth Gobalet**, Professor Department of Biology, California State University, Bakersfield; **John "Jack" Horner**, Regents Professor of Paleontology, Museum of the Rockies, Montana State University; **Bernie Housen**, Professor, Western Washington University; **Susanne Janecke**, Professor, Utah State University; **L. Louise Jee**, Cartographer, Colorado Desert District; **Charles Johnson**, Research Librarian, the Ventura County Museum of History and Art; **Mark Jorgenson**, Park Superintendent, Anza-Borrego Desert State Park; **Martin Kennedy**, Professor, University of California, Riverside; **Susan Kidwell**, Professor, University of Chicago; **Charles Lough**, retired hydrologist; **Jonathan Matti**, Research Scientist, U.S. Geological Survey; **Karen Paige**, Research Librarian, the California State Library; **Charles L. Powell II**, U.S. Geological Survey; **Tom Rockwell**, Professor, San Diego State University; **Peter D. Roopnarine**, Associate Curator, Department of Invertebrate Zoology and Geology, California Academy of Sciences; **Derek Ryter**, Program Manager, ASW Associates; **Susan Sheehan** and the staff at the Arizona Historical Society of Tucson; **David Van Cleve**, retired Colorado Desert District Superintendent; **Ray Weldon**, Professor, University of Oregon; and **Charles Winker**, Research Scientist, Shell International E&P.

From the San Diego Natural History Museum are: **Dee Parks**, Public Programs Manager; and **Michael W. Hager**, Executive Director.

From the Natural History Museum of Los Angeles County are: **Anne C. Cohen**, Research Associate, Crustacea Section; **George E. Davis**, Collection Manager, Crustacea Section; **Harry F. Filkorn**, Collection Manager, Invertebrate Paleontology Section; **Cathy L. Groves**, Curatorial Assistant, Echinoderms Section; and **Cathy McNassor**, Museum Archivist.

From the American Museum of Natural History are: **Robert Evander**, Senior Principal Fossil Preparator; **Barbara Mathe**, Museum Archivist; **Mark A. Norell**, Chairman and Curator, Division of Paleontology; and the Library Special Collections researchers.

The Life-Member Paleontology Volunteers for Anza-Borrego Desert State Park include: **Bob Anderson**, **Harley Garbani** (also a Field Collector for the Natural History Museum of Los Angeles County); **George McDaniel, Julie Parks**, and the late **Elizabeth "Betty" Stout**.

James F. Landers, author and President of Shannon River Systems, provided an essential and non-paleontological point of view in his editing of our chapters.

All of the Contributing Authors (see page *ix*) went above and beyond their own works, by contributing photographs for other chapters and by reviewing other chapters and providing useful comments and suggestions.

History of Fossil Collecting in the Anza-Borrego Desert Region

Most of the desert around Borrego Springs is literally alive with prehistory. It [Anza-Borrego Desert State Park] is a geological, paleontological and archaeological goldmine . . . Paleontologists from throughout the world are aware of this and it is the reason I have devoted the rest of my career to this particular area.

George J. Miller (1921 – 1989)
in the *Los Angeles Times*, June 2, 1981
Professor Emeritus of Paleontology,
Imperial Valley College

Barbara Marrs

History of Fossil Collecting in the Anza-Borrego Desert Region

Ancient Shoreline of Lake Cahuilla in the Colorado Desert. (Photograph by Barbara Marrs)

The year is 1853; the month, November. A tall gentleman, at first glance stalwart and imposing, sits by a desert campfire in the twilight hours, his piercing blue eyes intent as he recounts in his journal the day's events:

> *"17th. Started from camp at 4 AM before daylight and turned point of mountain, proceeding short distance observed shells of small varieties of fresh water specimens . . . I continued the opportunity of finding the shells so numerous and well preserved that I collected 6 or 7 species and observed a well defined water line of the rocks. I hardly know how to commence the notes of today, it has been so unusually interesting. Passing along over this desert looking plain and finding on the surface these relics of a former fresh-water lake and also on the bare and rugged rocks of the Sierras, a distinctly defined water line extending in a line for miles and miles. Since then are a thick calcerous [sic] coating like coral on the surface which recorded the former height of the water and gave an index of the inhabitants . . ."* (Blake, personal diary, 1853).

The person is 27-year-old William Phipps Blake (Figure 1.1), a geologist and mineralogist for the U.S. Pacific Railroad Survey team assigned to assess and determine a practical railway route through the passes of California's Sierra Nevada from the San Joaquin and Tulare Valleys (southeast of San Francisco) through the southern California desert to Yuma, Arizona.

Prior to the expedition's journey into the present parklands, Blake had recorded many important and significant geologic observations about the California terrain. But nothing prepared him for the remarkable data that he gathered on November 17, 1853, when he discerned concrete evidence of the existence of an enormous ancient freshwater lake in what was now before him, ". . . the wide Sahara bounded by rugged, barren mountains . . ." (Blake, personal diary, 1853). How amazed was the young scientist to witness prehistoric signs of an abundance of the very same natural resource, which, in their present sweltering and parched environment, was a matter of critical concern for the expedition members and their pack animals.

Blake was not the first to be astonished by the area's geologic phenomenon of the former incursion of a large body of water upon the land, which was now desert terrain. Similar observations were made 78 years earlier by Fray Pedro Font, official chaplain and recorder for Captain Juan Bautista de Anza, leader of the 1775 Spanish military expedition from Sonora, Mexico, which traveled across the Southern California desert, exploring the new territory for Spain. In his diary entry for December 9, 1775, Font forecasts Blake's journal:

Figure 1.1
Prof. William Phipps Blake,
Circa 1890s.
(Photograph courtesy of the
Arizona Historical Society)

> *"On account of the unfruitfulness of these lands, so level, and of the aspect of the sand dunes, and especially of the abundance of shells of mussels [oyster deposits] and sea snails [freshwater gastropods] which I saw today in piles in some places, and which are so old and ancient that they easily crumble on pressing them with the fingers, I have come to surmise that in the olden times the sea spread over all this land, and that in some of the great recessions which the histories tell us about, it left these salty and sandy wastes uncovered."*

In 1849, following the Mexican War, the U.S. Army deployed the first topographical boundary survey of the country's newly acquired lands, from Texas to California, led by Major W. H. Emory and Lieutenant C. J. Couts. Emory, a former Chief Topographical Officer, who had often traversed the Colorado Desert during the war, had observed and noted the area's oyster coquinas (limestone composed of shells and shell fragments). Couts, who commanded the military escort for the 1849 survey, and who had also become familiar with the desert region during the war, recorded, among his topographical accounts, his observations of the extensive arrays of invertebrate fossil deposits that he and the 1849 survey team encountered:

> *"The desert passed over has been described by many [i.e., Emory]. There is no doubt of its having once been the bed of the Gulph [sic] of California; Shells peculiar only to salt water [oysters] are spread over its surface in great quantities, scrubby chapparral [sic] bushes are strewed over its surface and [it] is nothing more or less than an immense sand plain."*

Though these early accounts of Anza-Borrego's fossiliferous sediments generated some interest, it was W. P. Blake's detailed and comprehensive notes, recorded from his observations and measurements in 1853, which became the first official scientific records to influence the geologic and paleontologic history of Anza-Borrego Desert State Park.

In his correspondence to the Smithsonian Institution in 1854, Blake wrote that he " . . . discovered that the *desert* or *a great part of it was formerly the bed of an immense fresh water lake* – now dry . . ." (Testa, 1996). It was proposed that this ancient body of water be called "Blake's Sea." However, Blake himself chose "Lake Cahuilla," derived from the name of the Indian tribe that occupied the valley.

Though only in the present parklands for less than one month, and working under the harsh conditions of the desert with the bulky tools of his trade, this astute geologist's lucid observations, meticulous field notes, and precise geologic sketches and wood engravings clearly demonstrated the area to be much more important and scientifically significant than anyone had previously imagined. By taking barometric elevation measurements as he traveled, Blake was the first to establish the fact that the region was an enclosed basin, the lowermost part of which was below sea level (today known as the Salton Trough). The fossil shell specimens collected by Blake were determined to be representative of three major types: freshwater specimens (of which three were new species); a new brackish water genus; and Tertiary marine fossil shells. Blake concluded that valley was originally occupied by seawater from the Gulf of California, as evidenced by the reefs of fossil oysters and other marine shells. He stated further that the influx of fresh water from the Colorado River eventually displaced the salt water, and that the valley was soon completely occupied by fresh water (Lake Cahuilla), which then disappeared very slowly through evaporation and loss of water flow from the river.

To merely credit this Renaissance scientist for the first significant invertebrate fossil discoveries of Anza-Borrego does not do justice to Blake's total efforts and the accomplishments which resulted from the 1853 survey (Schaeffer, 1857). His in-depth geologic analyses of the region's rocks and sediments and of the basin's transition from marine waters to fresh water to desert are major contributions to the scientific history of what is now Anza-Borrego Desert State Park. This "Yankee Gentleman and Pioneer Geologist of the Far West" is also credited for having named the Colorado Desert, since, Blake observed, the desert owed its origin to the Colorado River's deposition of sediment and displacement of its former marine waters.

Later California Geological Surveys into the area followed in the 1860s and 1880s. These were better publicized and better known than Blake's 1853 expedition, but no further reports on the region's fossil invertebrates were published as a result of these geologic excursions.

The first official publication on the geology of the Colorado Desert to follow Blake's account was by Charles R. Orcutt, who had traveled extensively throughout the area, collecting large numbers of invertebrate fossils, primarily oysters, corals, and sea urchins, from the southern part of the present parklands. His data was prepared for the California State Mining Bureau (today known as the California Geological Survey).

H. W. Fairbanks next collected in the Anza-Borrego desert, circa 1892. His invertebrate specimens were sent to the University of California at Berkeley, where they were studied by Dr. Thomas Wayland Vaughan, who described (identified) two new species and one new subspecies.

However, it was not until 1901, through the work of Dr. Stephen Bowers, that Anza-Borrego's fossil invertebrates began to garner their true scientific prominence.

The Reverend Dr. Stephen Bowers (Figure 1.2) hailed from the Midwest, and came to California in 1874 for "health reasons" and to accept pastorates in Napa and Santa Barbara. At age 42, he was already very well established as both a Methodist minister and an eminent geologist. While at Santa Barbara, the avid earth scientist wasted no time in establishing relations with the U.S. Geological Survey and with the Mineralogical and Geological Survey of California. Bowers later became the State Mine Examiner of California in 1899, " . . . a position which came to him on account of his superior scientific knowledge of geology in California" (*Los Angeles Times*, 1907). Always leaving a tumultuous trail of religious, political, and scientific polemical endeavors and controversial deeds in his wake, the itinerant Protestant preacher, Prohibitionist reformer, accomplished archaeologist (commissioned by the Smithsonian Institution and the U.S. Department of the Interior), and newspaper owner and publisher retained his passion for geology throughout his life.

Figure 1.2
Rev. Dr. Stephen Bowers, Circa 1870s.
(Photograph courtesy of the Ventura County Museum of History & Art)

In June and July of 1901, at age 69, the Reverend Dr. Bowers headed a Reconnaissance of the Colorado Desert Mining District for the California State Mining Bureau, the purpose of which was to scout for the existence of "liquid gold" (petroleum) in the Colorado Desert. Bowers divided the territory that he surveyed into four districts: Coyote Wells, Carrizo Creek, Fish Creek, and Seventeen Palms Springs; all of which are part of the Park today. Throughout his survey, the minister/geologist observed, as did Font, Emory, Couts, Blake, Orcutt, and Fairbanks before him, vast panoplies of fossil oyster shell and fossil coral reef deposits and many other invertebrate depositions. Though recording geologic data for petroleum locations was his first priority, Bowers recognized the significance of these marine and freshwater fossils and made collections of what he considered to be important specimens. His final report to the California State Mining Bureau includes much information on his observations of the Colorado Desert's fossil invertebrates and their sediments, and of the former incursion of seawater into the region. In his concluding notes, Dr. Bowers writes:

> *"The waters of the old Tertiary sea which once prevailed here must have been extremely favorable to the propagation and growth of mollusks, especially oysters. After the vast erosion that has taken place there are many square miles of fossil beds, especially of oyster shells, which in places are 200' thick, and may extend downward to a much greater depth. They existed*

*not only in vast numbers, but in many varieties, from the small
shell which is in evidence over so much of the territory forming
almost mountains in height and extent, to varieties nearly a foot
long, and to others weighing several pounds each. One variety
is nearly as round and as large as a dinner plate" (Bowers,
1901).*

Interestingly enough, invertebrate fossils were not the only paleontologic finds noted by the Reverend in his survey. In the Carrizo Creek area, he records that, "Sharks' teeth, a pecten [bivalve mollusk] 8″ or 10″ in diameter, ostrea [bivalve mollusks], large univalves [mollusks] . . . and other fossil forms are found here . . . " (Bowers, 1901). And, in his Seventeen Palms Springs notes, Bowers writes, "There are fossils also in Sec. 26 of this township. Among other forms the remains of the fossil horse (*Equus occidentalis*) were found here by M. C. S. Alverson . . . " (Bowers, 1901).

Despite their brevity, these 1901 records comprise the first official report of marine vertebrate and terrestrial vertebrate fossils found in what is now Anza-Borrego Desert State Park. The apparent nonchalant observations of these vertebrate finds perhaps reflected the mindset of the period. Oil was the top priority; therefore Bowers and his team surveyed primarily those areas with geologic evidence of this natural resource. Because oil is nearly always found in marine sedimentary rocks (petroleum is a product of the decomposition of organic matter deposited and trapped with marine sediment), the sediments surveyed also happened to be excellent sources for major invertebrate fossil deposits. The handful of sharks' teeth and horse bones and teeth found at that time probably seemed inconsequential and did not suggest the incredible profusion of vertebrate fossils and fascinating variety of deposits that later would be discovered in the area. Anza-Borrego's fossil vertebrates were not destined to come into their own until the 1930s, with expeditions from New York City's American Museum of Natural History.

After his report was published in August 1901, Dr. T. W. Vaughan (who had also examined and described Fairbanks' collection in 1900) studied Dr. Bowers' invertebrate collection at the University of California at Berkeley in 1903. Vaughan described five genera of corals from Bowers' specimens, which had also occurred in the fossil and recent fauna of the Antilles (chain of islands in the West Indies), and which had *never* occurred on the Pacific Coast. The unprecedented appearance of Atlantic Coast type invertebrates on the West Coast prompted the Berkeley professor to immediately conduct more detailed geologic studies in the area and he arranged for an expedition to return to Anza-Borrego in 1904. Mr. W. C. Mendenhall conducted the physiographic (physical geography) and stratigraphic (geology of stratified rocks) research for the scientific excursion, and the Reverend Dr. Bowers accompanied him and made a very extensive second collection of the invertebrate fossils found. As a result of this survey, Mendenhall published the most detailed account thus far of the geology of the Anza-Borrego region in 1910. Prior to this publication, W. P. Blake's report of 50 years earlier had remained the foremost scientifically complete geological report of the Colorado Desert lands.

During the next decade, the marine and freshwater fossil invertebrates of Anza-Borrego rose in visibility, esteem, and value through a number of scientific investigations, studies, and publications by researchers with the University of California at Berkeley, California Academy of Sciences, California State Mining Bureau, San Diego Society of Natural History, Stanford University, the Geological Society of America, and other prominent universities and institutions.

In 1919, G. Dallas Hanna assumed the duties of curator for the paleontology department of the California Academy of Sciences. There, he found two large collections of fossil mollusks from the Anza-Borrego region. Former curator Roy E. Dickerson had described a fair amount of these specimens, and in 1920 he turned the job over to Hanna " . . . to finish or return the collections to their owners" (Hanna, 1926). Hanna did complete the task, but withheld publication on his data until 1926. The new California Academy of Sciences curator realized that the unique specimens entrusted to him could not be accurately compared to either fossil or extant Pacific Coast or Atlantic Coast species. He found " . . . that the fauna needed for critical comparison was to be had only in the Gulf of California. Until the Academy sent its expedition there in 1921, no collection of consequence was available in any western museum for consultation" (Hanna, 1926). Upon examination of the newly acquired materials, Hanna observed that *the Anza-Borrego marine invertebrates,* though closely related to the Atlantic West Indies fauna, *were, beyond a doubt, most similar to the tropical fauna from the Gulf of California.* This significant observation revealed a substantial clue to the mysterious history of the Colorado Desert's extinct animals of the sea.

Hanna's extensive report described 80 species of gastropods (snails), pelecypods (mollusks), echinoderms (starfish), corals, and one shark. Some were new species identified by Hanna, and many more were new species previously described by earlier researchers. This added data corroborated the earlier reports of both Blake and Mendenhall: *Anza-Borrego's invertebrates of the past were a truly unparalleled paleofauna.*

Extensive collecting of the invertebrate fossils in the region tapered off in the 1930s with a simultaneous rise in the investigation, collection, and research of the Park's fossil vertebrates. But the dramatic importance and far-reaching significance of these North American invertebrate fossil faunas and their strata did not diminish. Researchers in the 1930s, '40s, '50s, and '60s continued to study the area's collections and regularly published their findings in hopes of unlocking more of the mysteries presented by this remarkable fossil evidence.

The development of new tools and technologies, specifically, the introduction of the science of taphonomy (the study of animal remains from death to fossilization) in 1940, and the steady increase of taphonomic studies in the years thereafter, led to a resurgence of investigations and published works on Anza-Borrego's invertebrates from the 1970s to the present. The past century and a half of the desert parklands' geologic and paleontologic investigations and research have yielded " . . . over 100 species of bivalves [mollusks], 72 species of gastropods [snails], 16 species of echinoids [starfish], and numerous ostracods [minute water crustaceans], corals, barnacles, trace fossils, bryozoans [moss water animals], and foraminiferans [marine protozoa]" (Remeika, 1998) (see Deméré and Rugh, this volume, *Invertebrates of the Imperial Sea*). Today, science professionals and graduate students from the University of California at Riverside, the University of Chicago, California State University at San Diego, the U.S.G.S., Anza-Borrego Desert State Park, and other renowned institutions continue the comprehensive research of both the area's sea and fresh water fossil collections as well as their sources: the Colorado Desert's numerous marine, estuarine (fresh and salt water environment), deltaic (river delta), and lacustrine (lake) localities, applying new techniques and theories to extract the answers to the region's extraordinary invertebrate fossil history.

With the exception of Dr. Stephen Bowers' 1901 report, there was no official word on the discovery of any vertebrate fossils in the Anza-Borrego region until 1928. In April of that year, a Mr. W.P. Van Derpoel from El Centro, California, found and collected fossil specimens from a site located east of the Vallecito Mountains (in the southern part of the Park). The only record of this discovery is Van Derpoel's correspondence to Dr. Chester Stock, a prominent vertebrate paleontologist at the California Institute of Technology (CIT), and a prolific author of scientific papers, particularly on fossil mammals. Van Derpoel had sent two of his specimens to the Smithsonian Institution whose staff scientists informed him that they were the foot bones of an extinct camel, probably *Camelops*, and that he should contact Dr. Stock concerning these paleontologic finds. Van Derpoel did just that. He also demonstrated an incredible discipline, insight, and appreciation of the site's scientific value because he left the majority of the fossil bones that he found in place. In his second letter to Dr. Stock, Van Derpoel writes, "Am not Geologist [sic] enough to tell you much about the immediate formation. In my opinion, it was once the bed of an ocean . . . Do not know the number or amount of bones as I did not disturb them at the time of discovery thinking perhaps they were of some scientific value."

Stock replied to Van Derpoel shortly thereafter and requested that arrangements be made for himself and/or one of his staff to visit the locality. Although there appears to be no more official records of their correspondence, it is presumed that Dr. Stock and/or staff did eventually study the camel site and found it to be significant, for in the mid-1930s, the California Institute of Technology sent expeditions and recorded two locations in the Borrego Badlands area of the Park. The specimens collected now reside at the Natural History Museum of Los Angeles County, though it appears that no scientific publications resulted from the recoveries.

To proceed further in our story, we must step back to the turn of the twentieth century, when much of the professional and scientific mindset was deeply captivated by the recent discoveries and dramatically publicized excavations of colossal dinosaurs. Finds of fossil mammals, for the most part, took a back seat to these titan vertebrate celebrities.

However, in the first few decades of this same century, Darwinism sprang back to life, and proponents of Darwinian gradualism sought concrete support for this long-term theory of evolution, sparking a profusion of expeditions, recoveries, and studies of Tertiary mammal fossils throughout North America.

One of the most notable personae of this paleontologic period was Childs Frick, a prominent industrialist and philanthropist, who was appointed a trustee of the American Museum of Natural History (AMNH) in 1919. Because Frick recognized the importance of fossil mammals as related to systematics (the study of systems and classification) and evolution, he personally planned the fieldwork and financed paleontologic expeditions to later Tertiary deposits throughout the continent.

Childs Frick had been a student at the University of California at Berkeley and may have been introduced to the Colorado Desert during his college years. (By this time, the university had acquired a significant invertebrate fossil collection from the Anza-Borrego region.) Later, Frick traveled extensively in the area and appears to have vacationed there on occasion. And so, the future AMNH fossil collector was indeed familiar with the parklands, and somehow he was convinced that this desert, hitherto known for its abundance of marine and fresh water invertebrates, was also a potentially fertile area for later Tertiary deposits of vertebrate fossils as well.

Paleontologist and gem rock collector, Guy Hazen, was hired by Frick to be his head field scout for the southwest region sometime in the late 1920s or early 1930s, after Frick's first southern California collector had passed away. It was reported that, "He [Hazen] travels with a super-camp wagon full of chests that all are cluttered up with prehistoric bones, gem rocks, petrified wood, ore – anything of interest. Hazen has an uncanny faculty for spotting areas where clay and sandstone outcroppings yield rich treasure for the bone-digging fraternity" (Henderson, 1941).

Circa 1935, an American Museum of Natural History expedition, led by Guy Hazen, began surveying for vertebrate fossils in the area later termed the Vallecito Creek-Fish Creek Badlands (VCFCB), in the southern part of the Park, and also in the Borrego Badlands, located in the northern end. Anza-Borrego was then a fledgling State Park (originally established in 1932 as Borego (sic) Palms Desert State Park) (Lindsay, 2001) and whether or not California State officials were cognizant of Hazen's activities is not known. At this time, there were no collecting policies in place nor State collecting permits issued; thus there appears to be no State documentation of AMNH's fieldwork in the parklands. However, shipping records from San Jacinto, California, clearly indicate

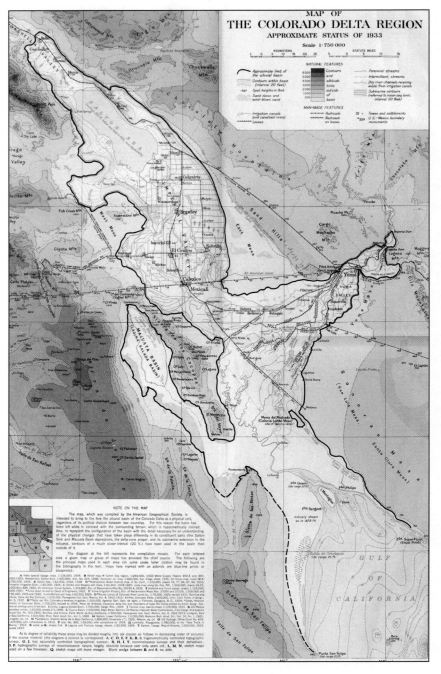

that Hazen's team was very productive and remained active in the region until 1938. Over 360 Anza-Borrego fossil vertebrate specimens were recovered and shipped back to New York City to their final destination at the Frick Laboratory. Mr. Childs Frick had inferred correctly; the Colorado Desert had revealed itself to be a literal goldmine for late Tertiary fossil mammals and the intuitive New York museum trustee had struck it rich.

At the American Museum of Natural History, the Anza-Borrego vertebrate collection, for reasons unknown, remained, for the most part, uncatalogued and the subject of little research. The New York fossil collector himself wrote the single publication on the area's paleontologic specimens. In his 1937 volume, *Horned Ruminants of North America*, Frick lists and discusses a few of the fossil deer antlers recovered by Hazen and his team during their Park surveys. Today, this collection still resides in the AMNH's Childs Frick Wing.

Following the New York museum's expeditions and the California Institute of Technology's field trips of the 1930s, there were no further official investigations, research, or publications on the region's fossil vertebrates. So it remained until December 1941. Though not as newsworthy as the December 7th attack on Pearl Harbor, a more subtle event occurred which initiated multiple repercussions, both vast and influential, for the fossil-collecting history of Anza-Borrego Desert State Park.

That momentous month, a 19-year-old southern California resident and some of his companions set out on a spontaneous adventure, which led them into the Park's Vallecito Creek-Fish Creek Badlands. Once there, the young man proceeded to do what came most natural to him; prospect for anything interesting that the terrain had to offer. His prospecting paid off and this remarkable

Figure 1.3
1933 Map of the Anza-Borrego Region Showing the Salton Sea and Gulf of California.
Note the high shoreline of ancient Lake Cahuilla (dark line).
(Sykes, 1937)

youth discovered several of what were to be the first of hundreds and hundreds of his Anza-Borrego fossil finds.

Harley James Garbani (Figure 1.4), who had been discovering and recovering fossils in the San Jacinto Mt. Eden Fossil Beds since he was a child, was a ranch hand/plumber by trade, and a self-educated fossil hunter. His passion for petrified bones, spawned by family members, enabled this young gentleman and his friends to find several vertebrate specimens on that fateful day, including horse teeth and a "piece of camel jaw with a tooth in it." (Garbani, personal communication, 2001). Garbani, who had met Guy Hazen in the early 1930s when the AMNH expedition leader was excavating in the Mt. Eden Fossil Beds, recollects that " . . . he [Hazen] told a friend of mine that there weren't any salable fossils there [VCFCB]. He'd get paid by the good skeletons . . . so he backed off that area" (Garbani, pers. comm., 2001). Garbani did not share this theory. He saw the region's potential and followed his instincts, exploring, surveying, and collecting in the southern part of the Park until he enlisted in the U.S. Army in 1944.

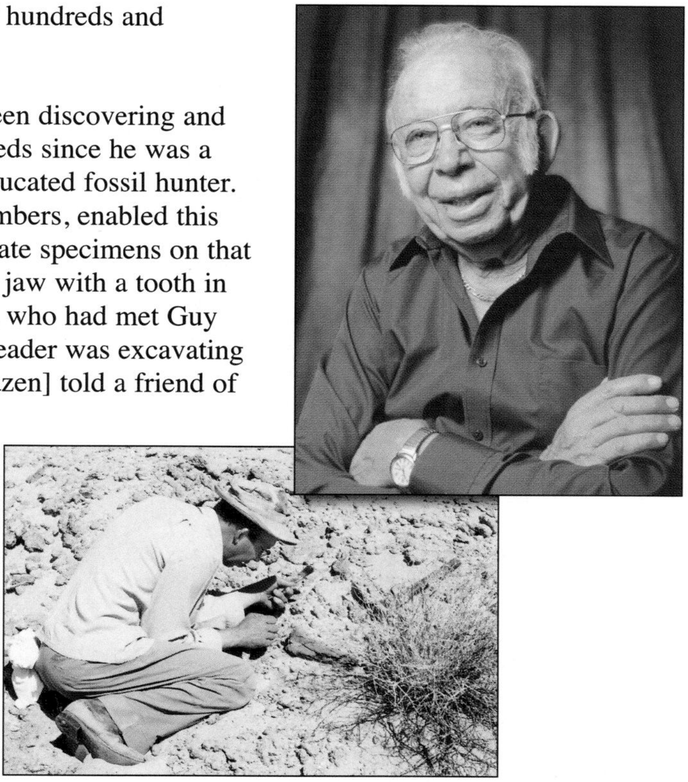

At this point in time, the parklands fossil hunter was not the only one to join Uncle Sam's armed forces. From 1941 through 1959, Anza-Borrego Desert State Park was the site for a variety of military operations conducted at various times by the U.S. Army, U.S. Navy, U.S. Marine Corps, and CIT rocket scientists, and much of the Park was officially closed to the public during this period.

Fortunately, the Park's military activities in the late 1940s and throughout the 1950s hampered neither Garbani nor his unremitting passion for fossil collecting in the Anza-Borrego desert. The postwar years found the tenacious bone hunter once again scouring the washes and gullies of the Vallecito Creek-Fish Creek Badlands in search of his "petrified booty."

Garbani eventually accrued a number of what he deemed to be significant vertebrate fossil specimens, which he wanted to have identified. After knocking on the doors of several California institutions (among them, the University of California at Berkeley, then the repository for most of the State's fossil finds) who showed little or no interest in his specimens, Garbani met with Dr. Hildegarde Howard, Senior Scientist at the Natural History Museum of Los Angeles County (LACM). Howard, in turn, directed Garbani and his collection to Dr. Theodore Downs (Figure 1.6), recently appointed Curator of the museum's Vertebrate Paleontology Department. Dr. Downs was very interested in both the Anza-Borrego fossil hunter's vertebrate finds, and in his description of the fossiliferous strata abounding in that section of the Park.

In January of 1954, Garbani and Downs, together in a borrowed jeep, visited the former's Anza-Borrego fossil localities in the VCFCB area. Here the

Figure 1.4
Top: Portrait of Los Angeles County Museum Field Collector Harley James Garbani, May, 2002. (Photograph by Barbara Marrs);
Bottom: Harley J. Garbani in the Vallecito Creek-Fish Creek Badlands, Anza-Borrego Desert State Park, 1958. (Photograph courtesy of the Natural History Museum of Los Angeles County)

curator observed specimens that Garbani had left *in situ* (in place) and was decidedly intrigued by what he saw. The two commenced prospecting and soon made several new vertebrate discoveries. In his field notes, Dr. Downs records findings of fossil horse, camel, mastodon (gomphothere), antelope, pocket gopher, rabbit, and turtle – an amazing variety of prehistoric animals for one day's work! Downs also proceeded to number and record all of Garbani's localities in that area, as well as note his own observations of the sediments and their lithology (the study of the composition and structure of rocks). It was this curator's constant penchant for meticulous field notes that made these and the museum's later State Park collections as valuable as they still are today.

Downs and Garbani returned to Anza-Borrego in June of 1954 accompanied by LACM preparator Leonard Bessom (Figure 1.5), for the first official museum field trip into the parklands. A second expedition followed in December of that year, and thus began the most productive phase of paleontologic research in the region's history.

During the years that followed, Garbani became an official Field Collector for the LACM and he was asked by Dr. Downs to explore and collect in Anza-Borrego Desert State Park whenever possible since the L.A. museum curator could not make himself available for such field trips on a regular basis. Garbani's fossil hunting expertise was enough to assure Downs that quality surveying would continue even in his absence.

When Theodore Downs did accompany Garbani, he began to make observations about certain enigmatic characteristics of the Vallecito Creek-Fish Creek Badlands strata. It soon appeared to him that in addition to the abundance and diversity of the vertebrate fossils discovered, there were other intriguing mysteries present. In his field notes he states, "This locality [VCFCB] adds a long chapter to the story of the history of ancient southern California. *It may provide insight into the interchange of mammals from Mexico.* It certainly suggests a different scene from what we know today" (Downs, field notes, 1954). Again and again, the LACM curator emphasized in his field notes that, " . . . something is definitely happening here . . . After walking the exposure out and taking several attributes, *I am convinced there is a new sequence of events revealed here*" (Downs, field notes, 1958).

Early in 1957, Dr. John Anderson White (Figure 1.6), a mammalogist and professor from California State University at Long Beach, joined the "LACM

Figure 1.5
The "LACM Party" in the Vallecito Creek-Fish Creek Badlands, Anza-Borrego Desert State Park.
Top: Dec. 1957, left to right is Dr. John A. White, Harley J. Garbani, Dr. Theodore Downs, and Park Ranger Dalton E. Merkel;
Bottom: Dec. 1958, left to right is Dr. Theodore Downs, Dr. John A. White, LACM preparator Leonard Bessom, James Garbani, and a photographer friend.
(Photographs courtesy of the Natural History Museum of Los Angeles County)

party." Because of the great numbers and varieties of microfossils being recovered (see White et al., this volume, *The Small Fossil Mammals*); Downs requested that White, a small mammal specialist, personally survey the VCFCB localities. Once on board, it did not take long for the distinguished university professor to catch the infectious enthusiasm shared by the LACM party members for Anza-Borrego's paleontologic riches. Dr. White proceeded to discover and collect many new species of rodents and lagomorphs (rabbits and hares) from the area's

sites, among them an important new species of pocket gopher, *Geomys garbanii,* which he named after Harley Garbani. In referring to the latter Dr. Downs states, "The 'paleo find' is *Geomys,* the pocket gopher, because of the quality and quantity of the remains for evolutionary study." By the end of 1957, over 75 specimens of this significant rodent were recovered.

The exceptionally high yield of many new microfossil species coupled with the discovery of several significant horizons in the VCFCB strata, convinced both Downs and White that the deposits from the south end of this section of the Park to its north end were in fact a strati-

graphic continuum of freshwater deposits. Both scientists realized that they were observing a true paleontological phenomenon, thus far unparalleled in North America: *thousands of feet of thick layers of fresh water sediments* that had been laid down in a conformable (unbroken) biochronologic sequence, *which contained a practically perfect record of vertebrate fossil fauna spanning a time of no less than 2.5 million years.* As Dr. White recalled:

> " . . . you've got about three million years of time represented and this stuff is all dipping . . . It's like a huge filing cabinet; at the upper end, about 9,000 feet . . . we have about 250 levels where we have vertebrate fossils occurring . . . It's a continuum and it's been dated . . . Downs and I did it at the LACM . . . Now remember, these are thick, thick sediments which were laid down in fresh water streams. We'll [often] get things such as stratigraphic sections which are very thick, but these are deposits in the ocean. But to find these things in a situation such as a floodplain where you have [a continuum] is very rare" (White, pers. comm., 2001).

In 1958, Dr. Downs invited Geoffrey Davidson Woodard, a former college acquaintance and a Teaching Assistant in Geology and Paleontology at the University of California at Berkeley, to examine all the Anza-Borrego sites that the Los Angeles curator believed to be geologically significant. Woodard hailed

Figure 1.6
Top: Dr. Theodore Downs in his LACM Office, Examining a Large Mammal Fossil from Anza-Borrego Desert State Park, March 1958;
Bottom: Dr. John Anderson White at the LACM, Examining Small Mammal Fossils from Anza-Borrego Desert State Park, Circa 1960s.
(Photographs courtesy of the Natural History Museum of Los Angeles County)

from Adelaide, South Australia, where he had made a name for himself in the field of geology; exploration mapping and stratigraphic test logging were his areas of expertise. In order for Drs. Downs and White to define the local faunas, which appeared throughout the Vallecito Creek-Fish Creek Badlands horizons, it was necessary for the region to be mapped geologically. Having a personal interest in the Plio-Pleistocene strata of southern California, Woodard joined the LACM party and eagerly applied himself to the task. Downs relied heavily on the "nice Australian's" observations and knowledge for the completion of his own lithologic and stratigraphic field notes.

By this point in time, Garbani's sons, James and David, had also joined the museum's Anza-Borrego team. Though only youths, they followed in their talented father's footsteps and both displayed an incredible knack for detecting and unearthing fossils. Said Dr. Downs of the Garbani trio, "We can be very thankful for the expert aid of Harley Garbani and his boys. They really have the true bone hunter desire and keep at it constantly" (Downs, field notes, 1958). The Garbanis made many significant fossil finds, among them the area's first porcupine and short-faced bear specimens, discovered by David; several large turtles and a giant tortoise found by James; an aiolorn fossil (the largest flying bird to have ever lived in North America), and an immense rare specimen of a giant ground sloth with the hide and dermal ossicles (small bones) attached, two of Harley's many major Anza-Borrego discoveries.

Promoted to Chief Curator of Earth Sciences at the Los Angeles County Museum in 1961, Dr. Downs continued to make the Colorado Desert's fossil resources a top priority, despite the additional duties of his new position. And, until his retirement in 1980, he considered himself the museum's principal investigator for the State Park. Downs personally accompanied the LACM party field trips through 1970, and during that time he initiated the first use of aerial photos for plotting the Park's fossil sites; he employed the first screen washing methods for extracting the area's microfossils (see White et al, this volume); he had Anza-Borrego's first pollen analyses performed; and it was he, together with J. A. White and G. D. Woodard, who first conducted an in-depth study of the lithology and stratigraphy of the Vallecito Creek-Fish Creek Badlands and Borrego Badlands regions.

The Anza-Borrego mammal, bird, and reptile fossils (approximately 6,000) recovered by the L.A. museum during the 1950s, '60s and '70s were from more than 450 individual sites and numbered over 100 different species, many of which were new. Numerous scientific papers were published on these remarkable fossil discoveries and the unique sedimentary formations that housed them, including several written by Drs. Downs and White, and a doctoral thesis by geologist G. D. Woodard. As with Anza-Borrego's invertebrate fossils and their particularly significant strata, discovered and researched years earlier, the Park's vertebrate fossils and extraordinary sediments were now on their rise to both national and international recognition in the global scientific community. By 1970, Anza-Borrego Desert State Park had been acknowledged, beyond a doubt, as a region of tremendous paleontologic and geologic importance, featuring several significant fossil faunas and prodigious depositional sequences unlike any other in North America. As stated by Dr. Downs, "The location of the region is particularly appropriate in terms of potential information concerning the migration of faunas to and from the Central and South Americas . . . [There is] little evidence of immigrants from Eurasia to North America in Anza-Borrego. Seven or more taxa are from Central and South American sources. [We] hope to evaluate the possibility of *in situ* evolution relative to the concepts of gradualism [theory of gradual evolution] and punctuated equilibrium [theory of punctuated evolution]."

In 1965, a notable newcomer joined the LACM party whose paleontologic endeavors in the years ahead would have a major impact on the future of Anza-Borrego's fossil collections. A student at California State University at Long Beach, George Jacob Miller (Figure 1.8) was a southern California contractor/journeyman plumber by trade. Inspired by his high school reading of Darwin's *The Origin of Species,* he decided to return to school in his early forties to study geology and paleontology. While completing his Bachelor's degree at Long Beach, he was introduced to Dr. White by a fellow student, and the professor took Miller under his wing. The latter's first visit to the State Park was circa 1955, and he returned in December of 1965 to work with the LACM's Anza-Borrego team. Though he always reported to Drs. Down and White, Miller usually prospected apart from the group and kept his own set of field notes. A serious bone hunter from the very start, he often made solo trips into the Park to survey, record data, and collect fossils for Downs and the LACM party.

In 1972, George Miller relocated both his home and his job, with a move to Canebrake (a community on the outskirts of Anza-Borrego) and a career transfer to Professor of Geology and Paleontology at the Imperial Valley College in Imperial, California. With the help of Dr. Downs, the new college professor was able to obtain a permit from the State of California to collect fossils in Anza-Borrego Desert State Park for the Imperial Valley College Museum (IVCM), which appointed him Curator of Paleontology in May of 1973. Still full of Anza-Borrego "fossil fever and fervor," the zealous paleontologist continued to discover and recover vertebrate fossils from the parklands, now bringing his collections to IVCM in El Centro, California, for preparation and curation.

Upon sharing his find of an ancestral Indian hearth with both State Park archaeologist, Bill Seidel, and Borrego Springs resident, Betty Stout (Figure 1.7), Miller observed that the latter was well educated and a hard worker. Stout and

Figure 1.7
Top: California State Park Volunteer Betty Stout in the Stout Paleontology Laboratory, Anza-Borrego Desert State Park, 1982;
Bottom, Betty Stout in the Borrego Badlands, Anza-Borrego Desert State Park, 1982.
(Photographs courtesy of the Colorado Desert District Stout Research Center Archives)

her husband, Charles, retired Nevada residents, had made Borrego Springs their winter home in 1969. Though proficient at both golf and bridge, the industrious Stout chose to spend the majority of her new leisure time taking geology and archaeology courses at the University of Nevada to educate herself about her new desert environment. Having been "appointed" by Park Superintendent Bud Getty after she had offered her time and knowledge (there was no official Park Volunteer Program, nor a Visitor Center at this time.), the volunteer aide and naturalist and her 4-wheel drive vehicle often accompanied the Anza-Borrego archaeologist and other Park rangers on various field trips and surveys throughout the parklands.

Miller wrote Stout and invited her and any interested parties to attend his two-week paleontology seminar at Imperial Valley College. The sagacious retiree replied, "Dear Mr. Miller: I don't have 3 or 4 friends *[pause]* who would be willing to go with me to [Imperial] three times a week for two weeks. But if you would come to Borrego Springs and teach your class, I could get you 50 people!" (Stout, pers. comm., 2001). Inspired by the college professor's positive response, Stout proceeded to gather and organize a large group of interested individuals (mostly Borrego Springs retirees), and on Sept. 26, 1974, at the Borrego Springs Youth Center, Professor Miller commenced teaching Anza-Borrego's first paleontology classes.

Through his lectures, which were described as "intense," "inspiring," "eye-opening," and "delightful," the passionate IVCM scientist greatly influenced his pupils and inculcated them with a true love of paleontology. With his eventual army of eager and newly educated paleontology volunteers, Miller led the next big wave of productive paleontologic surveying and collecting in the parklands, which continued throughout the 1980s. The Imperial Valley College professor and his crew surveyed and recovered vertebrate fossils from the southern part of the Park as well as from the northern end in the Borrego Badlands. Though previously "harvested" by both the American Museum of Natural History and the Natural History Museum of Los Angeles County, these areas continued to yield an astounding plethora of fossil vertebrates, particularly horses, camels, cervids (deer family), and Anza-Borrego's prehistoric celebrities, the mammoths. Because of the lack of proper facilities at the Park, the majority of these specimens were taken to the Imperial Valley College Museum for preparation and curation, and the new Colorado Desert repository began to amass a striking collection of the region's prehistoric treasures.

Figure 1.8
Prof. George Jacob Miller in the Borrego Badlands, Anza-Borrego Desert State Park, Circa 1970s.
(Photograph courtesy of the Colorado Desert District Stout Research Center Archives)

Throughout the 1970s and 80s, LACM party colleagues Theodore Downs, John White, George Miller, and Harley Garbani were in constant communication with each other, both professionally and personally, and the fossil resources of Anza-Borrego Desert State Park remained at the heart of their scientific concerns.

Dr. "Ted" Downs left his museum post in 1980, and received the honorary position of Chief Curator Emeritus, Earth Sciences Division, for the LACM in that same year. Although officially retired from his full time-job and despite ill health in later years, the former Chief Curator's primary interest and passion continued to be with the desert parklands that had become such an integral part of his life. With the aid of several colleagues, Downs and George Miller's paper on the fossil horses of Anza-Borrego was finally published in 1994, three years before his death in 1997. Ironically, in that same year (1997) the LACM's Anza-Borrego collection joined the State Park's fossil collection and the IVCM's Park collection (which came to the Park's facility in 1992) and all three were officially united under one roof (the present Colorado Desert District Stout Research Center). So, at the time of his death, Dr. Theodore Downs' dream and vision to consolidate Anza-Borrego Desert State Park's three major collections finally did materialize, accomplished under the auspices of the State Park and through the supervision of a former LACM staff member, George T. Jefferson.

Though his friend and colleague, Dr. Downs, had to minimize his Anza-Borrego research efforts throughout the 1980s because of the lack of funds and his failing health, Professor George Miller and his band of dedicated volunteers remained highly productive, primarily through the aid of a powerful and valuable ally, the Anza-Borrego Desert Natural History Association (ABDNHA). This dynamic local organization yearly appropriated funds for Miller and his Imperial Valley College class so they could purchase the supplies and equipment necessary to carry on their important work.

In March 1979, the new Anza-Borrego Desert State Park Visitor Center opened its doors to both the public and scientific community alike. A cooperative venture financed through State funds, a federal grant, and donations raised by ABDNHA members, this state-of-the-art structure included the first room designated for fossil curation and collections housing. The Daniel Laboratory, named after Harry Daniel and his wife, Julia, who were ABDNHA officers and also very active members of Miller's paleontology team, became the official classroom for all of the Park's scientific lectures and seminars. However, the burgeoning needs of the Park's paleontologic resources, as well as the use of the laboratory by the Park's archaeology department and naturalist programs soon made it evident to Miller and others that a larger, separate facility was necessary to properly accommodate the region's expanding fossil collections and their increased requirements for adequate preparatory space and housing.

With funds donated by volunteer Betty Stout and her husband, Charles, and under George Miller's guidance, the Stout Paleontology Laboratory was completed in May of 1982. This modern State Park facility met all the needs of Miller's volunteer paleontology team, who were now recovering, preparing, repairing, and curating the majority of the fossils collected in the Park. Prior to this, much of the preparatory work on large fossil specimens was done in a cramped, deteriorating wooden shanty, devoid of any ventilation or insulation, dubbed the "snake pit." Both the Stout Laboratory and the Daniel Laboratory were welcome additions, facilities critical to the continued success of Anza-Borrego's paleontology program.

George Miller's most acclaimed and publicized Anza-Borrego discovery occurred in December 1986 (Figure 1.9), during the group's last field trip for that year. The expert Borrego Badlands ridge-runner spotted an area dense with fossil shards and large bones. By the time his students had resumed classes the following January, a mammoth skull and tusk had been partially uncovered at the site by their frenetic professor. For approximately 11 months thereafter, Miller and his dedicated team of senior pupils excavated and scraped about 3.5 m (10 to 12 ft.) of earth from the mammoth's remains until the 667 kg (1,400 lb.) cranium and 3.4 m (11 ft.) tusk were ready to be jacketed and removed. On March 25, 1988, before an audience of anxious paleontology volunteers, State Park rangers, and several newspaper and television crews, the two elephantine specimens were separately airlifted by helicopter and flown to their new home at the Stout Laboratory.

Shortly after their recovery, Miller realized that these mammoth specimens were unlike any others previously recovered from the parklands. Five years later, after the excavation was completed and all the paleontologic and geologic evidence gathered, the "Miller mammoth" was determined to be the most complete skeleton of *Mammoths meridionalis* (southern mammoth) found in North America (see McDaniel, this volume, *Mammoths and Their Relatives*). Once again, Anza-Borrego's astonishing fossil treasures had placed the California State Park in the scientific world's national and international limelight.

Though credited for his numerous and extraordinary fossil finds, for his years of research, and for authoring many scientific publications on the Park's vertebrate fossil fauna, George J. Miller contributed to Anza-Borrego's paleontologic history in yet another significant way. He, along with park volunteer exemplar, Betty Stout, is credited with organizing the parklands first group of paleontology volunteers, and for actively recruiting these individuals from one of the area's most valuable resources: its senior citizens. It was the immense success of this volunteer venture that enabled Miller to survey and collect as extensively as he did in the 1970s and 1980s, and the Anza-Borrego paleontologist

Figure 1.9
Miller Mammoth Site.
Clockwise from top: Prof. George J. Miller and paleontology volunteers working at the Miller mammoth site, Feb., 1988.; Helicopter prepares to airlift jacketed mammoth tusk, March, 1988; Prof. George Miller supervises as paleontology volunteers prepare jacketed mammoth skull for the airlift, March, 1988; Prof. George J. Miller points to the Miller mammoth tusk, ready to be jacketed, Jan., 1988. (Photographs courtesy of the Colorado Desert District Stout Research Center Archives)

was indeed cognizant of the value of his distinguished students. Said Miller, in a *Los Angeles Times* interview: "This group of senior citizens is unlike any other group anywhere. What they have done here has even amazed the national Society of Vertebrate Paleontologists" (*Los Angeles Times*, June, 1981).

Following Miller's death in 1989, the professor's paleontologic novitiates found themselves without any leader and left standing completely on their own. Spurred on by their unremitting passion for Miller's work, and armed with the knowledge, experience, and paleontologic rudiments they had acquired from their former teacher, the close-knit band of volunteers chose to continue their professor's paleontology program as best they could.

Senior volunteer Betty Stout ultimately took the lead and for the next three years supervised the Park's volunteer paleontology program. Lacking a professional paleontologist for their State collecting permit, Stout approached Dr. John White and asked him to step in for his former student and colleague. The University of Arizona (at Tucson) professor consented by lending his name and title to the program so that the group's fieldwork and other paleontologic efforts could continue, and continue they did. And, while George Miller's determined apprentices worked in earnest to keep their professor's dream alive, another individual, under the auspices of California State Parks, came to the aid of Anza-Borrego's fossil future.

David Van Cleve (Figure 1.10) assumed his new duties at Anza-Borrego Desert State Park as the District Superintendent for the Anza-Borrego District in August 1989. Though he personally never met George Miller, Van Cleve soon became very aware of the Imperial Valley College professor's ongoing work and of the scope and significance of Anza-Borrego's paleontologic and geologic resources. A former agency scientist with the California Southern Region Office, the new District Superintendent was very familiar with the subject of science in the State Parks. However, paleontology was not his forte, nor was it an area of expertise for the California state government. Therefore, when Miller's death left behind a steadfast but struggling non-professional volunteer group responsible for all of Anza-Borrego's fossil collections and for the Park's entire paleontology program, Van Cleve knew that this skeletal organization had to have effective State support to achieve both stability and success.

Van Cleve sought help, and after conferring with several National Park paleontologists, he acted upon their consensus and proceeded to assemble a Paleontology Advisory Board to advise him and the State of California on the proper course of action to manage and conserve Anza-Borrego's paleontologic resources. Comprised of paleontologists and museum professionals from throughout the southwestern United States, the first official Advisory Board meeting convened in October 1991. Among the new board members were some faces very familiar to the parklands: Dr. John A. White, Betty Stout, and George T. Jefferson, Associate Curator for the Rancho La Brea George C. Page Museum in Los Angeles.

Figure 1.10
Colorado Desert District Superintendent Dave Van Cleve, Circa 1990s.
(Photograph courtesy of the Colorado Desert District Stout Research Center Archives)

Following the board's advice, Van Cleve immediately initiated and executed a number of important actions to lay down the foundation for a permanent Park paleontology program. Under Van Cleve's supervision both a Resource Management Plan and a Collections Management Policy were written and approved, which, for the first time, established formal goals, procedures and policies for managing Anza-Borrego's paleontologic resources. Also instituted was the Park's first State-sanctioned Certification Training Program in Paleontology. No longer to be left in "limbo," Miller's former students, as well as any interested parties, could now become official State Park Paleontology Volunteers and be recognized for their contributions.

One of Van Cleve's main concerns was to have the non-state agencies housing fossils excavated from the parklands transfer these collections to the Colorado Desert District's paleontologic facility. Working with Paleontology volunteer, Julie Parks, and Park ranger Paul Remeika, and with financial assistance from the ABDNHA, the District Supervisor adroitly executed the transfer of George Miller's Imperial Valley College Museum fossil collection to its rightful home at the Park's Stout Laboratory on March 31, 1992; only one day before the approximately 6,000 fossil specimens were to be relocated to the college's trailers, " . . . cargo containers outside on the desert floor" (Remeika, report, 1992).

In January 1993, George T. Jefferson (Figure 1.11), having retired from the Los Angeles County Museum, was hired by Van Cleve as a Park Environmental Services Intern, and by the ABDNHA as a Park Paleontologic Collections Manager, a temporary "joint" position so that the former Los Angeles curator could oversee the parklands collections until a permanent post was available. Jefferson, who for many reasons was perfectly suited for the task, became the State's first District Archaeologist/Paleontologist a year later in 1994.

Figure 1.11
Colorado Desert District Archaeologist/Paleontologist George Thomas Jefferson, March, 2002.
(Photograph courtesy of the Colorado Desert District Stout Research Center Archives)

Much like his District Superintendent, George Thomas Jefferson was a man of intense resolve. Because of his many years experience with the management of the LACM's vertebrate fossils, he knew what had to be done with Anza-Borrego's paleontologic resources and he knew how to do it. With funds received from the State, Jefferson hired Lyndon Murray, an MS graduate in Quaternary Studies from Northern Arizona University. Together, the two began to establish and organize a viable full-scale collections management program. Almost immediately, Jefferson revamped the Certification Training Program, taught the new lecture series, and succeeded in gathering what was once again a strong force of paleontology volunteers, comprised of both the experienced and the novitiate. Jefferson and Murray also tackled the colossal project of organizing Anza-Borrego's botanic, invertebrate, and vertebrate fossil collections into a professional working system, which included, among other standards, strict and accurate specimen curation and the creation of official Specimen Recovery, Laboratory, and Collections Guidelines.

One of Jefferson's many major achievements was the transfer of the Natural History Museum of Los Angeles County's Anza-Borrego collection back to its Colorado Desert homeland. After three years of countless trips to Los Angeles, much lengthy correspondence, and numerous phone calls to LACM officials, Jefferson's tenacity paid off, and the county museum's Park collections, totaling over 6,000 items, were transferred to Anza-Borrego Desert State Park in 1997.

As a response to the Advisory Board's recommendation for the State to provide an onsite district collections facility and research center, and with funds provided by State appropriations and grants and the Stout Foundation, Jefferson, Van Cleve, and other Anza-Borrego professionals worked together on plans for this major project. The proposed structure had to meet the requirements not only of the state park's paleontologic resources, but also those of Anza-Borrego's natural history, botanic, and archaeologic collections as well. Construction of the new Colorado Desert District Stout Research Center (built around the original Stout Laboratory) began in 1997, and the complex was completed in 1998. At the Center's dedication in 1999, it housed approximately 13,000 curated botanic, invertebrate, and vertebrate fossil specimens and featured one of the most state-of-the-art laboratories for fossil preparation in the country.

Equipped with the new State Park research facility, Jefferson and a seasonal team of 30 to 40 Certified Paleontology Volunteers thoroughly trained in the latest laboratory methods, curation procedures, and field technology, including the use of GPUs (Global Positioning Units), GIS (Geographic Information System) map imagery, and linked computer data bases, ushered in the new century with a continued wave of parklands discoveries. New cheetah-like cats, llamas, camels, and giant aiolorn bird specimens were "discovered" amongst the already collected and curated materials after the combined collections were reorganized and properly identified under Jefferson's direction and guidance. Also during this time, fossil wood deposits and fossil animal tracks were mapped and documented by Paul Remeika. Other "fossil firsts" included the discovery of *Gomphotherium* (elephant family) specimens, a possible new species of *Palaeolama* (found by none other than Harley J. Garbani) (see White et al., this volume), a fossil Gila monster, and important new discoveries in the VCFCB: a diverse and unusual mix of both marine and terrestrial fossil vertebrates, including sharks, sea turtles, walruses, ground sloths, gomphotheres, ancestral horses, camels, and llamas.

The discovery of two tusk ends and skull fragments of a subadult and a juvenile mammoth in the Borrego Badlands in February 2002, prompted the use of new recovery procedures by the Anza-Borrego Paleontology team

(Figure 1.12). Throughout that year, volunteers, under Jefferson's supervision, used gasoline-powered drills and pneumatic chisels connected to a gasoline-powered air compressor to break-up and remove layers of sediment, excavation methods unprecedented in Anza-Borrego's collection history. The mammoth fossils were then jacketed and the two plaster casts, weighing approximately 420 and 1,050 kg (880 and 2,200 lbs) respectively, were separately airlifted by helicopter in February, 2003, and safely flown to the Stout Research Center (Figure 1.12).

Regarding the future of the State Park's scientific status, Jefferson states, ". . . there's only one reason to do this whole business, you know, and that's because this [Anza-Borrego Desert State Park] is an important place paleontologically . . . this is an actively researched area and currently under study, and even with all of the work that Miller and Downs and Los Angeles County Museum and Imperial Valley College Museum have done, there are still big questions that this place can answer in terms of paleontologic history and the last couple of million years. No question about it" (Jefferson, pers. comm., 2001).

At present, Anza-Borrego Desert State Park is still the location for important research by visiting geologists, paleontologists, and other earth scientists and professionals from around the globe, as well as by paleontology and geology graduate students from universities throughout the U.S. Scientific publications on the region's internationally famous vertebrate and invertebrate paleofaunas and outstanding geologic features continue to prove the marked significance of Anza-Borrego and its incredible scope of fossil resources.

One wonders what William Phipps Blake would think if he could view this desert region today, over 150 years later, and share in all the paleontologic and geologic discoveries that have occurred since his 1853 expedition into what he described as a "wide Sahara bounded by rugged, barren mountains." If he was amazed by what he observed back in the nineteenth century, he would most certainly be awestruck by what he would witness today. The fossil treasures of

Figure 1.12
2002-03 Excavation and Airlift of Mammoth Specimens in the Western Borrego Badlands.
Bottom: George T. Jefferson supervises State Park paleontology volunteers as they use a gasoline powered drill to break-up and remove the sediment surrounding the fossils; center: State Park paleontology volunteers jacket the exposed tusk ends (covered with aluminum foil); top: jacketed fragments are airlifted to the Stout Research Center Laboratory. (Photographs courtesy of the Colorado Desert District Stout Research Center Archives)

Anza-Borrego Desert State Park, which once nestled serenely undisturbed among the region's washes, gullies, hills, and badlands for hundreds of thousands of years, have now traveled all the way from virtual obscurity to both national and international prominence in the scientific world, and the geological evidence abounds that this desert park will continue to proffer its petrified and stratified riches for generations to come.

Paleolandscape 1:
The Imperial Sea 5 Million Years Ago

INDEX
1. *Pelecanus* sp. (pelicans)
2. *Melanitta persipicillata* (surf scoter)
3. *Cheloniidae* (sea turtles)
4. *Valenictus imperialensis* (Imperial walrus, extinct)
5. *Mytilidae* (mussels)
6. *Clypeasteridae* (sea biscuits)
7. *Haliotis* sp. (abalones)
8. *Megabalanus tintinnabulum* (acorn barnacle)

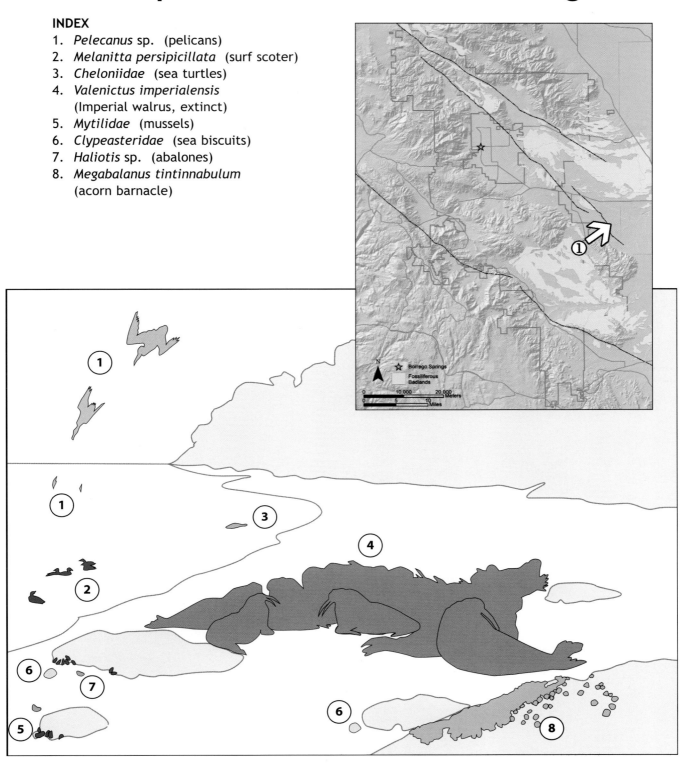

A warm, tropical sea, sandy intertidal zone, and beach are home to a group of basking *Valenictus imperialensis* (Imperial walrus) as pelicans dive (though today, only brown pelicans dive for their prey). The view today is to the northeast from north side of Fish Creek Campground (see map), the approximate location of the Miocene shore. (Picture by John Francis)

2 The Imperial Sea: Marine Geology and Paleontology

The old sea bed, where once rolled the headwaters of the Vermilion Sea, is still a ghostly memory of its former state. And a memory not too dim, either. At dawn all the hollows of the badlands swim with misty haze that startlingly suggest water. And when sunset flings the long blue shadow of Coyote peak far out across the dry reaches the effect is breath-taking. There they are again, all those ancient bays and winding gulfs and lagoons. And beyond them the purple grey of the great sea.

Marshal South
Desert Magazine **February 1940**

Thomas A. Deméré

The Imperial Sea:
Marine Geology and Paleontology

Bahia de Los Angeles — a View of Former Anza-Borrego. (Photograph by Paul Remeika)

On a recent warm spring day as I was hiking up the soft slopes of a mud hill in the badlands near the Coyote Mountains of Imperial County, I came across fossilized shells of ancient organisms. The shells were weathering out of the otherwise uniformly fine-grained claystones that underlie this portion of the Colorado Desert. Upon close inspection, I found that the shells belonged to an extinct species of marine snail first recognized from this area by G. Dallas Hanna in 1926 and named *Turritella imperialis*. Surprisingly, Hanna's fossil Turritella is closely related to another species of *Turritella* known from fossil deposits on the Caribbean coast of the Panamanian Isthmus. As it turns out, many fossil shells from the Imperial Valley share close relationships with both fossil and living Caribbean species. When viewed in a broader geologic context, these related fossils suggest an historical connection for the two regions, but we are getting ahead of ourselves.

As I examined the shells at my feet, I paused to survey the geologic setting of this place. I was standing on a distinct stratum of light brownish-yellow fossiliferous siltstone tilted rather steeply to the northeast. The deformed stratum was overlain and underlain by uniformly massive beds of similar colored claystone, the whole exposed stratigraphic sequence probably several hundred meters in thickness. To the east across the breadth of the Carrizo Badlands were identical mud hills exposing similar tilted sequences of ancient marine siltstones and claystones, while to the west I could see a more resistant sequence of light gray and pale brown sandstone strata lying well below the claystones and resting on the twisted and altered metamorphic rocks of the main mass of the Coyote Mountains. The claystones, siltstones, and sandstones form an easily recognized sequence of sedimentary rocks that geologists have named the Imperial Group after the tectonically active valley where they occur.

Today there is no surface water in this part of the Imperial Valley except for the small oasis at the old Carrizo stagecoach station along Carrizo

Creek. Yet this parched earth contains locally abundant fossil remains of marine organisms. The juxtaposition of these two conflicting realities – shells of marine animals and parched earth – begs for explanation. Thanks to earth scientists like Hanna and many others, the conflicting realities are easily seen as historical snapshots in a dynamic evolving system driven by plate tectonics and playing for all eternity. Of course both realities exist, however, the reality of the fossil shells is one based in a world 4 to 7 million years past, while

Figure 2.1
Deguynos Formation.
Marine strata of the Deguynos Formation are well exposed in the eroded Carrizo Badlands (note people for scale). (Photograph by Tom Demére)

parched earth is the current reality. What the future reality will be, we can only predict.

As a paleontologist, I have learned how to live vicariously in ancient realities using the abundant clues preserved in the crustal rocks of the Earth. It is always rewarding when I can take people unfamiliar with reading these clues on a journey back in time, if only in their imaginations. To begin a trip like this it is useful to look at an area from two very different perspectives or scales. On the one hand, you have to get down and literally touch the Earth, while at the same time you need to step back and take more of a bird's-eye-view, remembering that there are times when you "can't see the forest for the trees."

To touch the Earth I mean to look closely at the rocks at your feet. For the paleontologist this generally means looking at clastic sedimentary rocks; those sedimentary rocks formed by the accumulation of sedimentary particles (mud, silt, sand, and/or gravel) under the influence of running water, blowing wind, or grinding glaciers. Such rocks, much like sandpaper, have textures largely determined by grain size. Thus, we can recognize coarse-grained, medium-grained, or fine-grained sandstones. Sedimentary rocks formed of silt-sized particles are called siltstones and those composed of clay-sized particles are called claystones or, if finely layered, shales. A rock unit made up of gravel is called a conglomerate, with modifiers added to reflect more specific particle sizes like pebble conglomerate, cobble conglomerate, or boulder conglomerate. These textural distinctions become important when we try to interpret the conditions under which a particular sedimentary rock stratum was deposited. Generally speaking, the larger the grain or particle sizes the stronger (swifter) the current responsible for transporting and depositing those particles. For example, a storm-swollen stream can transport pebbles and cobbles (not to mention cows and houses), while a slow-moving creek may only transport silt. Of course, interpreting the details of ancient depositional environments requires a lot more information, but these simple textural clues can provide a glimpse of

Figure 2.2
Turritella imperialis.
Fossil shells of this extinct snail, weather out of a mudstone stratum in the Deguynos Formation, Carrizo Badlands. (Photograph by Tom Demére)

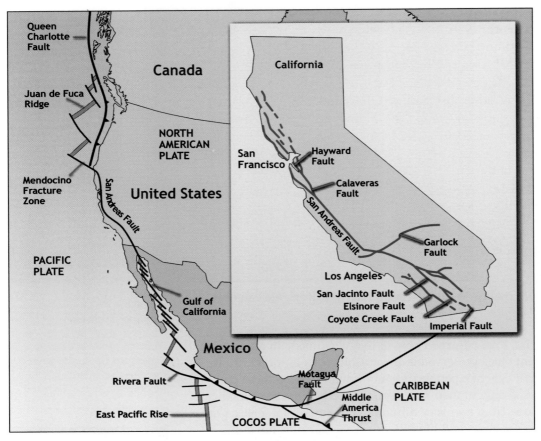

Figure 2.3
Plate Tectonic Setting of the Gulf of California.
Shown are the spreading centers, subduction zones, and transform faults.

the general conditions under which a particular rock stratum was formed. Fossil remains provide additional clues and at their coarsest level can indicate whether a stratum was deposited on land or on an ancient sea floor. Armed with these geological tools I have more than once been able to stand on an outcropping of fossil-rich marine sandstone and experience, in a virtual sense, tropical waters with their warm breezes and long vanished animal communities.

It is easy to get caught up in the minute details of sedimentary rocks and miss the bigger picture. This is especially true in the Colorado Desert where there are so many spectacular exposures of tilted, folded, and faulted strata. Stepping back and utilizing the broader perspective provided by maps and, these days, satellite images from space, we can see these sedimentary outcrops in their broader geologic context. The Colorado Desert is really a unique place, geologically. Like the East African Rift Valley of Kenya and Tanzania, the Colorado Desert lies in an area of active crustal thinning and continental rifting (see Dorsey, this volume, *Stratigraphy, Tectonics and Basin Evolution in the Anza-Borrego Desert Regions*). The southern portion of the Colorado Desert at the head of the Gulf of California is underlain at depth by a segment of the East Pacific Rise, a major crustal plate boundary that separates the largely oceanic Pacific Plate on the west from the largely continental American Plate on the east. Molten rock is welling up from deep inside the Earth along the axis of the East Pacific Rise, forcing the plates to spread away from each other. In places the central rift valley, where this spreading is concentrated, is broken and offset by transform faults with horizontal rather than vertical displacements. For most of its 8700 km (5437 mile) length the East Pacific Rise is a submerged volcanic mountain range on the ocean floor. This submerged mountain range enters the Gulf of California near Cabo San Lucas and extends along the length of the Gulf where it is broken into numerous short segments by transform faults. The combination of spreading and transform faulting is responsible for both the origin of the Gulf of California and the more than 250 km (156 mile) northward movement and clockwise rotation of Peninsular California (the landmass consisting of southern California and the Baja California Peninsula). At the head of the Gulf a major transform fault breaks through a short spreading segment of the East Pacific Rise and extends for over 1287 km (804 mile) northwestward to another spreading segment off the northern California coast at Cape Mendocino. This major transform fault is more popularly known as the San Andreas Fault and represents one of the few examples on Earth where transform faults deform continental rather than oceanic crustal rocks.

With these large-scale features of sea floor spreading and transform faulting in mind, it is easy to see why the Colorado Desert region of southern California is an area of high earthquake activity (transform faulting), as well as high geothermal activity (magmatic heat from sea floor spreading). These current conditions are nothing new for our region, which has been undergoing deformation since at least the middle Miocene, approximately 13 million years ago, when a proto-Gulf of California began to form in generally the same location as the modern Gulf. This proto-Gulf was flooded by ocean waters of the tropical eastern Pacific that entered the mouth of the proto-Gulf well to the south at about the latitude of present day Puerto Vallarta (20° N). As this narrow inland seaway filled, the shoreline transgressed northward, eventually reaching as far as San Gorgonio Pass in present day Riverside County. It was at this time, about 7 million years ago, that the oldest sandstone strata of the Imperial Group were being deposited in the shallow waters at the head of the proto-Gulf. The waters at the head of the proto-Gulf were warm and clear and supported diverse communities of marine organisms that lived in a variety of nearshore and intertidal paleoenvironments. Beginning about 5 million years ago, conditions changed dramatically when the ancestral Colorado River started depositing huge volumes of sediment into the northern end of the proto-Gulf. This tremendous influx of sediment was generated as the mighty river eroded headward into the Colorado Plateau of Arizona, excavating deep canyons including the Grand Canyon. The resulting sediment load caused increased turbidity of gulf waters and the eventual deposition of thousands of meters of clay and silt. Over time, the rapidly expanding river delta formed a massive "sediment dam" that forced the shoreline of the Gulf southward to near its present position, leaving the northern end of the proto-Gulf "high-and-dry." Actually, not that high and dry since the low-lying region north of the delta became first a series of Pliocene and Pleistocene fresh water and ephemeral playa lakes, followed by braided streams, and eventually coalesced alluvial fans. These later non-marine paleoenvironments and their fossil organisms are the subject of other chapters in this book.

Having presented this rather abbreviated overview of the late Cenozoic history of the Imperial Valley, we can now focus more closely on the ancient marine strata of the Imperial Group and its preserved biological record of fossils.

Latrania Formation

Geologists have named the older pre-delta marine sandstones at the base of the Imperial Group the Latrania Formation. In the Coyote Mountains, Fish Creek Mountains, and Vallecito Mountains the Latrania Formation (Figure 2.4) can be up to 100 meters (330 ft) thick and typically consists of a sequence of coarse-grained strata that include gray to red-brown, medium-grained, massive, micaceous sandstones; gray, fine-grained, laminated, and cross-stratified, micaceous sandstones; red-brown bioclastic sandstones; and pale yellow skeletal (shelly) limestones. In several areas, the basal marine sandstones of the Latrania Formation interfinger with coarse-grained alluvial fan conglomerates. These fanglomerates, as they are called, suggest that erosion and sedimentation rates were relatively high during the initial filling of the proto-Gulf in the late Miocene. A good place to see these basal deposits is at the mouth of Fossil Canyon on the south slopes of the Coyote Mountains near Ocotillo. As you walk up the canyon the sedimentary strata exposed in the canyon walls gradually change from coarse-grained fanglomerates with sharply angular cobbles and boulders to fine-grained marine sandstones. About 1000 meters (3300 ft) up the canyon from the Bureau of Land Management fence, a resistant one meter

Figure 2.4
Latrania Formation.
Wind eroded sandstones of the Latrania Formation are exposed in the Coyote Mountains. (Photograph by Tom Demére)

(3.3 ft) thick shelly limestone bed crosses the canyon floor as a distinctive, tilted stratum. A close inspection reveals that this limestone is packed with irregular rounded pieces of colonial corals, broken shells of oysters and scallops, and sandstone molds of marine snails. This limestone stratum, like most of the Imperial strata exposed in Fossil Canyon is tilted rather steeply to the east, the result of tectonic forces associated with the nearby Elsinore Fault. Overlying the limestone is a rather thick sequence of coarse-grained sandstones that contain local concentrations of fossil scallops, oysters, cones, conchs, whelks, sand dollars, and sea biscuits. For the most part, these fossils are not that well preserved and primarily consist of internal and external molds of marine clams and snails. The original calcium carbonate (in the form of the mineral aragonite) of the fossil shells has been leached away by slightly acidic groundwater percolating through the buried sandstone strata. As the shell material dissolves, some of the calcium carbonate is precipitated in the pore spaces between individual sand grains. When uplift and erosion eventually expose the fossil-bearing strata, they begin to crumble "releasing" the buried fossils, which now consist of hardened sandstone impressions of the internal or external surfaces of the shells. Typically, the internal molds are the most useful to paleontologists because, although the shell material is now lost, the mold preserves much of the form of the original shell allowing for taxonomic identification.

Not all of the fossils from the Latrania Formation are preserved as internal and external molds. For species that construct their shells with the mineral calcite, another form of calcium carbonate, the acidic groundwaters do not destroy the original shell material. Fossils of these species, which include scallops, oysters, certain snails, corals, and sand dollars, are preserved as original shell material and often retain a great deal of important surface morphology. Some even preserve remnant shell coloration (see Deméré and Rugh, this volume, *Invertebrates of the Imperial Sea*).

Figure 2.5
Sandstone Molds.
Internal molds of marine mollusks, note large cone snail left center, weather out of the Latrania Formation in the Coyote Mountains. (Photograph by Tom Deméré)

The basal marine sandstones of the Latrania Formation are not always associated with alluvial fan deposits and elsewhere in the Imperial Valley lie directly on older volcanic and/or metamorphic rocks that form the core of the Coyote and Fish Creek mountains. The contact between the Latrania Formation and these crystalline rocks is very irregular and represents an ancient eroded land surface that was rapidly submerged as marine waters spread northward with the advancing Miocene shoreline of the Imperial Sea. In some places, low cliff faces and benches carved during the Miocene into the volcanic and metamorphic crystalline rocks serve to mark the location of former shorelines and sea cliffs. Careful geologic studies of these ancient shorelines suggest that the Imperial coastline was quite irregular and consisted of a mixture of prominent headlands, steep and linear sea cliffs, eroded embayments, and rocky islands. Such a description could also be made for long stretches of the modern Gulf coast of Baja California. Good exposures of the Miocene shoreline features occur on the south slopes of the Coyote Mountains between Fossil Canyon and Painted Gorge.

The rarity of clay and silt in the Latrania Formation suggests that deposition generally occurred in relatively clear marine waters at intertidal to subtidal depths. This hypothesis is supported by the occurrence of locally diverse and abundant assemblages of fossil marine invertebrates dominated by mollusks, echinoderms, and importantly, colonial corals. As already mentioned, many of the fossil invertebrate organisms found in the Latrania Formation are closely related to species found in the Caribbean region, either as fossils or as still living organisms. Their occurrence in the Latrania Formation suggests some degree of faunal interchange between the two regions. Compelling geologic evidence from Central America further indicates that a direct marine connection formerly existed between the Caribbean and the tropical eastern Pacific via seaways across southern Costa Rica, central Panama, and western Colombia. Today this area is occupied by the Isthmus of Panama, which was uplifted above sea level during the Pliocene Epoch approximately 3.5 to 3.1 million years ago. Prior to that time the Central American Seaway formed the southern shoreline of the North American continent and separated North America from the island continent of South America. Given this very different paleogeography, it is possible to visualize a former unbroken tropical region encompassing the Caribbean Sea, western Gulf of Mexico, and equatorial eastern Pacific that supported a shared community of marine organisms. Paleontologists refer to this region as the Tertiary Caribbean province. Surface ocean currents of the eastern Pacific (Equatorial Countercurrent) and Caribbean (Caribbean Current) presumably flowed west to east and east to west, respectively, through the Central American Seaway and provided a means for dispersing larvae of various marine invertebrate phyla. Countercurrents flowing north along the west coast of Mexico, in turn, allowed dispersal of Tertiary Caribbean species into the proto-Gulf of California. This was especially the case for species of colonial corals. Paleontologists have recognized at least nine different species of colonial corals from the Latrania Formation (Vaughan, 1917), only one of which belongs to a genus that still lives in the eastern Pacific. All of the other fossil species belong to genera that evolved in the Caribbean and only survive there today.

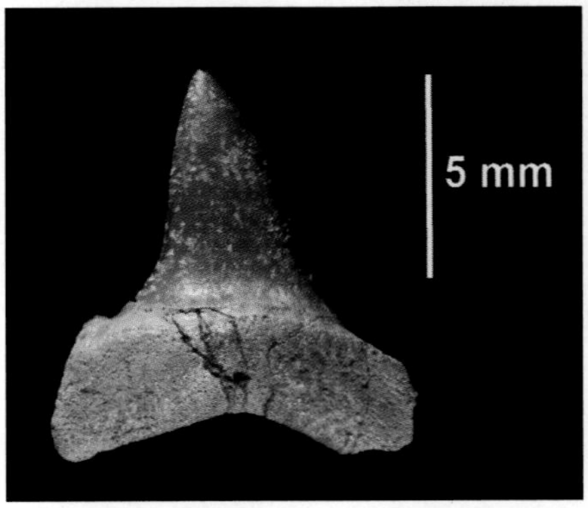

Figure 2.6
Isurus sp.
Fossil tooth of this mako shark (SDSNH 86003) was collected from the Latrania Formation in the Coyote Mountains. (Photograph by Barbara Marrs)

Besides preserving a record of tropical Tertiary Caribbean affinities, the Latrania Formation also preserves interesting records of ancient marine ecological and environmental conditions. Paleontologist Rodney Watkins (1990a, 1990b) has discovered rare rocky intertidal deposits of skeletal (shelly) limestone in the Latrania Formation as exposed in the Coyote Mountains. These rather unique deposits lie directly on marble of the ancient metamorphic bedrock complex and preserve a rocky shore trace fossil community consisting of burrows and borings of rock-boring bivalves, sponges, worms, and echinoids, as well as body fossils of encrusting barnacles and corals (see Appendix, Table 2). Outcrops of intensely bivalve-bored marble in this area occur in association with wave-eroded sea cliffs and serve to define former shorelines around prehistoric rocky islands of the Miocene Imperial Sea. In this same area, local concentrations of fossil corals occurring directly on wave eroded bedrock platforms appear to represent ancient coral patch reefs that formed in shallow nearshore waters. Walking along this ancient shoreline, it is almost possible to hear the Miocene waves breaking against the low sea cliffs and to imagine the currents swirling around the margins of slightly submerged coral reefs.

Figure 2.7
Myliobatis sp.
Tooth plates of the bat ray *Myliobatis* sp. (SDSNH 86005) were collected from the Latrania Formation in the Coyote Mountains. (Photographs by Barbara Marrs)

Susan Kidwell is another scientist who has studied the fossiliferous marine deposits of the Imperial Valley. Her studies (Kidwell, 1988; Winker and Kidwell, 1996) have focused on major fossil shell beds in the Latrania Formation. Some of these shell beds are as much as 7 meters (23 ft) thick and contain a complex variety of fossil concentrations including sandy and muddy layers with low diversity assemblages of oysters and snails, gritty limestone layers with low diversity oyster and coral assemblages, and limey sandstone layers with very diverse clam, snail, and echinoid assemblages. Typically, the fossils occur as jumbled concentrations of whole and partial shells suggesting that the shell beds represent long-term accumulations of dead skeletal debris on subtidal sea floors, probably below the level reached by normal wave action.

Although the bulk of fossils known from the Latrania Formation represent species of invertebrate organisms like corals, molluscs, crustaceans, and echinoderms, recent field work by staff of the San Diego Natural History Museum has resulted in the recovery of partial remains of marine vertebrates including teeth of sharks, rays, and bony fishes, as well as bones of sea turtle and marine mammals (Deméré, 1993). The fossil shark and ray assemblage currently consists of the bat ray *Myliobatis* sp. (Figure 2.7), the extinct giant shark *Carcharocles megalodon*, the mako shark *Isurus* sp. (Figure 2.6), and the sand shark *Odontaspis* sp. Fossil bony fish from the Latrania Formation include a giant barracuda *Sphyraena* sp., a sheeps head *Semicossyphus* sp., and a triggerfish family Balistidae. A single carapace plate of a sea turtle belonging to the family Cheloniidae has been recovered from sandstones of the Latrania Formation. Fossil remains of marine mammals from this rock unit include a partial rib of a sea cow and several partial lower jaws and ribs of a baleen-type whale. All of these vertebrate fossils occur in the shell-rich strata of the Latrania Formation and probably represent scattered remains that were transported and deposited by localized currents. Unfortunately, complete or even partial skeletons of marine vertebrates have so far eluded the dogged attempts by field parties of vertebrate paleontologists to find them. Rather than being seen as a "half empty glass," this meager vertebrate fossil assemblage suggests that there is still a great deal of paleontological work to be done in these marine rocks and that many more fossils are to be found.

Deguynos Formation

As mentioned earlier, beginning about 5 million years ago the ancestral Colorado River began building its massive delta at the head of the proto-Gulf of California. This Pliocene event is recorded in the sedimentary rocks of the Imperial Group by a dramatic change from the coarse-grained sandstones of the Latrania Formation in the lower portion of the Imperial Group to the fine claystones of the Deguynos Formation in the upper portion of this rock unit. The Deguynos Formation is at least 1000 meters (3300 ft) thick in the Carrizo Badlands and consists primarily of greenish-gray to grayish-olive massive claystones, yellowish-gray laminated siltstones, light gray silty very fine-grained sandstones, and gray shell coquinas. Resistant skeletal

30 cm

Figure 2.8
Baleen Whale.
This fragmentary rib of a baleen whale (SDSNH 46015) was collected from the Latrania Formation in the Coyote Mountains. (Photograph by Barbara Marrs)

Figure 2.9
Deguynos Formation.
Cross-bedded deltaic sandstones and pro-delta mudstones of the Deguynos Formation are well exposed in Painted Canyon, Coyote Mountains (note person in foreground for scale). (Photograph by Tom Deméré)

Figure 2.10
Coquina.

This massive bed of oyster shells is in the Deguynos Formation in Painted Canyon, Coyote Mountains. (Photograph by Tom Deméré)

15 cm

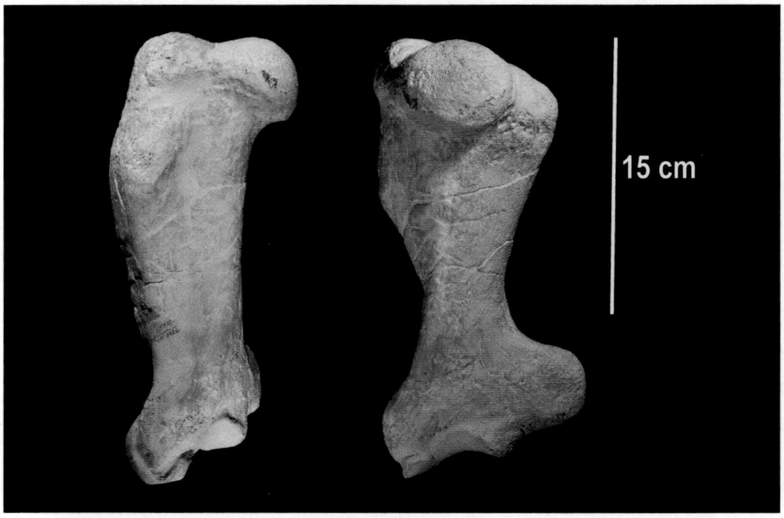

Figure 2.11
Valenictus imperialensis.
Left humerus (upper arm bone) of the extinct Imperial walrus (side and rear views of a cast of LACM 3926). (Photograph by Barbara Marrs)

limestone and bioclastic sandstone strata typically weather to a dark brown color, while the softer claystone and siltstone strata weather to a pale yellow color and often develop a blistered expansion surface littered with white, platy crystals of Gypsum, calcium sulfate. Because of relatively high rates of sediment accumulation in and around the massive delta of the ancestral Colorado River, fossils are generally rare in the Deguynos Formation. Highly fossiliferous strata, however, do occur in the middle and upper portions of the formation, which is characterized by resistant oyster-shell coquina strata varying in thickness from less than one meter (3.3 ft) to over four meters (13 ft). The coquinas are often cross-bedded and contain tightly packed concentrations of whole and fragmentary shells in a very hard calcite cemented sandstone matrix. Fossil assemblages in these coquinas are typically dominated by shells of the small oyster, *Dendrostrea vespertina*. Some coquinas also contain common shells of the bivalve, *Anomia subcostata*, and rarer shells of the small scallop, *Argopecten deserti,* and the snail, *Turritella imperialis*. Rodney Watkins has closely studied these coquinas in the Coyote Mountains and Fish Creek Mountains and concluded that they formed in large submerged distributary channels on the shallow marine portions of the ancestral Colorado River delta (Watkins, 1992). The dense concentrations of oyster shells in these coquinas are the result of swift currents transporting the large shells as bedload (suspended in the water mass) particles while flushing the finer-grained sand and silt offshore into deeper water. Fossil assemblages in the mudstone strata underlying many of the coquinas include widely dispersed fossils of the pholads (rock piddock clam), *Cyrtopleura costata,* and the pen shell, *Pinna latrania,* often in life position. By life position, I mean that the fossil shells are upright in their original burrows rather than lying on their sides in a transported death pose. These occurrences suggest that the infaunal (living in mud or sand bottom) pholads and pen shells were buried alive by deltaic sediments and not ripped up by storm currents and transported to a common site of deposition. There is a difference between fossil concentrations that represent life assemblages versus those that represent death assemblages. With the former paleontologists are provided a glimpse of ancient organisms as they lived with their associated community members and actual population densities. The latter preserves more of a jumble of organisms from a wider range of communities and habitats that were only brought together in death.

Some of the shell beds in the Deguynos Formation have produced rare remains of marine vertebrates including teeth of sharks and rays, as well as bones of marine mammals. Unfortunately, as was the case in the Latrania Formation, these vertebrate remains are fragmentary and do not include even partial skeletons. The fossils however, do give us a hint of the types of animals that did live in the turbid waters of the Pliocene Imperial Sea (see note below). Fossil bones of an extinct species of walrus, *Valenictus imperialensis*, recovered in the Painted Gorge and Fish Creek areas of the Carrizo and Vallecito Creek-Fish Creek Badlands are the only records of walrus from the proto-Gulf (Figures 2.11, 2.12, 2.13). Nearly complete skeletons of another species of *Valenictus* (*V. chulavistensis*) have been recovered from Pliocene-age sandstones near San Diego and preserve close anatomical similarities with the modern Arctic walrus, *Odobenus rosmarus*. Similarities include large tusks in the upper jaw, an arched and elongate palate, heavy muscle attachments at the back of the skull, and fused bony "chin." The recovery of these temperate to subtropical walruses in Pliocene rocks of San Diego and the Imperial Valley supports the hypothesis that the ice-loving aspect of modern *Odobenus rosmarus* is the result of fairly recent evolutionary events of the Pleistocene.

Isolated fossil vertebrae of small dolphins have recently been recovered from the Fish Creek exposures of the Deguynos Formation and represent the first record of Pliocene odontocete (toothed whale) cetaceans from the Proto-Gulf (Figures 2.14, 2.15). These new discoveries underscore the fact that there is still much to learn about the marine vertebrate fauna of the Imperial Sea.

The marine sedimentary rocks of the Deguynos Formation are overlain by and interfinger with mid-Pliocene age (3.5 to 4.0 Ma) playa lake and river deposits of the Palm Spring Group. The transition between these two geological units records the final period of silting of the head of the Imperial Sea and of regression of the proto-Gulf shoreline to near its modern position south of the international boundary. The present-day Laguna Salada in northeastern Baja California south of Mexicali serves as a useful model of the depositional setting of the final sediments of the Imperial Group and the initial sediments of the Palm Spring Group (see Dorsey, this volume).

Today the Imperial Sea is only a memory, a memory preserved in the Miocene and Pliocene sedimentary rocks of the

Figure 2.12
Valenictus imperialensis.
Fragmentary femur (upper leg bone) of the extinct Imperial walrus collected from the Deguynos Formation, East Mesa (ABDSP 2441/V6747.01). (Photograph by Barbara Marrs)

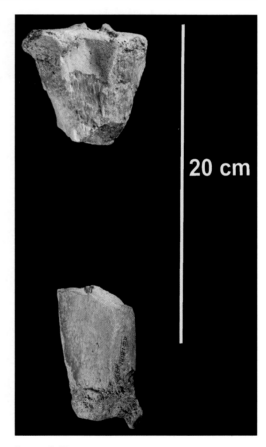

Figure 2.13
Valenictus imperialensis.
Left tibia (lower leg bone) of the extinct Imperial walrus (ABDSP 2441/V6747.02) collected from the Deguynos Formation, East Mesa. (Photograph by Barbara Marrs)

Anza-Borrego Desert State Park and adjoining federal and private lands. The opportunity to tap into this memory through a firsthand encounter with the exposed geological and paleontological record is worth the trip. The experience can be life altering if you come away with some sense of the vastness of geological time and with the realization that the current world is just a snapshot of a dynamic and evolving sequence of past and future worlds.

Figure 2.14
Dolphin Vertebra.
Side view of a weathered lumbar (lower back) vertebra from a fossil dolphin (ABDSP 2442/V6746) collected from the Deguynos Formation, East Mesa. (Photograph by Barbara Marrs)

Figure 2.15
Dolphin Vertebra.
Front and side views of a caudal (tail) vertebra from a fossil dolphin (ABDSP 2441/V6745) collected from the Deguynos Formation, East Mesa. (Photograph by Barbara Marrs)

The Ethics of Fossil Collecting

The fossil-bearing deposits of our local desert generally occur on public lands, either under U.S. Bureau of Land Management or California State Parks control. Collection of fossils on these lands is strictly regulated and research permits are required.

In this age of explosive human populations, we need to realize that the pioneer mentality of our forebears has to be tempered by thoughts of the needs of future generations. We can no longer lay claim to all we see and in the case of fossils, we should be content to observe their occurrence without requiring a souvenir.

I have seen areas where fossils were formerly abundant almost completely stripped of them by curious visitors. Unfortunately, in most cases the collected fossils were probably quickly forgotten or later thrown out when the collector grew tired of the clutter. Even museums need to be content in some cases to "only take pictures" and leave an area intact for future visitors. Please keep these thoughts in mind during your next visit to the desert and enjoy the geologic experience.

3 Invertebrates of the Imperial Sea

One could not ignore a granite monolith in the path of the waves. Such a rock, breaking the rushing waters, would have an effect on animal distribution radiating in circles like a dropped stone in a pool.

John Steinbeck and Edward F. Ricketts, in *Sea of Cortez*, 1941

Thomas A. Deméré
N. Scott Rugh

Invertebrates of the Imperial Sea

Tilted Shell Beds in the Deguynos Formation, Painted Gorge Area. (Photograph by Tom Deméré)

Introduction

The fossil marine invertebrate assemblages recovered from sedimentary rocks of the late Miocene Imperial Group (see Deméré, this volume, *The Imperial Sea*) have, for many years, attracted the attention of professional and avocational paleontologists, as well as casual collectors. Perhaps it is the stark contrast between the dry and rugged desert setting and the tropical looking, "petrified" shells that makes the Imperial Group fossils so appealing. In any event, the fossils from this rock unit are locally abundant and provide an excellent opportunity to study the life of the ancestral Gulf of California.

The last time that marine waters flooded this part of southern California was during the early part of the Pliocene Epoch four to five million years ago. Ironically, it was the Colorado River with its tremendous volume of water and high sediment load that spelled the death of the northern portion of the ancestral Gulf of California (see Deméré, this volume; Dorsey, this volume, *Stratigraphy, Tectonics, and Basin Evolution in the Anza-Borrego Desert Region*). However, during the late Miocene and early Pliocene, before this portion of the proto-Gulf dried up, its waters supported diverse assemblages of marine invertebrates including an impressive variety of corals, molluscs (clams and snails), crustaceans (barnacles, crabs, and shrimp), and echinoderms (sand dollars, sea urchins, and sea stars).

Although many of these fossils represent species that still live today in the Gulf of California, a surprising number also represent species or genera surviving only in the Caribbean Sea. This strong Caribbean component is a historical artifact of a time when the equatorial eastern Pacific Ocean, the western Gulf of Mexico, and the Caribbean Sea were all linked to form an unbroken tropical region – the Tertiary Caribbean faunal province. At this time during the late Miocene and early Pliocene, the Isthmus of Panama did not exist and, instead, the Central American Seaway occupied the region of Panama and Costa Rica (Figure 3.1). Surface ocean currents of the eastern Pacific (Equatorial Countercurrent) and Caribbean (Caribbean Current) presumably flowed west to east and east to west, respectively, through the Central American Seaway and provided a means for dispersing larvae of various groups of marine invertebrates. Counter-

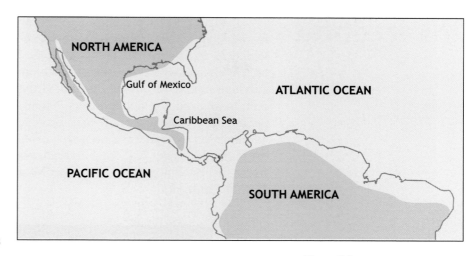

Figure 3.1
Central American Seaway During the Late Miocene.
The gold depicts the North and South American continents before they were joined forming the Isthmus of Panama about 3.5 million years ago (see McDonald, this volume, *Anza-Borrego and the Great American Biotic Interchange*).

currents flowing north along the west coast of Mexico, in turn allowed dispersal of Tertiary Caribbean species into the proto-Gulf and vice versa (see Deméré, this volume).

A thick sequence of sedimentary rocks accumulated during this time period in the region of the modern day Imperial Valley. Geologists call these rocks the Imperial Group and have subdivided them into an older series of coarse-grained sandstones and conglomerates called the Latrania Formation, and an upper series of claystones and fine-grained sandstones called the Deguynos Formation (see Dorsey, this volume). The Latrania Formation was deposited during the late Miocene, prior to the beginning of deposition of delta sediments of the ancestral Colorado River. The fine-grained delta deposits of the Deguynos Formation overlie the sandstones of the Latrania Formation.

Mode of preservation is an important aspect of the invertebrate fossils from the Imperial Group, with the majority of specimens represented not as intact original crystalline shell material, but as hardened sandstone molds preserving either the internal or external form of the shell. The original mineralogy of the shells and/or tests of the fossils is largely responsible for determining their mode of preservation. Fossil remains consisting of original shell/test material typically represent organisms with a higher percentage of $MgCO_3$ (magnesium carbonate) in their shells/tests. Organisms that fall into this group include oysters, spiny oysters, scallops, barnacles, and sand dollars. In contrast, shells/tests constructed with little or no $MgCO_3$ are more likely to be leached away by percolating ground waters and/or under the harsh physical and chemical conditions of the desert region. The result is that the original shell/test material is selectively removed leaving a void in the hardened sandstone. If a fossil specimen is the impression of the outside of the shell it is called an *external mold*. If instead the fossil is an impression of the inside of the shell it is called an *internal mold* (Figure 3.2). The German word "steinkern" is often used for three-dimensional internal molds, such as the internal sandstone mold formed within a snail shell or articulated pair of clam shells.

This chapter presents an overview of the more common Imperial Group marine invertebrates (corals, molluscs, and echinoderms) and discusses their temporal, environmental, and geographic distributions. Other groups of invertebrates (e.g., sponges, bryozoans, brachiopods, foraminifers, and crustaceans), although reported from the Imperial Group (Appendix, Table 2), are not common fossils and will not be treated herein. Thus, this chapter is not meant as a comprehensive treatment of Anza-Borrego's invertebrate fossils but will provide a foundation for a future monograph of this marvelous fauna. It is hoped, however, that the reader will come to realize some measure of the diversity of life forms that once flourished in the lost Miocene and Pliocene seas of the proto-Gulf of California.

Figure 3.2
Malea ringens.
An example of a fossil gastropod preserved as an internal mold, or steinkern (SDSNH 12234). (Photograph by Barbara Marrs)

Figure 3.3
Solenosteira sp.
An example of a fossil gastropod preserved with original crystalline shell material (ABDSP 1327). (Photograph by Barbara Marrs)

Cnidaria (corals)

The Imperial Group fossil coral assemblage consists of seven to nine colonial species representing both encrusting and branching forms. All belong to the Order Scleractinia or stony corals, which today are important reef builders in tropical seas. The majority of these corals are known for their symbiotic relationship with photosynthetic zooxanthellae (unicellular yellow-brown algae) that live within the body tissues of the coral.

Typically, the Imperial Group coral specimens occur as fragmentary coral heads (coralla) that were transported by ocean currents to a common site of deposition and concentrated together to form dense fossil shell beds. There are, however, rare instances where specimens are preserved in life position as *in situ* coral heads or small patch reefs still firmly attached to the ancient sea floor bedrock. The vast majority of confirmed records of Imperial fossil corals have come from the Latrania Formation, with only a single species reported from the Deguynos Formation. Among the genera known from the Latrania Formation (Vaughan, 1917), only one, *Porites*, is represented by a species that still lives today in the Pacific Ocean region. All of the other genera are either extinct today or have been extirpated (localized extinction) from the region since at least the early Pleistocene. In the latter case, the fossil taxa are primarily represented by species living today only in the western Atlantic Ocean and adjacent regions (i.e., Caribbean Sea).

Figure 3.4
Examples of Fossil Corals from the Imperial Group.
A. *Dichocoenia merriami* (SDSNH 11953),
B. *Siderastrea mendenhalli* (SDSNH 11964),
C. *Astrangia haimei* (ABDSP 1325),
D. cf. *Meandrina* sp. (SDSNH 11935), and
E. *Solenastrea fairbanksi* (SDSNH 11966).
(Photographs by Barbara Marrs)

Meandrinidae (brain and flower corals)

There are four species of meandrinid corals from the Latrania Formation. Perhaps the most beautiful of the Imperial fossil corals, *Dichocoenia merriami* (Figure 3.4A) is relatively rare and characterized by large irregular corallites (skeleton of individual polyp) up to

10 mm (0.4 in) across and distinctly separated from neighboring calices by complex winding valleys. The fossil species is similar to the extant *D. stokesi* from the Caribbean Sea. A related fossil species from the Latrania Formation, *Meandrina bowersi* is a small extinct brain coral similar to the extant *D. labyrinthiformis* from the Caribbean Sea. In brain corals the individual corallites are fused into long meandering skeletal rows resembling wrinkles of a human brain. The Latrania Formation species of *Meandrina* (Figure 3.4D) has coarse septa (radially positioned walls within a corallite) and irregular, elongate corallites measuring up to 15 mm (0.6 in) in length.

Eusmilia carrizensis is an extinct species of branching coral originally named from the Imperial fossil beds of the Coyote Mountains. Specimens of this species are relatively rare and often consist of single branches with paired, terminal corallites 13-17 mm (0.5-0.7 in) in size. Septa are discontinuous and there is no central columella (central crystalline column within a corallite). The Latrania Formation fossil species resembles *E. fastigiata* from the Caribbean Sea.

Siderastreidae (starlet corals)

Siderastrea mendenhalli (Figure 3.4B) is a relatively common fossil coral from the Latrania Formation with massive coralla up to 300 mm (12 in) across; although most are much smaller. This coral is distinguished by its closely crowded polygonal corallites (4-7 mm in diameter; 0.2-0.5 in), which often are deformed because of their tight packing. The septa in each corallite are very thin and form a delicate star-like pattern. Although Vaughan (1917) named two fossil species of *Siderastrea* and one variety from the Imperial fossil beds, most workers today only recognize the one species, *S. mendenhalli*. This extinct species is similar to the extant *S. siderea* from the Caribbean Sea.

Rhizangiidae (cup corals)

The Deguynos Formation typically is devoid of fossil coral specimens, a pattern apparently reflecting the more turbid nature of the deltaic portion of the Imperial Group deposits. However, there is one species of coral, *Astrangia haimei* (Figure 3.4C), which has been reported from the Deguynos Formation as exposed in the East Mesa area of the Anza-Borrego Desert State Park. This coral is distinguished by its separate and elongate, cylindrical corallites, which typically extend above the main mass of the corallum.

The occurrence of this more typically temperate genus of stony coral in the Pliocene portion of the Imperial Sea indicates that paleoenvironmental conditions for the Deguynos Formation were different from those of the Latrania Formation. Today, most members of this family lack symbiotic zooxanthellae perhaps indicating that their tolerance for lower light levels is greater than for the true zooxanthellate corals, which occur in the clear water sandstone facies of the Latrania Formation.

Faviidae (star corals)

Solenastrea fairbanksi (Figure 3.4E) is by far the most common fossil coral from the Latrania Formation. Specimens typically occur as large fragmentary coral heads up to 400 mm (16 in) across with distinctly separate polygonal corallites only 2-3 mm (0.1 in) in diameter. Specimens often occur with corallites covering all exposed surfaces, indicating that the coralla were periodically dislodged from the sea floor and rolled about by currents to expose new growth surfaces. This is in contrast to having a fixed coral head growing continuously upwards from its base. The Latrania Formation fossil species is similar to *S. hyades* from Pleistocene rocks of Florida and the living *S. bournoni* from the West Indies.

Poritidae (finger corals)

Six species of *Porites* are still living in the Gulf of California today. Fossil specimens from the Latrania Formation are assigned to *P. carrizensis* and typically occur as small rounded bioclasts (40-60 mm, 1.6-2.4 in) in sandy limestone strata. The surface of the coralla appear smooth because the polygonal corallites of this species are almost microscopic (1-1.5 mm in diameter, 0.06 in) and are very closely crowded together. The septa are relatively thick for such a small corallite. The fossil species from the Coyote Mountains is similar to the extant *P. astreoides* from the Caribbean Sea.

Mollusca (bivalves, clams, and snails)

Molluscs are by far the most diverse group of marine invertebrate fossils from the Imperial Group and were first described by Hanna (1926). Currently there are over 65 species of gastropods and over 60 species of bivalves reported as fossils from these rocks (see Appendix, Table 2). The composite molluscan assemblage from the Latrania Formation is more diverse than that from the Deguynos Formation. The former contains nearly equal numbers of gastropods and bivalves, while the latter is dominated by species of bivalves. Fossil molluscs in the Latrania Formation often occur in well-cemented sandstones as dense concentrations of shells and molds. Sometimes fossils occur in sandy limestones, while other occurrences consist of thick massive sandstones with widely dispersed shells and/or molds. Fossil molluscs in the Deguynos Formation generally occur in well-cemented shell coquinas, which due to their high calcium carbonate volume (shells) often form resistant shelly strata capping hogbacks and ridgelines in the badlands of the Anza-Borrego Desert State Park and adjacent Bureau of Land Management lands.

Fossil preservation in the Latrania Formation is variable and includes fossil shells with original calcite or recrystallized calcite and in some cases even original aragonite (carbonate mineral). In rare cases some specimens even retain original coloration (e.g., cones, scallops, and oysters). The most common style of preservation consists of internal and external sandstone molds with all original

Figure 3.5
Gastropods.
Preserved with original
crystalline shell material.
A. *Architectonica nobilis*
(SDSNH 23148),
B. *Nerita scabricosta*
(SDSNH 53865),
C. *Turritella imperialis*
(SDSNH 97265),
D. *Cancellaria obesa*
(SDSNH 12142),
E. *Oliva spicata*
(SDSNH 12332), and
F. *Vasum pufferi*
(SDSNH 23112).
(Photographs by Barbara Marrs)

shell material leached away (e.g., the venerid and lucinid clams and strombid and cypraeid snails).

Gastropoda (snails)

Most gastropods reported from the Imperial Group have been collected from sandstone strata in the Latrania Formation. The majority of these fossil snails are species that lived in and on a sandy substrate at shallow marine depths. A very few, including *Nerita scabricosta,* lived in rocky areas. Species of the Family Calyptraeidae lived attached to solid objects, such as the surfaces of living and/or dead clam and snail shells lying on the sandy sea floor.

Although a number of the Imperial Group fossil snails represent species still found in the Gulf region today, many are actually very similar or nearly identical to living species from the Caribbean Sea. A few examples include *Melongena patula, Pleuroploca princeps,* and *Vasum pufferi.*

While some gastropod species are typically preserved as sandstone steinkerns, including species of *Strombus* and *Macrocypraea,* many of the gastropod species are frequently preserved with some or all of the original crystalline shell material intact. Species that are most commonly preserved as steinkerns are thin shelled taxa, such as *Malea ringens* and species of *Bulla.* Specimens that have been preserved with the entire original shell intact can be very beautiful. A few examples of these are: *Turritella imperialis, Oliva spicata,* and several of the *Conus* species with original shell ornamentation and even color patterns preserved.

The discussions below focus on those gastropod families best represented by fossils in sedimentary rocks of the Imperial Group. Several families have been omitted from this listing because species attributed to them are either very rare or the original identifications are incorrect. Omitted families include Haliotidae, Lottiidae, Turbinidae, Potamididae, Cerithiidae, Modulidae, Littorinidae, Hipponicidae, Personidae, Muricidae, Columbellidae, Mitridae, Turridae, Fissurellidae, Trochidae, Calyptraeidae, Ficidae, Terebridae, and Bullidae (see Appendix, Table 2).

Neritidae (nerites)

Fossil nerites collected from the Latrania Formation include relatively large (for the Family) specimens 38-45 mm (1.5-1.8 in) in diameter, preserved as original shell material with numerous fine spiral lines. These large specimens have been identified as *Nerita scabricosta* (Figure 3.5B), an extant species living today along the Pacific coast of Baja California Sur throughout the Gulf of California and south to Ecuador. Other nerites reported as fossils from the Latrania Formation include *N. funiculata*, an extant species living from the west coast of Baja California throughout the Gulf of California and south to Peru, and *Theodoxus luteofasciatus*, an extant species living in the Gulf of California and south to Peru.

Turritellidae (turret shells)

Of the eleven species of turret shells living in the Gulf of California today, only *Turritella gonostoma* has been reported as a fossil in the Imperial Group. This record, however, cannot be confirmed. Additionally, the only other fossil turritellid reported from these rocks is the extinct species, *T. imperialis* (Figure 3.5C), first described from the Coyote Mountains. Shells of *T. imperialis* have been primarily collected from the sandstone and limestone strata of the Latrania Formation. However, certain strata within the mudstone beds of the Deguynos Formation contain dense concentrations of this distinctive fossil snail. *T. imperialis* is characterized by its two heavy spiral ridges, one at the shoulder, and one at the base of each whorl (one complete turn of the spiraled shell). Adult specimens are long and slender and can measure up to 145 mm (5.7 in) in length with a diameter of only 25 mm (1 in). *T. imperialis* is similar to *T. altilira* from the Miocene Gatun Formation of Panama.

A partially uncoiled member of the family Turritellidae, *Vermicularia pellucida eburnea*, is locally common in sandy limestone beds of the Latrania Formation. Fossil specimens cannot be separated from modern shells of the extant species, which lives today from southern California throughout the Gulf of California and south to Panama.

Strombidae (conch shells)

Fossils representing at least three different species of this group of large gastropods occur in the Latrania Formation. Although occasionally specimens are found that preserve original crystalline shell material, most fossil strombids from this region occur as internal and external sandstone molds. Two fossil forms are referable to extant taxa still living in the Gulf of California. Fossils of *Strombus gracilor* measure 70 to 85 mm (2.8-3.4 in) in length with a short flaring lip and small pointed nodes on the shoulder of the final whorl. This species still lives from the Gulf of California south to Peru. *Strombus pugilis* is a closely

Figure 3.6
Conch Species.
A. *Strombus obliteratus.* Juvenile steinkern (SDSNH 12155),
B. *S. obliteratus.* Front view of specimen shown in A., and
C. *Strombus galeatus.* Steinkern (SDSNH 12148).
(Photographs by Barbara Marrs)

related Caribbean species. The largest strombid from the Latrania Formation is *S. galeatus* (Figure 3.6C), a species that still lives today in the Gulf of California. Some fossil specimens measure up to 150 mm (6 in) in length with a large flaring outer lip that is often only partially preserved. It should be noted that the steinkern of juvenile fossil specimens is conical in shape, lacking the flaring lip of the adult, and may be mistaken as a fossil of a large *Conus*. *Strombus goliath* is a closely related Atlantic species. A third species, *S. obliteratus* (Figures 3.6A, B), is extinct and was originally named from fossils collected from the Latrania Formation. This fossil species is closely related to the extant *S. granulatus* and is characterized by its relatively small size (50-60 mm, 2-2.4 in) and robust rounded nodes on the shoulder of the final whorl.

Cypraeidae (cowries)

Fossil cowries are uncommon from the Latrania Formation and are unknown from the Deguynos Formation. In the Latrania Formation, specimens typically occur as sandstone steinkerns with minor amounts of crystallized shell material preserved. For this reason, identification to species level is difficult. However, there are at least two different fossil taxa recognized. One is tentatively identified as *Macrocypraea* sp. cf. *M. cervinetta*, a species living today from Sonora, Mexico, to Peru and Islas Galapagos. Known fossil specimens of this species may reach 50 mm (2 in) long, but living specimens may reach 115 mm (4.5 in) in length. The second type of fossil cowrie from the Latrania Formation is represented by large, globose sandstone steinkerns that reach 70 mm (2.8 in) in length and 50 mm (2 in) in diameter. These cowries are assigned to the genus *Muracypraea* (Figure 3.7), but specific identification is hampered by the lack of original shell material. This genus is extinct in the Gulf of California today. However, *Muracypraea* is known throughout the Caribbean Basin, Panama, Ecuador, Peru, Costa Rica, and Baja California Sur, Mexico, from the early Miocene to recent. *Muracypraea* specimens from the Latrania Formation may represent a new species (Arnold, 1998).

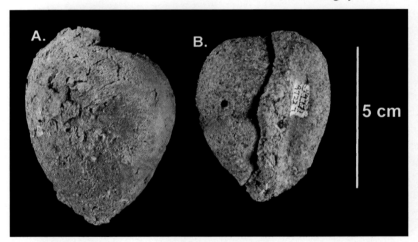

Figure 3.7
Muracypraea sp.
Steinkerns:
A. SDSNH 88085 and
B. SDSNH 97227.
(Photographs by Barbara Marrs)

Naticidae (moon shells)

Fossil naticids are locally common in the Latrania Formation and are usually preserved with crystalline shell material. The two dominant genera are *Polinices* and *Natica*. Specimens of *Natica* are more globular in shape than those of *Polinices*. However, reliable identification is only possible if the umbilical callus (thickened shelly area) is present and this feature may be missing if the crystalline material has crumbled or if only the sandstone steinkern has been preserved. Although *N. chemntizii* is the most common living *Natica* found in the Gulf of California today, it is rare as a fossil in the Imperial Group. The two most common species of fossil *Polinices* can be difficult to distinguish (even when the callus is present). *Polinices uber* is an extant species living along the west coast of Baja California throughout the Gulf of California and south to Peru. This species may be related to *P. stanislasmeunieri* from the Miocene of the Caribbean (Woodring, 1957). The generally larger *P. bifasciatus* is extant and lives from the Gulf of California to Panama. Fossil naticids from the Latrania Formation are small, generally not exceeding 30 mm (1.2 in), but specimens of *P. bifasciatus* may reach 40 mm (1.8 in) in length.

Cassididae (helmet shells)

Specimens of the extinct species *Cassis subtuberosa* are occasionally found in the Latrania Formation as sandstone internal molds. These fossils are conical in shape and easily may be confused with internal molds of *Conus fergusoni*, the largest fossil cone from the Imperial Group. *Cassis* may be distinguished from *Conus* by the presence of a few widely spaced vertical rows of shallow depressions on the fossil, which correspond to the "teeth" that run along the inside edge of the outer lip of the aperture (opening in shell from which body protrudes). The fossil species is small, about 60 mm (2.4 in) long by 70 mm (2.8 in) in diameter. Most of the closest living relatives are found in the Indo-Pacific, Atlantic Ocean, and the Caribbean Sea, none survive today in the eastern Pacific Ocean. These species have a thick outer lip, and most reach impressively large sizes. The largest, *Cassis cornuta* of the Indo-Pacific, can reach 387 mm (15 in) in length.

Tonnidae (tun shells)

Malea ringens (Figure 3.8) is a locally common fossil gastropod in the Latrania Formation and is typically found as sandstone internal molds (steinkerns) up to 70 mm (2.8 in) in length and 50-65 mm (2-2.6 in) in diameter. This is considerably smaller than the maximum shell size (up to 240 mm, 9.5 in) of living *M. ringens,* which today occurs from the Gulf of California south to Peru. Modern shells are relatively thin, which probably contributes to its loss by dissolution in fossil forms. Steinkerns typically preserve distinctly separated fine spiral lines and some specimens even preserve impressions of the thickened and toothed outer aperture lip.

Figure 3.8
Malea ringens.
Steinkerns:
A. SDSNH 103648 and
B. SDSNH 12234.
(Photographs by Barbara Marrs)

Turbinellidae (vase shells)

Fossil shells referable to this group of tropical neogastopods are rare from the sandstone strata of the Imperial Sea. They are represented by only a single species, *Vasum pufferi* (Figure 3.5F), characterized by its heavily built shell with broad nodous shoulder, thick spiral ridges, strong columellar folds, and flattened spire. This extinct species was originally named from the Latrania Formation as exposed in the Coyote Mountains. Specimens of this fossil species were formerly misidentified as *V. caestum* by Hanna (1926), an extant species living today in the Gulf of California and south to Ecuador. The fossil species, however, more closely resembles the living *V. muricatum* from the Caribbean Sea.

Buccinidae (whelks)

There are at least two species of buccinid snails found as fossils in the Deguynos Formation. Specimens typically are preserved with crystalline shell material intact. *Solenosteira anomala* is an extant species living along the Pacific coast of Baja California, throughout the Gulf of California south to Ecuador. This species has a coarse sculpture of axial ribs crossed by strong spiral chords, and grows to about 40 mm (1.6 in) in length. By contrast, the second species has a relatively smooth shell, with no distinct axial ribs, and evenly spaced spiral cords (small ridges). This species, which reaches 50 mm (2 in) in length, may represent a new and undescribed form (Figure 3.9).

Melongenidae (crown conchs)

The only species from this family found as a fossil in the Imperial Group is *Melongena patula*. This very large species – fossils over 100 mm (4 in) in length are common – may be found living from the Gulf of California south to Panama. Because this species is so large and shaped much like a large conch with a wide, open aperture, fossils of *M. patula* are often mistaken for fossils of *Strombus galeatus*. By comparison, *M. patula* has a lower spire than *S. galeatus,* and has a row of knobs at the shoulder. In contrast *S. galeatus* has a continuous, very low-lying bulge at the shoulder.

Fasciolariidae (horse conchs and tulip shells)

Representative taxa of this family are characterized by a very long siphonal canal (narrow channel near aperture), particularly in *Pleuroploca* and *Fusinus,* the most common genera that occur in the Latrania Formation. *Pleuroploca princeps* (Figure 3.10A, B) is the largest fossil gastropod from the

Figure 3.9
Solenosteira sp.
Possibly a new species
(ABDSP I326).
(Photograph by Barbara Marrs)

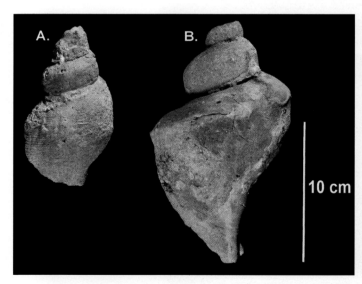

Figure 3.10
Pleuroploca princeps.
Steinkerns:
A. Smooth form
(SDSNH 12174) and
B. Shouldered form
(SDSNH 12183).
(Photographs by Barbara Marrs)

Imperial Group; the large sandstone steinkerns can be up to 162 mm (6.4 in) long and 60-90 mm (2.4-3.5 in) in diameter. This species lives in the Gulf of California today and is closely related to the Caribbean species *P. gigantea*. Although several extant species of *Fusinus* are common in the Gulf of California today, only one is common as a fossil in the Latrania Formation, *F. dupetitthouarsii*. Internal sandstone molds of this species are usually missing most of the siphonal canal. Incomplete specimens can be 80 mm (3.2 in) in length or larger.

Olividae (olive shells)

A number of extant species of *Oliva* and *Olivella* are found as fossils in the Latrania Formation, usually with the crystalline shell present. *Oliva spicata* is the most common fossil olivid and is today found living in the Gulf of California south to Panama. Fossils of this species are often beautifully preserved and reach up to 42 mm (1.7 in) in length. Two other species, *O. porphyria*, living today from the Gulf of California to Panama, and *O. incrassata*, found today from the Gulf of California to Peru, are less common as fossils in the Imperial Group.

It might seem that identification of species of *Oliva* would be difficult, because as fossils they appear to be very similar. However, if shell material has been preserved, some characteristics may be very helpful in identifying the species. In *O. spicata* (Figure 3.5E) the spire is elevated and the sides of the shell are nearly parallel. The spire of *O. incrassata* is also elevated, but the outer aperture lip angles out from the top edge, and is greatly thickened. *Oliva porhyria* has a very low spire with only the very earliest part of the whorls elevated. The latter two species are larger than *O. spicata*. Fossils of *Olivella* have also been reported from the Imperial Group, but are uncommon. Compared to *Oliva* species, shells of *Olivella* are usually much smaller and have a more elevated spire.

Cancellariidae (nutmeg shells)

Characteristic features of shells of this family include strong folds on the collumella and ornate decoration on the outer surface, often with knobs on the shoulders and strong reticulate, net-like sculpture. *Cancellaria obesa* (Figure 3.5D) is a relatively large species common today from Magdelena Bay, Baja California Sur throughout the Gulf of California and south to Ecuador. This species is also locally common as a fossil in the Latrania Formation and is usually preserved with the crystalline shell intact. Individual fossils are up to 30 mm (1.2 in) in length, while living specimens may grow to be nearly 60 mm (2.4 in) in length. Many species of *Cancellaria* live in the Gulf of California, and in addition to *C. obesa*, several species may be found as fossils from the Imperial Group. Identification, however, may be difficult due to the close similarity of surface sculpture between some species.

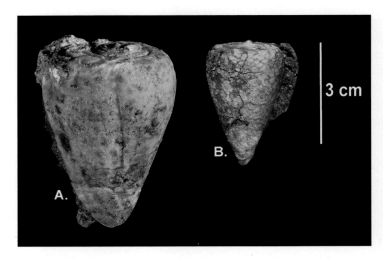

3 cm

Figure 3.11
Extinct Cone Shell Species.
Preserved with crystalline shell
and color patterns:
A. *Conus bramkampi*
    (SDSNH 23257), and
B. *Conus durhami*
    (SDSNH 103738).
(Photographs by Barbara Marrs)

Conidae (cone shells)

There are several species of fossil cones reported from the Latrania Formation, including extant and extinct forms. It is not uncommon to find specimens that are preserved with original crystalline shell material. However, this shell layer is fairly thin in relation to the underlying sandstone steinkern, and frequently, much of this fragile crystalline material breaks away. Specimens preserved only as steinkerns or with poorly preserved shell are extremely difficult to identify to species level because all cones have a superficially similar form. Therefore, the rare fossil specimens with all, or virtually all, shell material preserved are extremely important because they preserve such features as spire shape and surface sculpture, and even color patterns in some specimens.

The most common fossil species from the Latrania Formation are *Conus fergusoni,* and *C. regularis,* both still living today from the Gulf of California to Panama. These two species are very similar in form with elevated spires, so it is possible to misidentify one as the other. The less common *C. ximenes, C. arcuatus,* and *C. patricius* are also reported as fossils from the Imperial Group, however, these identifications need to be confirmed by detailed study. At least two extinct fossil cones have been named from the Latrania Formation. One is *C. bramkampi,* with a beautiful shell up to 53 mm (2.1 in) long and 35 mm (1.4 in) in diameter. This species is often preserved with a color pattern of small dark spots on a white shell. Unless the color pattern is present, this species could easily be confused with *C. fergusoni* or *C. regularis.* Another extinct fossil cone named from the Latrania Formation is *C. durhami.* This rare, low-spired fossil cone can be up to 50 mm (2 in) in length and specimens have been found preserved with a color pattern of white spots on a dark colored background.

Architectonicidae (sundial shells)

A rare, but beautifully ornamented fossil gastropod from the Latrania Formation, *Architectonica nobilis discus* (Figure 3.5A) represents an extinct subspecies of the living species, *A. nobilis,* which lives today from the Gulf of California south to Panama. This species also survives in the Gulf of Mexico. Fossil shells of this snail are generally preserved as original shell material and can measure up to 35 mm (1.4 in) in diameter. They are characterized by their very low, disk-like, flattened spire.

Bivalvia (oysters, scallops, and clams)

Most of the 60 or more species of bivalves known from the Imperial Group, are from the Latrania Formation. Like the gastropods, most of the bivalves are species that lived in or on a sandy substrate. Only a few species lived attached to solid objects. Attached species include *Spondylus bostrychites, Arcinella californica,* and a few species of oysters (Family Ostreidae and Family Gryphaeidae). While these species might seem to indicate the presence of a rocky habitat, it is more likely that attaching forms settled on empty mollusk shells lying on the ancient sea floor.

Dendostrea vespertina and *Anomia subcostata* are the dominant species from the Deguynos Formation, occurring in strata of substantial thickness and incredible numbers of individuals. Other species, also characteristic of tropical estuaries are *Pinna* and *Atrina,* and the pholad *Cyrtopleura costata.* While *C. costata* is extinct in the Gulf of California today, it still exists along the Atlantic coast in estuaries from Massachusetts to Brazil.

Another fossil bivalve with an Atlantic coast relative is *Arcinella californica,* which is nearly identical to *A. arcinella,* a species common in the Caribbean Sea today. *Arcinella californica* also may be found living in the Gulf of California.

The discussions below focus on those bivalve families best represented by fossils in sedimentary rocks of the Imperial Group. Several families have been omitted from this listing because fossils attributed to them are either very rare or the original identifications are incorrect. These omitted families include Nuculidae, Nuculanidae, Glycymeridae, Limidae, Corbiculidae, Semelidae, Donacidae, Corbulidae, Carditidae, Cardiidae, and Myidae (see Appendix, Table 2).

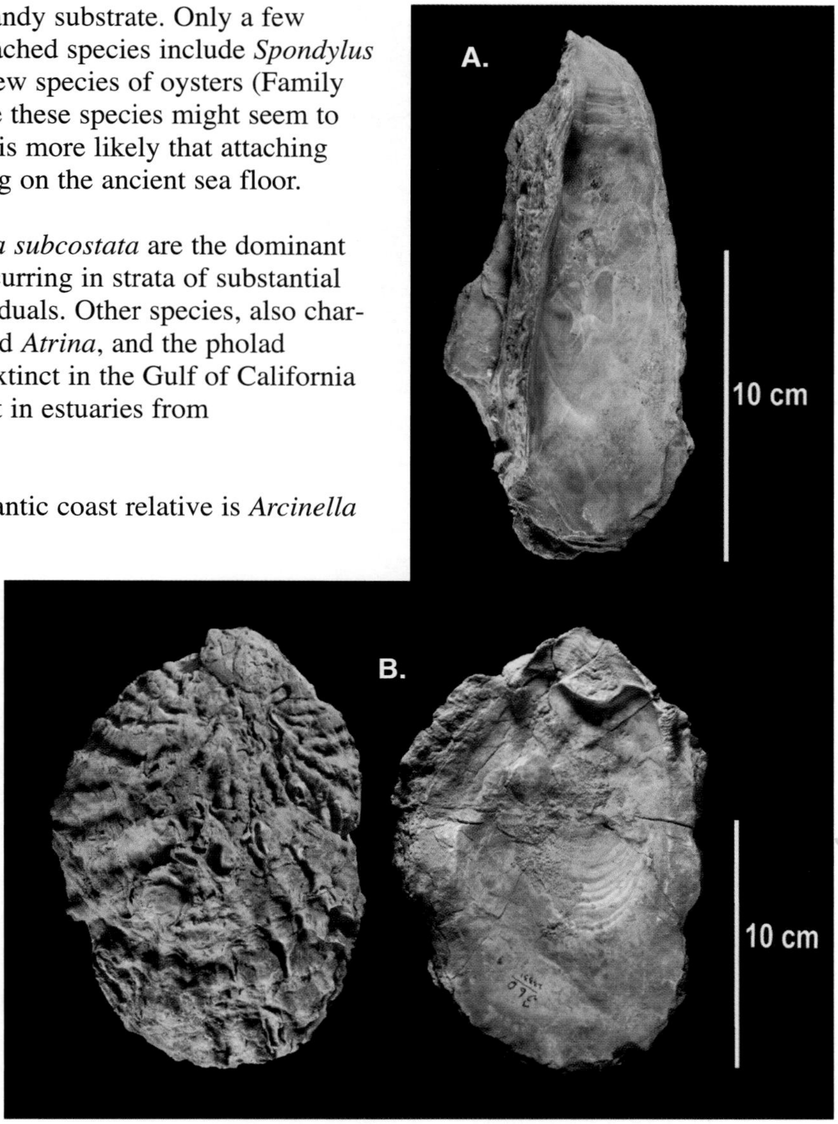

Figure 3.12
Fossil Oyster Species.
Both oyster species are preserved with original crystalline shell.
A. *Ostrea iradescens,* left valve (SDSNH 23101), and
B. *Pycnodonte heermani,* left valve and right valve of a pair (SDSNH 23331).
(Photographs by Barbara Marrs)

Arcidae (ark shells)

Ark shells identified as fossils from the Imperial Group include several species that still live today in the warm waters of the Gulf of California and south to Peru. *Arca pacifica* and *Barbatia reeveana* are intertidal species, while *Anadara formosa* (Figure 3.16C) is a subtidal species. *Anadara carrizoensis* Reinhart is an extinct species similar to *A. formosa,* but more closely resembling *A. secticostata* from the western Atlantic Ocean and Caribbean Sea. Sandstone steinkerns of these species are characterized by sharply angular hinge lines (top edge of valve where hinge teeth are located) and numerous tiny hinge teeth. *Arca pacifica* is an extant species from the Gulf of California similar to the extant *A. zebra* from Caribbean Sea. Fossil specimens from the Latrania Formation are large, measuring 52 mm (2 in) high by 72 mm (2.8 in) long.

Mytilidae (mussels)

Of the many species of mussels living in the Gulf of California, only one species, *Lithophaga* sp.cf. *L. plumula* has been reported as a fossil from the Imperial Group. This taxon is usually preserved as small (20-30 mm long, 0.8-1.2 in) cylindrical sandstone steinkerns embedded within fossil coral heads. The fossils occur in corals because in life *Lithophaga* bores into thick mollusc shells, corals, and/or rock. A species similar to *L.* sp. cf. *L. plumula* is *L. antillarum* from the Caribbean Sea.

Pinnidae (pen shells)

The largest bivalves known from the Imperial Group belong to the family Pinnidae. Shells of *Pinna latrania* measure up to 264 mm (10.4 in) in length and are typically found as large sandstone steinkerns of articulated valves (both valves still attached at the hinge). Often only the narrow apical portions (tip of hinge) (apical angle of 35°) of the original prismatic crystalline shell is preserved. These fossils are smaller than the extant *P. rugosa* which lives intertidally on protected mudflats from the Gulf of California south to Panama. A second species of extinct pinnid from the Imperial Group is *Atrina stephensi,* a smaller form with a broader apical angle (55°) first described from the mudstones of the Deguynos Formation. Fossil pinnids from the Latrania Formation typically occur as transported shells lying parallel to bedding, while specimens from the Deguynos Formation often occur in life position. A third species, *P. (P.) mendenhalli,* named from the Imperial Formation may be conspecific with *P. latrania.*

Ostreidae (true oysters)

One species of oyster is an exceedingly common fossil in the Imperial Group. *Dendostrea vespertina,* an extinct species originally named from exposures of this rock unit in Carrizo Creek, is a small oyster generally not exceeding 40 mm (1.6 in) in length. In the Deguynos Formation, *D. vespertina* is so abundant in some places it forms beds of solid shells 10-100 cm (3.9 in-3.3 ft) thick. Shells of this species may be recognized by the presence of a few widely spaced, distinct radial ribs. *Ostrea iridescens* (Figure 3.12A) is uncommon as a fossil but has been found in the Latrania Formation. Today it lives from La Paz, Baja California Sur, in the Gulf of California, to Peru. Living and fossil specimens on the average are 75-100 mm (3- 3.9 in) long. Fossils of this species are characterized by their long, narrow shells, many with a dark, metallic sheen inside the valves.

Gryphaeidae (pycnodont oysters)

Shells of gryphaeid oysters closely resemble the true oysters (Ostreidae) but tend to have upper and lower valves of equal size, whereas the true oysters generally have a flatter upper valve, and a more cupped lower valve. Also, gryphaeids have a flat area on the hinges with a U-shaped furrow in the center on both valves. In ostreids, the hinge on the lower valve has a concave furrow, and the hinge on the upper valve has a convex form. The one representative of this family from the Imperial Group, *Pycnodonte heermani* (Figure 3.12B) is by far the most common large bivalve in both the Deguynos and Latrania formations. This large oyster – specimens can be 150 mm (6 in) long – is locally abundant at some outcrops of the Deguynos Formation (e.g. Yuha Buttes). The shell texture, similar on both upper and lower valves, is a beautiful branching pattern of heavy, scaly ribs. The valves are very thick and heavy, some as thick as 40 mm (1.6 in).

Pectinidae (scallops)

Species of scallops make up the most diverse group of bivalves from the Imperial Group. Although most represent extinct species, the fossil scallops can be assigned to genera still found in the Gulf of California today. Some shell beds in the lower part of the Latrania Formation are nearly monospecific (composed of a single species) containing abundant shells of *Euvola keepi* (Figure 3.13F). This extinct species was first described from the Latrania Formation and is similar to the living *E. ziczag* from the West Indies. Shells of *E. keepi* measure up to 80 mm (3.2 in) in height and are characterized by their inflated upper valve and flat lower valve. The radial ornamentation consists of broad flattened ribs. The large scallop *Lyropecten tiburonensis* is an extinct form originally described from Miocene rocks on Isla Tiburon in the Gulf of California. Earlier workers assigned this Imperial Group pectinid to either *L. modulatus* or *L. subnodosus*. Distinguishing features include its size (up to 100 mm in length and 90 mm in height) and 11-12 broad, flattened, radiating ribs with fine ridges in the interspaces. *Flabellipecten carrizonensis* is another common fossil scallop from the Latrania Formation and is recognized by its inflated right valve with 18-19 radial ribs and flat left valve with 16-17 ribs. In contrast, both valves of *Argopecten mendenhalli* (Figure 3.13E) are only slightly convex and possess 19-20 radiating ribs. The *Anomia-Ostrea* shell beds of the Deguynos Formation often contain shells of a small extinct scallop, *Argopecten deserti* (Figure 3.13A-D). Individual shells of this species measure

Figure 3.13
Various Scallop Species.
A.-D. Valves of
    Argopecten deserti
    (SDSNH 12003),
E. *Argopecten mendenhalli* pair
    of valves (SDSNH 97281),
    and
F. *Euvola keepi* pair of valves
    (SDSNH 12030).
(Photographs by Barbara Marrs)

Figure 3.14
Paired Bivalves Preserved with
Original Crystalline Shell.
A. *Codakia distinguenda*
(SDSNH 12120),
B. *Eucrassatella subgibbosa*
(SDSNH 23122), and
C. *Spondylus bostrychites*
(SDSNH 23240).
(Photographs by Barbara Marrs)

Figure 3.15
Anomia subcostata.
(SDSNH 97260)
A. Left valve, and
B. Right valve.
(Photographs by Barbara Marrs)

only 30-40 mm (1.2-1.6 in) in height and are characterized by their equally inflated valves with 22-23 radiating ribs.

Plicatulidae (kitten's paw)

While a number of species of plicatulids exist today in the Gulf of California, only one species, *Plicatula inezana* occurs as a fossil in the Imperial Group. This species lives today from the southern Gulf of California to southern Mexico. The exterior sculpture of the valves, with numerous, scaly ribs, is remarkably similar to that of *Myrakeena angelica*, a medium-sized oyster (average length of 75 mm, 3 in) not known from the Imperial Group but extant in the Gulf of California, and is often misidentified as such. However, *P. inezana* has a very different "ball-and-socket" style hinge like that of the spondylid scallops. The size of the living species and fossils is up to 50 mm (2 in) in length.

Spondylidae (thorny oysters)

Probably one of the more unusual fossil bivalves from the Latrania Formation, *Spondylus bostrychites* (Figure 3.14C) represents an extinct species named from the Caribbean Sea. Well-preserved shells can measure up to 150 mm (6 in) in height and are often ornamented with large flattened spines. Occasionally, specimens are found with fossil acorn barnacles attached to the outer shell surface. The fossil species for the Latrania Formation is similar in size to *S. calcifer* Carpenter an extant species living from the Gulf of California south to Ecuador.

Anomiidae (jingle shells)

One member of this family, *Anomia subcostata* (Figure 3.15), is common in the Latrania and Deguynos formations. Shells of this extinct species can be up to 40 mm (1.6 in) in length. The upper valve is the one most commonly found. However, the lower (right) valve, with a hole near the hinge through which the attaching byssus apparatus (horny material secreted by a gland and used to attach bivalves to objects) passed, is much thicker than most modern and fossil relatives, so it is not unusual to also find fossils of whole right valves. The outer surface of the valves of these species is characterized by a series of fine radiating threads.

Although the color of most fossils is usually brownish, specimens may be found with an attractive iridescent red or orange-brown color.

Lucinidae (lucines)

A number of species of lucinids assigned to different genera occur as fossils in strata of the Imperial Group. All are still living today. The diversity of forms exhibited by shells of these species ranges from *Miltha xantusi,* a large and very flat lucinid that lives in the southern part of the Gulf of California, to the highly inflated *Pegophysema edentuloides* (Figure 3.16B) that lives from Cedros Island throughout the Gulf of California to Tenacatitia Bay, Jalisco, Mexico. *Miltha xantusi* may be found as sandstone steinkerns 70 mm (2.8 in) high by 80 mm (3.2 in) wide, occasionally as original shell with fine concentric lines. *Pegophysema edentuloides* is locally common in the Latrania Formation and typically occurs as paired valve sandstone steinkerns, 51 mm (2 in) high by 47 mm (1.88 in) wide. *Pegophysema schrammi* is a closely related Caribbean species. *Codakia distinguenda* (Figure 3.14A) is another warm water lucinid from the Latrania Formation. Fossils of this species typically occur as beautiful flattened circular shells with a reticulate sculpture of fine radial lines crossed by concentric ridges. This species lives today from Magdalena Bay, Baja California Sur south to Panama. Fossil specimens with preserved shell may be up to 60 mm (2.4 in) in diameter. *Divulinga eburnea* is a small lucinid (25 mm in diameter, 1 in) found in the sandstone strata of the Latrania Formation. Fossil specimens are usually preserved with original shell material and are ornamented with a series of fine, chevron-shaped ridges on the outer valve surface. *Divulinga quadrisulcata* is a closely related Caribbean species.

Crassatellidae (crassatellas)

Three species of crassatellids are found as fossils in the Imperial Group, *Eucrassatella digueti, E. subgibbosa* (Figure 3.14B)*,* and *Crassinella mexicana.* Fossils are preserved either as steinkerns or as specimens with the crystalline shell material intact. The most common species, *E. digueti* and *E. subgibbosa* are about the same size, up to 60 mm (2.4 in) in length. *Eucrassatella subgibbosa* is an extinct species named from the Latrania Formation exposed in the Coyote Mountains (similar to the extant *E. gibbosa*). This species has a distinct siphonal beak (posterior edge of shell extended to accommodate the incurrent and excurrent siphons). *Eucrassatella digueti* is an extant species now living from the Gulf of California to Colombia, usually dredged from depths of 13-64 mm (0.5-2.5 in). Shells of this species are relatively flat, whereas *E. subgibbosa* has a fairly well inflated shell.

Chamidae (jewel boxes)

Of the several species of *Chama* and *Pseudochama* living in the Gulf of California today, any could have been preserved as fossils in the Imperial Group. However, unless the spines characteristic of these genera were somehow preserved, identification to species would be virtually impossible. One species

however, *Arcinella californica* (Figure 3.16A) may be recognized by the unusual paired steinkern, triangular in shape, with a few strong radiating ribs. This oddly shaped fossil does not resemble at all the beautiful, spiny shell from which it was formed. Living specimens, found from Cedros Island, Baja California to Panama, and fossils from the Latrania Formation, are uncommon. This species very closely resembles *A. arcinella,* a species common in the Caribbean today.

Veneridae (Venus clams)

The Venus clams represent the most diverse living bivalve family, with over 400 living species worldwide, many of these in the Gulf of California. Several venerid species have been reported as fossils in the Imperial Group. An unidentified species of *Megapitaria* is often found as sandstone steinkerns of articulated valves measuring up to 50 mm (2 in) in length. Two species (*M. squalida* and *M. aurantica*) occur today from the Gulf of California to northern South America. Living specimens of both species grow to a length of a little more than 100 mm (3.9 in) and are very similar in appearance with smooth, compressed, and oval shells. This similarity in general form makes steinkerns of *Megapitaria* difficult to identify. The same can be said for sandstone steinkerns of another venerid identified as either *Dosinia ponderosa* (Figure 3.16E) or *D. dunkeri.* Both species have nearly circular shells and live today in the Gulf of California and south to Peru. Steinkern fossils of *Dosinia* from the Latrania Formation are generally under 100 mm (3.9 in) in length.

The most massive venerid fossil found in the Latrania Formation is referred to the extant species, *Periglypta multicostata.* Usually, these fossils are found with the original crystalline shell material intact. Living specimens grow up to 118 mm (4.7 in) in height and have a thick, well-inflated shell. This species lives from the Gulf of California to Peru. Another species of venerid reported from the Latrania Formation is *Ventricolaria isocardia,* which lives today from the Gulf of California to Colombia. Fossils of this species generally preserve original shell material and are characterized by their strong concentric growth lines on distinctly inflated valves. Fossils can be rather large and often attain the same size as the living species, which can grow to a length of 87 mm (3.4 in), a height of 81 mm (3.2 in), and a diameter of 64 mm (2.5 in). The

Figure 3.16
Bivalve Species Preserved as Steinkern Pairs.

A. *Arcinella californica*
   (SDSNH 12115),
B. *Pegophysema edentuloides*
   (SDSNH 17701),
C. *Anadara formosa*
   (SDSNH 12026),
D. *Panopea abrupta*
   (SDSNH 12101),
E. *Dosinia ponderosa*
   (SDSNH 53797), and
F. *Megapitaria* sp.
   (SDSNH 23150).
(Photographs by Barbara Marrs)

extant Caribbean species *V. rigida* is very similar in appearance to fossils recovered from the Latrania Formation.

Tellinidae (tellins)

This family of thin-shelled bivalves includes the genera *Tellina, Macoma,* and *Florimetis,* with many more species in the first two genera. Fossil tellinids are almost always preserved as steinkerns, generally as an articulated pair of valves. A very high diversity of tellinids exists in the Gulf of California today, as in other tropical areas of the world. However, few species are known from the Imperial Group. This is largely due to the difficulty in distinguishing between the steinkerns of different species of tellinids. It can even be difficult to determine if a specimen belongs to the genus *Macoma* or *Tellina*.

One tellin from the Latrania Formation is *Tellina lyra*, a small species represented by fossils up to 40 mm (1.6 in) in length. Today it is found from Baja California, Mexico to Tumbes, northern Peru. Internal sandstone molds of this tellin may be recognized by the neat, oval shape and the evenly spaced, sharply defined growth lines pressed onto the specimen from the external mold. Other species reported from the Imperial Group include *T. ulloana*, an extant species living from Magdalena Bay, Baja California Sur, south to Panama (*T. martinicensis* is a closely related Atlantic species), *T. ochracea*, an extant species living throughout the Gulf of California (*T. laevigata* is a closely related Atlantic species), and *Florimetis dombei*, an extant species living from Panama to Peru.

Solecurtidae (tagelus clams)

A number of very similar species from this family occur in the Gulf of California today, and precise identification of steinkern fossils may not always be possible. Specimens identified as *Tagelus californianus*, an extant species living along the coast of southern California, throughout the Gulf of California, and south to Panama, are known. These long and narrow steinkerns measure from 30-50 mm (1.2-2 in).

Hiatellidae (geoducks)

The one representative of this family that may be found as sandstone internal molds in the Latrania Formation is *Panopea abrupta* (Figure 3.16D). These steinkerns usually represent articulated valves, with the characteristic gaping of the posterior and anterior ends visible. The fossils are up to about 130 mm (5 in) in length, which is smaller than the living species, which grows to a length of 160 mm (6.3 in). Today it lives from Alaska to the Gulf of California.

Pholadidae (rock piddocks)

The one example of this family known as a fossil from the Imperial Group is *Cyrtopleura costata*. It is found today living along the western Atlantic from Massachusetts south to Brazil, but is extinct in the Gulf of California. Local extinctions (extirpation) like this also characterize many Imperial Group corals. Fossils of this thin-shelled species are more common in the Deguynos Formation where they occur as sandstone internal molds. Specimens may be up to 60 mm (2.4 in) in length. An extant relative living in the Gulf of California today is *C. crucigera*. This species is rare throughout its range from Guaymas, Sonora, Mexico to Ecuador. The living species of *Cyrtopleura* may be found burrowing in silty mud.

Thraciidae (thracias)

This family typically includes species of very thin-shelled clams that generally live offshore. *Cyathodonta undulata* is the only thraciid reported from the Latrania Formation and usually is found as sandstone steinkerns. *Cyathodonta undulata* lives today from Sonora, Mexico, to Peru in water depths of 3-110 m (10-215 ft). The squarish, paired steinkerns can be up to 50 mm (2 in) long and are characterized by undulating, low-lying concentric ribs.

Echinodermata (sand dollars, sea biscuits, urchins, and sea stars)

Although not as diverse as the fossil molluscs, the echinoderm assemblage from the Latrania Formation is none-the-less a conspicuous element of most fossil collections from this rock unit (Kew, 1914) (see Appendix, Table 2). Specimens represent members of all three main echinoderm groups including sea stars, sea urchins, and brittle stars. Fossils are typically preserved with their original crystalline tests intact and many specimens retain even the most delicate structures of the ambulacral petals (specialized zones of pores that support respiratory tube feet) on their aboral or upper surfaces. Sand dollars and heart urchins possess a characteristic five-pointed petal system on their aboral surfaces, while their oral surfaces have a more-or-less centrally placed mouth (peristome) and a marginal, posteriorly placed anus (periproct). In sea urchins the peristome is on the oral surface of the roughly spherical test, while the periproct is centrally placed on the aboral surface. The test is constructed from ten columns of plates on which are located the tubercles (tiny circular structures on test) to which the spines attach. Heart urchins (spatangoids) are also constructed of ten columns of plates, but have a bilaterally symmetrical, generally oval test with sunken ambulacral petals.

In his review of the Imperial Group echinoderms, Powell (1995) noted that the echinoderm assemblage primarily consists of subtropical to tropical taxa, many with affinities to species living today in the Caribbean Sea.

The discussions below focus on those echinoid families best represented by fossils in sedimentary rocks of the Imperial Group. Brittle stars have been omitted from this listing because they are very rare, and have not been studied.

Asteroidea (sea stars)

Astropectinidae (sand stars)

The general absence of hardparts makes it unlikely that sea stars will be preserved in the fossil record. However, fragmentary specimens of a robust sea star have been collected from the Latrania Formation. These specimens consist of portions of "arms" preserved with the relatively thick crystalline plates intact and have been referred to the genus *Astropecten* by Powell (1995).

Echinoidea (sea urchins and sand dollars)

Cidaridae (club-spined urchins)

Fragmentary remains (spines and partial test plates) of a regular (i.e., symmetrical) urchin tentatively referred to *Eucidaris thouarsii* have been reported from the Imperial Group (Latrania Formation) as exposed in the Coyote Mountains. Fossil specimens preserve the characteristic large tubercles of this urchin. This species lives today in the eastern Pacific Ocean from southern California south to Panama and Islas Galapagos.

Arbaciidae (regular urchins)

Small tests of a regular sea urchin similar to the living purple urchin, *Arbacia* sp., have been collected from the Latrania Formation as exposed in the Coyote Mountains. These fossils have not yet been described and may represent a new fossil species (Powell, 1995). Today, species of *Arbacia* live from southern California to Peru.

Toxopneustidae (white urchins)

Tripneustes californicus (Figure 3.18C) is an extinct species of regular sea urchin originally named from the Latrania Formation. Typical specimens measure 85 mm (3.4 in) in diameter and have fewer tubercles and a more distinctly conical shape than the extant *T. depressus* from the Gulf of California and tropical eastern Pacific Ocean. The fossil species has tiny tubercles and resembles *T. esculentus* from the Caribbean Sea. These three species may represent a closely related group.

Powell (1995) reported fragmentary specimens of a second member of this urchin family from Imperial Group deposits in the Coyote Mountains. These specimens are tentatively referred to *Toxopneustes* sp. cf. *T. roseus,* a species living today from Guaymas, Mexico, south to Isla La Plata, Ecuador.

Clypeasteridae (sea biscuits)

Clypeaster bowersi (Figure 3.17A) is a large extinct species of sea biscuit originally described from exposures of the Latrania Formation in the Coyote

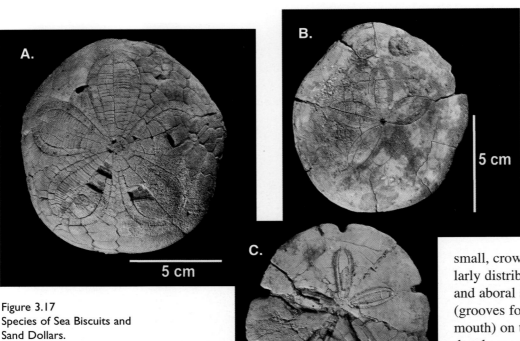

Figure 3.17
Species of Sea Biscuits and Sand Dollars.
A. *Clypeaster bowersi* (SDSNH 103642),
B. *Clypeaster deserti* (SDSNH 12260), and
C. *Encope tenuis* (ABDSP specimen).
(Photographs by Barbara Marrs)

Mountains. Large specimens measure 135-145 mm (5.3-5.7 in) in length, 120 mm (4.7 in) in width, and up to 50 mm (2 in) in height. The general outline is suboval to slightly pentagonal. On the aboral (upper) surface, the ambulacral petals are broad and ovate and not elevated. Tubercles are small, crowded together, and irregularly distributed over both the oral and aboral surfaces. Food grooves (grooves for transport of food to the mouth) on the oral surface are deeply recessed and straight, without any branching. The peristome (mouth) is centrally located and the periproct (anus) is on the posterior margin of the oral surface.

Clypeaster bowersi resembles *C. caudatus* from Miocene rocks in the Dominican Republic and is also similar to *C. pallidus,* a recent species from Barbados Island.

Clypeaster carrizoensis is a rare small fossil species originally named from the Latrania Formation as exposed in the Coyote Mountains. Typical specimens have a flattened oval outline and measure only 24 mm (1 in) in length, 21 mm (.8 in) in width, and 7 mm (0.3 in) in height. The ambulacral petals are open at their lateral ends. Tubercles are small and of uniform size. *Clypeaster carrizoensis* is similar to *C. cotteaui* from the Miocene of Jamaica. The fossil species also resembles *C. subdepressus* a living clypeasterid from the Caribbean Sea.

Clypeaster deserti (Figure 3.17B) is another extinct species of sea biscuit originally named from sandstones of the Latrania Formation. The general outline is distinctly pentagonal with an elevated central region at the ambulacral apex. Larger specimens measure 93 mm (3.7 in) in length, 81 mm (3.2 in) in width, and 15 mm (0.6 in) in height at the center. The height of the margin is only 8 mm (0.3 in). Food grooves on the oral surface are recessed and straight, without any branching. The peristome is more strongly recessed. *Clypeaster deserti* is similar in size to the extant *C. rotundus* from the modern Gulf of California.

Mellitidae (key-hole sand dollars)

Encope tenuis (Figure 3.17C) is a large extinct sand dollar originally named from the Latrania Formation as exposed in the Coyote Mountains. Large individuals measure up to 110 mm (4.3 in) in diameter, but only 8-11 mm (0.3-0.4 in) in height. Many specimens are slightly wider than long. Distinctive features include 5 marginal notches each aligned with one of the delicate narrow abulacral petals. The notches tend to close in older individuals. A distinct oval lunule (opening) occurs between the posterior two petals. The food grooves on the oral surface have an elaborate branching pattern radiating from the centrally placed peristome (mouth). The oval periproct (anus) is located between the peristome and the lunule. Tests of *E. tenuis* are locally common to abundant in certain strata within the Latrania Formation and their well-preserved test fragments often litter the ground. Some workers have identified other species of *Encope* from the Latrania Formation (*E. sverdupi* and *E. arcensis*), however, the amount of morphological variation observed in large samples of *Encope* tests collected from a single locality suggests only a single highly morphologically variable species. *E. tenuis* is the oldest valid name applied to this group.

Figure 3.18
Species of Heart Urchins and Sea Urchins.
A. *Schizaster morlini* (SDSNH 12196),
B. *Meoma* sp. (SDSNH 103644), and
C. *Tripneustes californicus* (SDSNH 12195).
(Photographs by Barbara Marrs)

Echinoneidae (irregular heart urchin)

Echinoneus burgeri is an extinct species of irregular echinoid originally named from rocks of the Latrania Formation exposed in the Coyote Mountains. The test of this rare urchin is walnut-sized, measuring 40-50 mm (1.6-2 in) in length, 30-40 mm (1.2-1.6 in) in width, and 20-30 mm (0.8-1.2 in) in height.

The test outline is oval with delicate and elongated ambulacral petals on the aboral surface. The oral surface is characterized by a horizontal grouping of tubercles and a very large oval periproct placed relatively close to the peristome. Species of *Echinoneus* are common today in the Atlantic Ocean and Caribbean Sea, as well as in the Indo-Pacific. Its absence from the eastern tropical Pacific Ocean suggests that this genus has been extirpated from the region.

Schizasteridae (puffball heart urchins)

Schizaster morlini (Figure 3.18A) is an extinct species of heart urchin originally named from specimens collected from the Latrania Formation exposed in the Coyote Mountains. Specimens measure 42-55 mm (1.7-2.2 in) in length, 35-49 mm (1.4-2 in) in width, and 24-29 mm (1-1.1 in) in height. The ambulacral petals are sharply depressed into the aboral surface and consist of a posterior pair of short petals, an anterolateral pair of longer petals, and a single, centrally placed, elongate anterior petal. The large oval periproct occurs high on the posterior margin of the test.

Brissidae (heart urchins)

Powell (1995) reports three species of brissids from the Latrania Formation. These include poorly preserved specimens referable to *Brissus* sp., *Metalia spatagus*, and *Meoma* sp. (Figure 3.18B). Species of all three genera still live in the Gulf of California and range south to at least Panama. One fossil specimen of *Meoma* sp. measures 120 mm (4.7 in) in length and 95 mm (3.7 in) in width. This specimen preserves delicate aboral spines measuring over 25 mm (1 in) in length.

Loveniidae (porcupine heart urchins)

Specimens of a gracefully heart-shaped urchin found in the Imperial Group of the Coyote Mountains have been tentatively identified as *Lovenia* sp. cf. *L. hemphilli*. This extinct species is similar to living *L. cordiformis* from the temperate and subtropical eastern Pacific Ocean. The fossil taxon is also similar to *L. dumblei* from the Miocene of northeastern Mexico.

Conclusions

Marine invertebrate fossils from the Latrania and Deguynos formations are locally abundant in the Imperial Valley (e.g., Loop Wash, Shell Reef, East Mesa, Fossil Canyon, Painted Gorge, and Yuha Buttes) and, consequently, are the only fossils that most people directly observe and enjoy in this region. As discussed in this chapter, the fossils provide critical evidence for understanding the evolutionary, biogeographic, ecological, and environmental history of the Imperial Sea. From an evolutionary perspective the Imperial fossils show a general pattern of extinct species mixed with extant species. Nearly 100% of the corals represent extinct species, while approximately 20% of the gastropods, 25% of the bivalves, and almost 50% of the echinoderms are extinct. A careful study of this pattern eventually may lead to a clearer understanding of the timing of particular speciation and extinction events for these groups.

The discovery of both western Caribbean and tropical eastern Pacific species in the Imperial fossil beds suggests a former biotic interchange between the two biogeographic regions and serves as evidence that the modern separation of these two regions only occurred within the past three to five million years. Some 15-20% of the species (or closely related species) survive today in the Caribbean, while 60-70% still live in the tropical eastern Pacific. It is noteworthy that an inverse relationship of interchange, termed complementarity, often exists between terrestrial and marine environments and their biotas. A classic case of complementarity occurred during formation of the Isthmus of Panama in the mid-Pliocene. Elevation of the isthmus created a corridor for terrestrial interchange between North and South America (see McDonald, this volume, *The Great American Biotic Interchange*), but it also created a barrier to marine interchange between the tropical Pacific Ocean and the Caribbean Sea.

Fossil molluscs from the Latrania Formation comprise species that played a variety of ecological roles including epifaunal (living on or above sea floor) predatory snails (e.g., *Conus* spp., *Malea ringens*, and *Natica* spp.), epifaunal grazing snails (e.g. *Nerita scabricosta*, *Strombus* spp., *Macrocypraea* sp., and *Plueroploca princeps*), infaunal (living within sediments of sea floor) suspension feeding bivalves (e.g. *Miltha xantusi*, *Divulinga eburnea*, *Eucrassatella* spp., and *Megapitaria* spp.), and attached epifaunal suspension-feeding bivalves (*Dendostrea vespertina*, *Pycnodonte heermani*, *Spondylus* spp., *Anomia subcostata*, and *Arcinella californica*). The species diversity in the Deguynos Formation is much lower, as is the complexity of paleoecological communities.

Although approximately 3% of the Imperial species live today in temperate waters along the California coast and another 15% live in subtropical waters off Baja California Sur, the majority of species (60%) survive today in the tropical Panamic region of the eastern Pacific between 25° north latitude and 5° south latitude (including the Gulf of California). The warm, tropical Imperial fossil assemblages clearly define a dichotomy between clear water, shallow marine paleoenvironments preserved in the Miocene-age Latrania Formation and turbid water, marine-dominated deltaic paleoenvironments preserved in the Pliocene-age Deguynos Formation. This transition correlates with the initial growth of the Colorado River delta during the early Pliocene. The buildup of this huge sediment dam eventually lead to the southward regression of the Imperial shoreline and the development of the Salton Trough, at over 5000 square kilometers (2000 square miles) the largest below-sea-level dry basin in the Western hemisphere. The fossil biotas of the various terrestrial and fresh-water habitats that existed in this basin from the later Pliocene through the Pleistocene are discussed in subsequent chapters of this book.

Paleolandscape 2: Colorado River Delta 3.5 Million Years Ago

INDEX

1. *Anser* sp. (geese)
2. *Charadrius vociferous* (killdeer)
3. *Dinohippus* sp. (Pliocene horse, extinct)
4. *Xyrauchen texanus* (razorback sucker)
5. *Thamnophis* sp. (garter snake)
6. *Gomphotherium* sp.
   (gomphothere proboscidean, extinct)
7. *Washingtonia* sp. (fan-palm)

Borrego Springs
Fossiliferous
Badlands

N

0   10,000   20,000
                    Meters
0   5   10
           Miles

A low, flat topography with shallow river distributary channels and wide tidal beaches is inhabited by *Gomphotherium* (Gomphothere) and *Dinohippus* (Pliocene horse). The view today is southeast from the west end of Loop Wash. (Picture by John Francis)

Ancestral Woodlands of the Colorado River Delta Plain

Specimens of silicified wood were found to be abundant and of various sizes . . . generally of a brown color, and retain all the appearance of wood; the grain and knots show distinctly, and resemble the wood of the mezquit.

Lieutenant R.S. Williamson in
***Report of Explorations in California* (1856)**

Paul Remeika

Ancestral Woodlands of the Colorado River Delta Plain

Petrified Wood in Blackwood Basin Fish Creek. (Photograph by Paul Remeika)

Introduction

It is difficult to imagine the now-rugged and arid Anza-Borrego region covered by a lush, watery floodplain, or a dense, hardwood forest. Yet the presence of petrified wood scattered about the desert floor tells a vivid story of prehistoric woodlands in a landscape of the distant past. During the early to middle Pliocene Epoch, between 4.0 and 2.6 million years ago, the Anza-Borrego region was near sea level. It was a receiving flood basin for the ancestral Colorado River as it flowed across the Colorado Plateau and carved out the Grand Canyon. There was no hint of the desert to come. The area was one of subdued relief on which a riverine system grew and matured, part of the vast deltaic depositional apron that prograded ("progressively graded" or spread out) across the Salton Trough at the northern end of the Gulf of California (Paleolandscape 2).

A temperate woodland plant community, known as the Carrizo Local Flora (Remeika, 1994), occupied this ancient delta (see Table 4.1, this chapter). Angiosperm (flowering) trees, up to a meter (3 ft) in diameter at their base and reaching over 25 meters (80 ft) in height, dominated the canopy. These trees were uprooted and toppled by recurrent floodwaters. Many were carried some distance along the meandering rivers and streams across the delta. As they were tumbled and jostled in the muddy waters, many fallen logs lost their crowns, branches, bark, and root systems. Eventually, they came to rest and were buried in clay and silt flood plain sediments. This sediment cover was thick enough to cut off oxygen and slow the woods' decomposition. The deposits are exposed in parts of the Vallecito, Fish Creek, and Carrizo Badlands. Today, ancient riverbanks and beaches are upended into multi-terraced ridges and sparsely vegetated bluffs along Fish Creek Wash. Broken petrified tree trunks, logs, limbs, branches, roots and bark, differentially weathering out of cliff edges, are the only haunting reminders and evidence of tall trees in the now desert landscape.

Much of the petrified wood in Anza-Borrego occurs in cross-bedded, arenitic (well-sorted and well-cemented) sandstones of the Palm Spring Formation (Woodring, 1931; see Appendix, Table 4). This formation consists of resistant, cream-colored layers of hard-packed siltstones and sandstones that alternate with less obvious red-brown bands of softer claystones to form a geo-logical layer cake. These deposits are exposed throughout the many arroyos of the Vallecito-Fish Creek Badlands. Over hundreds of thousands of years, the fine-grained and concretionary sediments, with well-developed cross-bedding features of natural levee, point bar, channel sands, and overbank silts, were deposited by high-energy rivers and braided streams of the ancestral Colorado River. These sediments spread across a broad flood plain extending south of the Mexican border and eventually accumulated to a thickness of approximately 2,500 m (8200 ft) (Dibblee, 1954). The deposits yield a mixed assortment of fossil woods (Remeika et al., 1988; Nations and Gauna, 1998), eroded and redeposited pollen assemblages (Fleming, 1994; Fleming and Remeika, 1997), vertebrate footprints (Remeika, 1999, 2001, this volume, *Fossil Footprints of Anza-Borrego*), and a lesser amount of mammalian fossil bones (Downs and White, 1968; McDaniel, this volume, *Mammoths and Their Relatives*; Scott, this volume, *Extinct Horses and Their Relatives*).

Throughout the graveyard of the formation are exposures containing such large accumulations of fossil wood that they are termed petrified forests. Most of the wood is brown or tan colored on weathered surfaces and dark brown on fresh surfaces. It is not semiprecious or museum-quality. Many speci-mens were erroneously identified as desert ironwood because of an earlier mis-conception that Anza-Borrego has always been a desert. There is no typical mode of preservation. The fossil wood varies from well-silicified (formed into rock with silicate cement) to splintery and unidentifiable. Often, the grain on a specimen surface appears to be bark but is actually the degraded weathering pat-tern of the now-mineralized (permineralized) wood, which, of course, began as trees. Petrifaction, a process by which the wood tissues, which were buried in sediments, were gradually replaced by minerals, was so precise in some speci-mens that the cellular structure of the wood can still be recognized. This is the basis for microscopic studies, in which researchers are able to identify the dif-ferent wood species.

These remains represent seven tree families in two classes, angiosperms (flowering plants) and gymnosperms (cone-bearing plants). Angiosperms are further divided into two subclasses: monocotyledons (the seedling has one seed leaf and usually parallel-veined narrow leaves) and dicotyledons (the seedling has two seed leaves and usually reticulated or network-veined broad leaves). Dicot tree family members include fossils representing buckeye (Hippocastanaceae), walnut (Juglandaceae), bay laurel and avocado (Lauraceae), cottonwood and willow (Salicaceae), and ash (Oleaceae). There is evidence of one fossil monocot angiosperm, a palm (Arecaceae), and one indeterminate fossil gymnosperm (Cupressaceae *Pineoxylon* sp.) (see Table 4.1 this chapter).

In the Borrego Badlands, some 900-1,900 m (2,900-6,200 ft) of fine-grained sediments record the presence of a long-lived lake, because of a protracted subsidence of the Salton Trough. This late Pliocene age lake originated from the Colorado River and the sediments are referred to as the Borrego Formation (Tarbet and Holman, 1944), which yields incidental petrified woods plus leaf impressions similar to the wood assemblage from the Palm Spring Formation (Woodring, 1931; see Appendix, Table 4).

What is Petrification?

The word "petrified" is derived from the Greek word *petros,* meaning stone. Petrified wood is wood that has literally turned into stone. The process of petrifying (petrifaction) requires three key elements: the wood, water, and sediment.

Anza-Borrego's wood petrifaction began during the Pliocene Epoch when tree branches, limbs, and trunks fell into the flooding waterways of the ancestral Colorado River. These ended up beached downstream as driftwood. Rivers swelled and overflowed their banks, rapidly covering the flood plain and its woody treasure with sediment-laden waters. As these beached, waterlogged woods were buried under clays, silts, and sands, they were compressed. This contributed a high content of dissolved silica to the slightly acidic groundwater to form a solution that gradually infiltrated the woody tissue's pores, and cellular cavities. During this process, lignin and cellulose were replaced by minerals, usually quartz or silica (silicon dioxide, SiO_2 or SiO_2 (H_2O), or calcite (calcium carbonate, $CaCO_3$), and also by gypsum (calcium sulfate, $CaSO_4$). Other factors, such as temperature and compression from the weight of overlying sediments, also played an important role in the process, as did the presence of various inorganic compounds and trace mineral impurities in the sediments, which created a spectrum of color in the silicified wood. For example, manganese dioxides stained the wood blue, black, and purple, while iron oxides produced orange, rust, red, or yellow shadings. On the blander end of the spectrum, quartz crystals are usually colorless, or gray; silica is tan-gray, white, or brown in appearance, and calcite and gypsum are usually white to tan.

How to Identify Petrified Wood

In order to identify a three-million-year-old piece of wood that has turned into stone, it must be large enough to reveal the desired structural details on the rough cross section when viewed macroscopically, or with the naked eye. A three-dimensional study usually requires tedious and time-consuming examination of both cross (transverse) and longitudinal (lengthwise) wood sections. This begins by securing the specimen in a vice. Then comes the sectioning of the specimen at right angles to the grain with a vertically mounted, metal-core, diamond-bladed rock saw (Figure 4.1). This type of cross cut produces a surface that can be polished on a lapidary wheel to reveal, in detail, the internal annual growth ring structure of the wood. No prior laboratory preparation of the material is required to identify petrified wood, except removal of the encompassing matrix. Next, samples with adequately preserved details are wet-sanded on an abrasive lapidary grinding wheel (16-inch diameter, horizontal, tapered iron plate). A silicon carbide grit slurry is used to remove all saw cut scratches and to obtain a smooth, transverse, optical surface. Additional finer grit sandings may be used to polish the surface so that the sample can be viewed with a 10-power hand lens (10X magnification) or high-powered microscope.

Figure 4.1
Sectioning Fossil Wood.
This section of wood, clamped in the rock saw vice, is being cut across the grain for polishing and examination. (Photograph by Paul Remeika)

The microscopic wood features include the wood's annual growth rings, rays, pores, and cellular tissue structure, all of which can be characterized as to size, shape, and distribution. In transverse sections, the annual rings appear as concentric bands, with rays extending outward like the spokes of a wheel. These rings may be counted to derive an age for the tree. Rays are ribbons of tissue that run perpendicular to the grain and radiate outward from the central core of the tree. Pores are water-conducting tubes. Springwood portions of the annual growth ring, the earlier, softer, and more porous wood, and its opposite, the summerwood, reveal their characteristics in round or oval ring outlines.

Figure 4.2
Palm Frond.
Impression of a fossil palm leaf *Washingtonia* sp. in sandstone, 6 inch ruler for scale.
(ABDSP[IVCM]1315/P58)
(Photograph by Paul Remeika)

The petrified wood identification process then becomes even more precise for purposes of examination and characterization. If the wood is well preserved, it is possible to accurately describe and identify it to species level. For example, Anza-Borrego is unique for having a fossil record containing palm wood and permineralized leaf impressions of palm fronds (Figure 4.2). The fragmentary leaf impressions show affinities to both sabal- and brahea-type palms. Palms contain prominent rod-like and cellular structures within the regular grain of the wood. The rod-like structures show up as tapering rods or fibrovascular bundles that comprise part of the woody tissue that give the wood its vertical strength. Unfortunately, Anza-Borrego specimens are too incomplete to permit accurate description and assignment to species.

Well-preserved wood (Figure 4.3) may also reveal whether it is a gymnosperm (softwood) or angiosperm (hardwood) by examining it in cross-section with a hand lens. Softwoods, such as conifers (polycotyledons), are composed mainly of narrow, elongate cells (tracheids) that function in both structural support and conductance of water. Conifers do not possess vessels (pores), thus the wood has a smooth, even appearance. Hardwoods, such as bay laurels (a dicotyledon) are composed of both vessels and fibers. The latter are narrow, elongate, thick-walled cells functioning in structural support. Typically, woods of temperate softwoods and hardwoods show concentric growth rings in cross-section.

Figure 4.3
Tree Section.
A well-preserved 3.5 million-year-old fossil tree fragment.

The majority of Anza-Borrego wood samples, based on what the eye can see and under a microscope, are identified as angiosperm hardwoods, both dicotyledons and monocotyledons. The dicotyledon samples may be further divided into three well-defined groups based on cell structure across the rings on the polished transverse sections: semi-ring porous, diffuse porous, and ring-porous. Wood sample observations revealed that the dicotyledons closely resemble extant broad-leaved riparian forest trees of the western U.S. The monocotyledon (no-subdivisions) samples include palm frond impressions, seeds, and wood. Wood sample identifications are refined even more by observing detailed anatomical features at high magnification and these are generally accurate to family or genus, and in some cases, species.

Carrizo Local Flora

The Carrizo Local Flora (Table 4.1, this chapter), which existed as part of the Pliocene landscape, is a mixture of three distinct tree associations, and is part of the regionally occurring extinct Madro-Tertiary Geoflora (Axelrod, 1950a, 1958). The three associations are:

- Fossil mixed-evergreen hardwoods of the California Woodland Element (Axelrod, 1950a) include bay laurel (*Umbellularia salicifolia*), walnut (*Juglans pseudomorpha*), ash (*Fraxinus caudata*), black cottonwood (*Populus alexanderi*), willow (*Salix gooddingii*), and buckeye (*Aesculus* sp.), descendants of which today make up the understory element of the redwood community along the central California coastline (e.g., Big Sur),

- Fossil members of the Sierra Madrean Woodland Element (Axelrod, 1950a) include the fan palm (*Washingtonia* sp.), avocado (*Persea coalingensis*), and black cottonwood (*Populus alexanderi*), descendants of which today survive in the semi-arid and arid lands of southwestern U.S. and subtropical and tropical Mexico,

- The Conifer Woodland Element (Axelrod, 1950a) consists of one representative *Pineoxylon* sp.; whether it is cedar or juniper remains indeterminate.

Members of the Carrizo Local Flora testify to environmental changes over the past three million years. Each member listed below is referable to extant genera by leaf impressions, seeds, nuts, and wood.

Anza-Borrego's fossil bay laurel, *Umbellularia salicifolia,* is one of the most abundant petrified woods and is commonly mistaken for desert ironwood. Typically, the wood in this riparian species is diffuse porous, characterized by distinct annual growth rings, each delineated by a dense band of summerwood. Pores are small, uniform in size, and barely visible to the naked eye. They are sparse but are evenly distributed throughout the growth rings. Occasionally, pores may string together in sets of two, three, or more running together in a radial direction. These fossil are numerous and widely distributed throughout the Anza-Borrego region. They occur in the Mio-Pliocene Remington Hill and the Table Mountain Floras (Condit, 1944) as well as with the lower Pliocene Mulholland Flora (Axelrod, 1944a) of northern California. In addition, they occur with the Pliocene Esmeralda Flora (Axelrod, 1940) of Nevada, the Pliocene Troutdale Flora (Chaney, 1944) of Oregon, the upper Pliocene Sonoma Flora (Axelrod, 1944b) of northern California, and other paleofloras in California, Oregon, and Nevada.

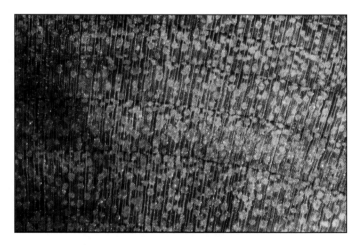

Figure 4.4
Fossil Bay Laurel.
Microphotograph of a diffuse-porous growth ring pattern typical of the fossil wood *Umbellularia salicifalia.* Magnified 80X. Under the microscope, the pores and pore multiples are seen to be encircled by a whitish sheath of parenchyma.
(Photograph by Paul Remeika)

The California laurel, *Umbellularia californica,* is a living descendent of fossil bay laurel. It is known as both the California bay and the Oregon myrtle and is the only member of the Laurel family found in the western U.S. In northern California, the California laurel is associated with the coniferous forests, an area of moderate rainfall (mesic). It is the principal undergrowth species of the coastal redwood forests along the Big Sur coastline south of Monterey, and in the Muir Woods and Armstrong Woods north of San Francisco. In southern California, this tree forms pure stands in Santa Barbara. A relict outpost of California laurel survives on the eastern slopes of the Laguna Mountains (Oriflamme Canyon) in Anza-Borrego Desert State Park. During the Pliocene Epoch, the bay laurel thrived here in wet soils along riverbanks, streams, lakes, and floodplains. It usually grew as a small-stature tree with a short trunk forked into several large, spreading branches and rarely grew to 25

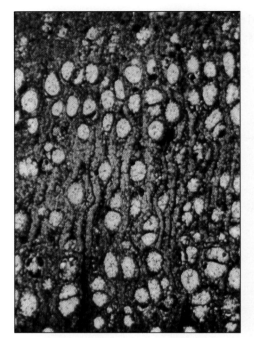

Figure 4.5
Fossil California Walnut.
Microphotograph of a semi-ring-porous growth ring pattern typical of the fossil wood *Juglans pseudomorpha.* Magnified 80X. Note deformed ray structures and zones of partially collapsed parenchyma cells and vessels. With a hand lens (10X), short bands of diffuse-in-aggregates parenchyma are visible. (Photograph by Paul Remeika)

meters (80 ft) tall. The bay laurel is universally known as an indicator of a mild, temperate climate subject, principally, to winter rainfall.

The California walnut tree, *Juglans californica,* is a direct descendant of Anza-Borrego's fossil walnut, *Juglans pseudomorpha.* It is characterized by having distinct growth rings, delineated by a dramatic difference in pore size between early springwood pores and late summerwood pores. Early springwood pores are easily visible to the naked eye, but decrease in size through the summerwood. Walnut wood is semi-diffuse porous, meaning there are more pores at the beginning of a ring than at the end of the summerwood of a ring. In addition, walnut trees have larger pores at the beginning of a ring and only show an occasional doubled pore. Fossil specimens occur in the Mio-Pliocene Remington Hill Flora (Condit, 1944) of northern California, and in the Pliocene Mount Eden Flora (Axelrod, 1937) of southern California. Today, the California walnut tree occurs in the southwestern U.S., and generally shows a preference for moist soils along larger streams or on fertile floodplains. In southern California, it is included in cooler coastal areas (the fog belt) on the western slopes of the Santa Ana Mountains and in the Puente Hills and in widespread savanna associations along with bay laurel, cottonwood, willow and many other species. It is a round-crowned, bushy tree, standing between 3 and 16 meters (15-50 ft) tall, often with several stems.

Regionally the fossil walnut and bay laurel trees occur together. This suggests that these two dicotyledonous species were the dominant trees in a paleoecological association that may have been restricted to the flood-plain, perhaps riparian, because the sediments in which they are found as fossils clearly represent a deltaic setting (Paleolandscape 2). In riparian flood-plain forests, the canopy may be dominated by a few taxa (groups) that have adapted to the clayey substrate. Periods of flooding, high water table, and low soil oxygen levels characterize these settings.

The fossil avocado tree, *Persea coalingensis,* is one of the most common Pliocene species in California. Typically, the wood is diffuse porous, with small pores scattered throughout the growth ring. Rays are abundant and undistorted. The Mio-Pliocene Table Mountain Flora (Condit, 1944) of northern California contains fossil avocado as well as the Pliocene Piru Gorge Flora (Axelrod, 1950b) and Anaverde Flora (Axelrod, 1950c) of southern California, the Pliocene of Chula Vista on the San Diego County coast (Axelrod and Demére, 1984), and other paleofloras in California. The Pliocene fossil avocado specimen is indistinguishable from its living equivalent, *Persea podadenia.* Today, this small stature riparian form thrives in large numbers along stream banks and lake borders of the Sierra Madre Occidental in Sonora and Durango, Mexico.

In the Willow family, both the black cottonwood tree, *Populus alexanderi*, and the willow tree, *Salix gooddingii,* are represented by leaf impressions and wood. Both fossil species are characterized by distinct growth rings delineated by a difference in pore size and number between the late summerwood and early springwood. Their pores are numerous and small with the largest springwood pores barely visible to the naked eye. The wood varies from semi-ring porous to diffuse porous.

Today, the black cottonwood tree, *Populus trichocarpa,* is widely distributed along riparian woodlands in California, ranging north into Alaska and the northern Rocky Mountains. Along the Pacific coast mountain ranges of Washington and Oregon, its extensive bottomland forests follow stream courses. In southern California and Baja California, it is most abundant along coastal stream banks at low to moderate elevations. The fossil willow tree has a wide Quaternary distribution along the Central Valley of California southeastward and into southern Arizona, New Mexico, and northern Mexico. The living Descendant ranges from Alaska and Canada to southern California on the Pacific coast, and eastward into the southern Rocky Mountains and on high plateaus of the southwestern U.S. It is common in riparian habitats and in association with woodland vegetation.

Additionally, tall western cottonwoods and thicket-forming willows are found in the Lower Pliocene Mulholland Flora (Axelrod, 1944a), and the Upper Pliocene Sonoma Flora (Axelrod, 1944b) of northern California, in the Pliocene of Chula Vista (Axelrod and Demére, 1984), and locally in the Pleistocene Soboba Flora from the San Jacinto Mountains (Axelrod, 1966). Today, as in the past, cottonwood trees and willow trees have an extravagant desire for water, and generally grow in riparian environments where the soil is moist. Both forms are abundant along the Colorado River drainage, preferring relatively low-elevation habitats. During the Pliocene Epoch, their presence along the Colorado River reflects a preference for an ever-moist, river environment, and compared with their modern distribution suggests a temperate climate and predominantly winter rainfall, between 37 and 62 centimeters (15-25 in) per year.

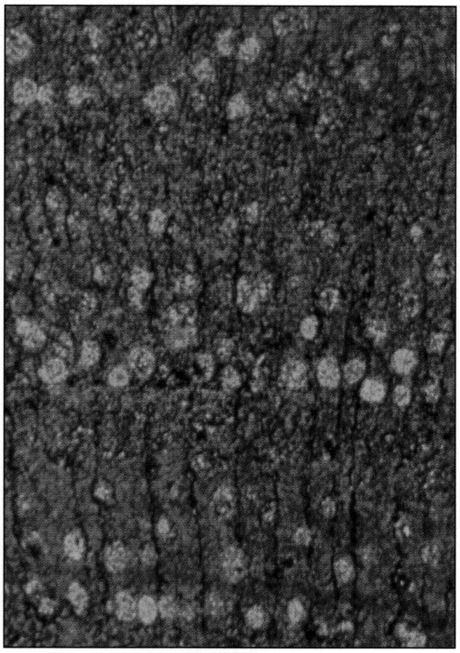

The present-day Oregon ash tree, *Fraxinus oregona,* is a descendant of the fossil ash tree, *Fraxinus caudata,* which is a typical riparian tree. Its growth ring pattern is ring-porous with earlywood (springwood) pores abundant and latewood (summerwood) pores widely scattered and small. The Oregon ash tree is widely distributed, ranging from southern British Columbia through western Washington and Oregon to the Coast Ranges, Sierra Nevada foothills, and interior foothills of southern California. It inhabits stream-border areas throughout oak woodlands to the lower margins of the yellow (Ponderosa) pine forests. Although rare in the fossil record, the fossil ash tree occurs in the Pliocene Troutdale Flora of Oregon (Chaney, 1944), the Wildcat, Orinda, and Etchegoin Floras of northern California (Dorf, 1930), and in the Pleistocene Soboba Flora of southern California (Axelrod, 1966).

Figure 4.6
Fossil Oregon Ash.
Microphotograph of a ring-porous growth ring pattern typical of the fossil wood *Fraxinus caudata.* Magnified 80X. Note that the growth rings are narrow and the latewood pores are seldom connected. (Photograph by Paul Remeika)

Also, the modern California buckeye, *Aesculus californica*, which is the only species of buckeye endemic to California, is related to the fossil buckeye, *Aesculus* sp., which is another typical riparian tree common in Anza-Borrego deposits but rare elsewhere. Its growth ring pattern is diffuse-porous, with numerous small pores scattered throughout the finer-celled tissues, which are not visible to the eye. The fossil buckeye is reported only from the upper Pliocene Santa Clara Flora of northern California. Present-day buckeyes appear along both streambeds and open dry slopes to wooded canyons of the California Central Valley, and Coast Ranges of Monterey County, California, and are members of the mixed evergreen and redwood forest communities.

Finally, permineralized monocot wood, morphologically and anatomically identical to the modern California fan palm, *Washingtonia filifera,* has recently been found in Anza-Borrego. This may represent the first documented fossil *Washingtonia* sp. within California. Previously (Axelrod, 1939, 1950b), specimens of this taxon have been assigned erroneously to the genus *Sabal* based on palm frond impressions. This leads to paleobotanic confusion, since it is very difficult to identify fossil specimens accurately from their leaf stem impressions alone. The leaf and associated fossil wood material from Anza-Borrego, however, is indistinguishable from the genus *Washingtonia* sp. (based on anatomical pore structure morphology) and does not resemble *Sabal* palm. It is unlikely that *Sabal* sp. (usually referred to the species *S. miocenica*) and *Washingtonia* sp. co-existed in Anza-Borrego during the Pliocene. *Washingtonia* sp. favored temperate conditions and had a much wider distribution. Today, the presumed relict species *W. filifera* is widely distributed throughout the interior dry regions of southern California where water is available. It ranges from Death Valley National Park to the Colorado River and includes all of the Colorado Desert extending south into the San Felipe Desert of Baja California, Mexico, along the western margin of the Gulf of California.

Figure 4.7
The Colorado Delta.
During the Pliocene Epoch the Anza-Borrego region was a lowland of rivers and streams. (Picture by John Francis)

What Does Petrified Wood Tell Us?

During the Pliocene Epoch, petrified wood evidence suggests, the Anza-Borrego region was not a barren desert landscape, but a lowland of meandering rivers and riparian streams associated with the ancestral Colorado River drainage (see Paleolandscape 2). Mixed woodland flora of western temperate hardwoods, namely bay laurels, walnuts, avocados, cottonwoods, willows, ashes, buckeyes, and palms occupied this setting. The hardwood flora is an excellent indicator of wet soil and permanent water, which suggests that during this time the region was cooler and wetter than now, with four seasons per year. The Pliocene climate of Anza-Borrego was likely blessed with warmer winters and wetter summers than now (see Sussman et al., this volume, *Paleoclimates and Environmental Change in the Anza-Borrego Desert Region*). Temperatures would have ranged from -9.4° to 26.6° C (15° to 80° F) with annual rainfall of 38-63 cm (15-25 in) or more, mostly through winter and spring. This explains the fast growth rate measured in most of the wood samples.

Microscopic study of fossil wood, including its structure, ring counts, worm damage, and distortion, yields important clues to the climate and sedimentation processes of 3 to 4 million years ago. Bay laurel ring counts, for example, show growth rates ranging from 1 to 5 rings per cm (3/8 in), with an average range of 2-5 rings. This reinforces the hypothesis that the Pliocene Anza-Borrego region received mainly winter rainfall and was, in addition, subject to the maritime influence (coastal fog and overcast skies) of both the ancestral northern Gulf of California and the Pacific Ocean. The paleoclimate, more temperate than today's, clearly predates the dramatic uplift of the Peninsular Ranges and the severe consequences of the rain shadow subsequently created by the mountains.

Upon additional microscopic study, some fossil wood shows small tunnels and burrows stuffed with partly digested wood, evidence of attack by insects or their larvae. Much of this "worm" damage was caused by the larvae of the powder-post beetle (*Lyctus* sp.), which attacks bay laurel and walnut on the Pacific coast. Other wood specimens contain burrows filled with pellets of partly digested wood, probably the work of other wood-boring insects.

After the wood was buried, it experienced very high pressure from the weight of the overlying sediment. This caused it to be compressed down to two-thirds of its normal volume and it was held in this compressed and distorted state during petrifaction. In thoroughly water-saturated wood, distortion and deformation of the rays occurred without the tissue splintering or fracturing. However, some walnut wood specimens exhibit partial replacement with massive calcite crystals, which obliterate the structure of the tissue, while in adjoining tissue only minor distortion of the cell structure is observed. This phenomenon – deformation of much of the wood structure within an undeformed crystalline matrix – indicates that petrifaction occurred very slowly while the wood was under continuous pressure.

In the Anza-Borrego region, we have an astonishing record of plant community dynamics; a striking contrast between fossil and present-day plant species. There are plants that range from deciduous hardwoods that flourished on a Pliocene floodplain to the modern xerophytic "dry plant" community (almost exclusively composed of low-elevation, drought-tolerant shrubs, and an entourage of spiny succulents) typical of the hot and dry Colorado Desert. The history of Anza-Borrego's flora is another testament of the dramatic climatic and topographic changes that have occurred here over the past few million years, a story that echoes throughout this book.

Time Travelers
(microfossils from the age of dinosaurs)

A host of freshwater invertebrates (Taylor, 1966), reworked marine shellfish (Watkins, 1990) (see Appendix, Table 2), and other marine organisms are entombed in the Palm Spring Formation (Woodring, 1931) and in the underlying Yuha Formation of the Imperial Group (Remeika, 1998; see Appendix, Table 4).

Among the fossil specimens from these formations are bizarre, calcite-filled shells (tests) of tiny late Cretaceous age foraminifers (single-celled marine organisms) (Merriam and Bandy, 1965) and plant pollens, that were eroded from the Colorado Plateau and redeposited in the Salton Trough. Many of these forms became extinct at the Cretaceous-Tertiary time boundary (approximately 65 million years ago). The most abundant forms, including the pollens *Proteacidites, Mancicorpus,* and *Aquilapollenites* (Fleming, 1994; Remeika and Fleming, 1995), have restricted biostratigraphic ranges and are distributed in the western interior of North America. In total, eleven types of alien extralimital (outside the region) fossils have been found in sediments in the Anza-Borrego region. They come primarily from the late Cretaceous marine Mancos Shale, a geologic formation that crops out in northern Arizona, Utah, and southwestern Colorado. Additional microfossils of the Eocene age pollen *Pistillipollenites* were derived from the Green River Formation of Utah-Wyoming. Rapid floodwaters of the ancestral Colorado River eroded, transported, and redeposited the fossils.

Stratigraphic distribution of these fossils within the Anza-Borrego sediments indicates that erosion of the Mancos Shale in the southern part of the Colorado Plateau that contains *Proteacidites*, but lacks *Mancicorpus* and *Aquilapollenites*, began about 4.5 Ma. *Proteacidites* first appears in the marine-deltaic Yuha Formation (Remeika, 1998) during this time. *Mancicorpus* and *Aquilapollenites*, from the northern part of the Colorado Plateau, occur in the overlying Palm Spring Formation (Woodring, 1931; see Appendix, Table 4) at about 3.9 Ma. Thus, erosion of Cretaceous rocks from the northern part of the plateau began during the middle part of the Pliocene (sometime after 5 million years ago). Rapid erosion and transport of a large volume of sediment to the northernmost Gulf of California implies higher precipitation levels in the southwestern U.S. than today. This wetter climate (Thompson, 1991; Fleming and Remeika, 1997) is also evidenced by plants of the Carrizo Local Flora. The presence of microscopic foraminifers and pollen in the stratigraphic package of Anza-Borrego, testifies that the erosional landscape on the Colorado Plateau, and extensive down cutting of the Grand Canyon, are relatively recent phenomena.

FOSSIL WOODS	ALLIED LIVING WOODS	MADRO-TERTIARY GEOFLORA		
		1	2	3
Arecaceae				
Gen. et. sp. indet.	monocot (palm)		■□	
Washingtonia sp.	*Washingtonia filifera* (California fan palm)		■□	
Hippocastanacea				
Aesculus sp.	*Aesculus californica* (California buckeye)	■□		
Juglandaceae				
Juglans pseudomorpha	*Juglans californica* (California walnut)	■		
Lauraceae				
Persea coalingensis	*Persea podadenia* (black avocado)		■□	
Umbellularia salicifolia	*Umbellularia californica* (California bay laurel)	■		
Oleaceae				
Fraxinus caudata	*Fraxinus oregona* (Oregon ash)	■□		
Salicaceae				
Populus sp. indet.	*Populus* sp. (cottonwood)	■		
Populus sp. cf. P. alexanderi	*Populus trichocarpa* (black cottonwood)	■□		
Salix sp. indet.	*Salix* sp. (willow)	■	■□	
Salix gooddingii	*Salix gooddingii* (Dudley willow)	■□		
Cupressaceae				
Pineoxylon sp.	softwood (cedar or juniper)?			■□

Table 4.1
Carrizo Local Flora.
Distribution of Anza-Borrego fossil taxa compared to related living species and occurrence in the Madro-Tertiary Geoflora.

Explanation:
1 = California Woodland Element;
2 = Sierra Madrean Woodland Element;
3 = Conifer Woodland Element;
■ = Presence in Madro-Tertiary Geoflora;
□ = New to Anza-Borrego.

5 Stratigraphy, Tectonics, and Basin Evolution in the Anza-Borrego Desert Region

*The most beautiful thing
we can experience
is the mysterious;
It is the source of all true art
and all science.*

Albert Einstein

Rebecca Dorsey

Stratigraphy, Tectonics, and Basin Evolution in the Anza-Borrego Desert Region

Southern Borrego Badlands Looking West. (Photograph by Rebecca Dorsey)

Introduction

The fossil record of past life is commonly preserved in ancient sediments and sedimentary rocks. Sediments accumulate in subsiding basins that contain different kinds of depositional environments such as rivers, lakes, deltas, and marine seaways. These environments are friendly to life, and often support assemblages of plants and animals. Through integrative studies of stratigraphy, sedimentology, and paleontology, we can reconstruct ancient life communities and the environments in which they lived.

Plate tectonic forces determine where sedimentary basins form, how long and how fast sediments accumulate, and how they may later be faulted, uplifted, and eroded at the surface. Climate also affects basins and sediments; precipitation, wind, and temperature variation affect surface processes such as erosion and soil formation. In the Salton Trough region of southern California, styles, rates, and environments of basin formation have evolved through time in response to complex changes in driving tectonic forces, fault interactions, and climate change. Because of the rich history of geologic research in Anza-Borrego Desert State Park and adjacent areas, it is impossible to summarize all of the knowledge on this subject in a few pages. This chapter presents a brief overview of existing knowledge about the regional stratigraphy, tectonic evolution, and major sedimentary basins preserved in the Park, which have supported a great diversity of plants and animals during the past ~10 million years.

Anza-Borrego Desert State Park is located within a complex zone of strike-slip faulting and oblique crustal extension and compression that defines the tectonically active boundary between the North American plate and the

Pacific plate in southern California (Figures 5.1, 5.2). The southern San Andreas fault system, which includes the San Andreas, San Jacinto, and Elsinore faults, is a broad zone of past and ongoing seismic activity that separates areas belonging to the Pacific plate (Baja California and southern California) from areas located on North America (mainland Mexico and the U.S.). Long-term northwesterly movement of the Pacific plate relative to North America has resulted in progressive right-lateral fault displacements and related crustal deformation during the past ~25 to 30 million years, producing a complicated network of faults, rotating crustal blocks, mountain ranges, and sedimentary basins. The two plates are also diverging slightly in some areas, which has caused the Salton Trough and Gulf of California to open up by oblique rifting and extension during the past 10 to 15 million years. These aspects of relative plate motion and the development of geologic structures on a regional scale are known from landmark studies by Atwater (1970), Lonsdale (1989), Stock and Hodges (1989), Powell et al. (1993), DeMets (1995), Dickinson (1996), Atwater and Stock (1998), Axen and Fletcher (1998), Oskin and Stock (2003), and others.

As described below, we now know that regional subsidence related to crustal extension and transtension (a combination of strike-slip movement and oblique extension of a fault) produced a number of fault-bounded basins that filled with sediments from Miocene to Pleistocene time (Figures 5.2, 5.3, 5.4). In the recent geologic past (last 1 to 2 million years) many of these basins have been uplifted and eroded to reveal the diverse stratigraphic record of their tectonic, climatic, and paleontologic evolution. In the Salton Sea, fault-controlled subsidence has continued to the present day, accumulating a thick section of young sediments that are buried in the modern basin beneath the surface.

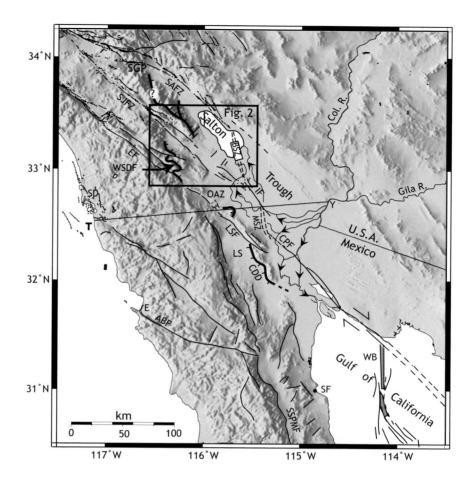

Figure 5.1
Faults and Topography of the Northern Gulf of California and Salton Trough Region.
Decorated thicker lines are detachment faults, tick marks on upper plate; plain lines are high-angle normal and strike-slip faults. Explanation: ABF, Agua Blanca fault; BSZ, Brawley Spreading zone; CDD, Cañada David detachment; CPF, Cerro Prieto fault; E, Ensenada; IF, Imperial fault; SAFZ, San Andreas fault zone; SD, San Diego; SGP, San Gorgonio Pass; SF, San Felipe; SJFZ, San Jacinto fault zone; SSPMF, Sierra San Pedro Martir fault; T, Tijuana; and WB, Wagner basin. (Shaded-relief map base courtesy of H. Magistrale)

Our understanding of geological events summarized below is based on decades of research by many scientists. This chapter does not present new data, it is simply an attempt to synthesize a vast body of existing knowledge and make it accessible to a broad audience. Some of the more influential studies of regional stratigraphy, basin evolution, and related structures in the western Salton Trough appear in papers and theses by: Axen and Fletcher (1998), Bartholomew (1968, 1970), Brown et al. (1991), Dean (1988, 1996), Dibblee (1954, 1984, 1996a, 1996b), Dronyk (1977), Feragen (1986), Frost et al. (1996a, 1996b), Girty and Armitage (1989), Ingle (1974), Johnson et al. (1983), Kerr (1982, 1984), Kerr and Kidwell (1991), Lough (1993, 1998), Merriam and Brady (1965), Muffler and Doe (1968), Opdyke et al. (1977), Quinn and Cronin (1984), Remeika (1995), Remeika and Beske-Diehl (1996), Schultejann (1984), Sharp (1982), Stinson (1990), Stinson and Gastil (1996), Tarbet and Holman (1944), Wells (1987), Winker (1987), Winker and Kidwell (1986, 1996), Woodard (1963, 1974). This list is only a partial sampling of many theses, papers, and abstracts that have contributed to our understanding of this fascinating region. Interested readers are encouraged to explore the original literature upon which the following summary is based.

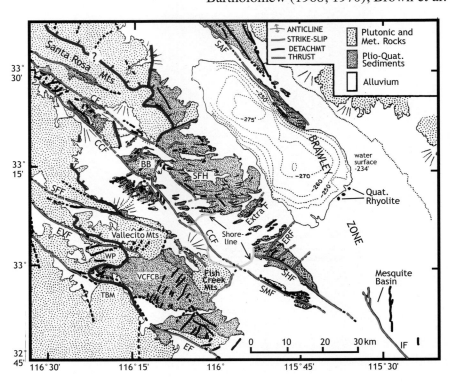

Figure 5.2
Geologic Map of the Salton Trough Region.
Explanation: BB, Borrego Badlands; CF, Clark fault; CCF, Coyote Creek fault; EF, Elsinore fault; ERF, Elmore Ranch fault; EVF, Earthquake Valley fault; VCFCB, Vallecito Creek-Fish Creek basin; IF, Imperial fault; SFF, San Felipe fault; SFH, San Felipe Hills; SHF, Superstition Hills fault; SMF, Superstition Mtn fault; TBM, Tierra Blanca Mts; and WP, Whale Peak. (Map compilation courtesy of L. Seeber)

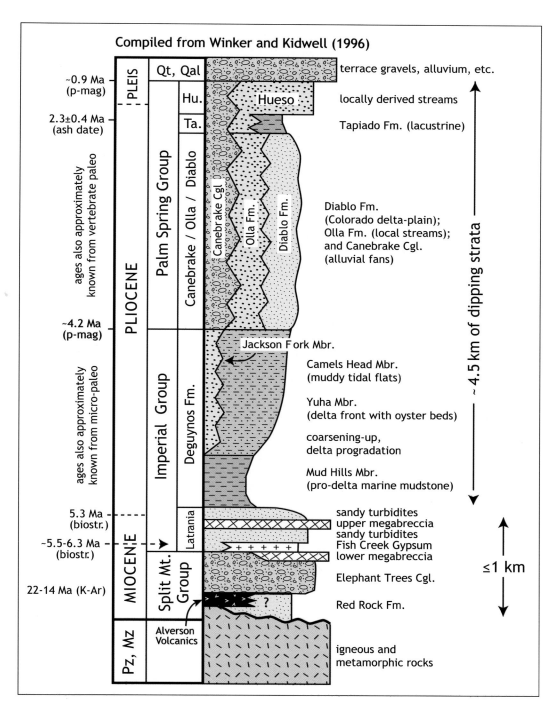

Figure 5.3
Generalized Stratigraphic Column for the Vallecito Creek-Fish Creek Stratigraphic Section.
Paleomagnetic and ash dates are from Opdyke et al. (1977) and Johnson et al. (1983). Biostratigraphic age controls are from studies of Stump (1972), Downs and White (1968), Ingle (1974), Pappajohn (1980), Dean (1988), and McDougall (personal communication, cited in Winker and Kidwell, 1996). K-Ar (potassium-argon) ages in the Alverson Volcanics are from Ruisaard (1979) and Gjerde (1982), summarized by Kerr (1982). Adapted from Winker and Kidwell (1996).

Tectonics, Stratigraphy, and Basin Evolution

Overview

Changing patterns of faulting and subsidence in the San Andreas system have exerted a primary control on sedimentation and stratigraphy in the western Salton Trough through time. Although significant uncertainties and questions remain, the Miocene to Pleistocene tectonic evolution of this region can generally be divided into three stages: (1) early (?) to late Miocene continental sedimentation, volcanism, and formation of fault-bounded nonmarine rift basins; (2) Pliocene to early Pleistocene extension and transtension on a system of regional detachment faults (low-angle normal faults) and formation of a large basin that filled first with marine and later with terrestrial sediments; and (3) Pleistocene to modern strike-slip faulting and related folding in the San Jacinto and Elsinore fault zones, which results in uplift and erosion of the older deposits. Much of the evidence for these tectonic stages is contained in deposits that accumulated in ancient sedimentary basins. Thus, the stratigraphy can be regarded as both an integral component of the dynamic fault-basin system, and a natural record, not always easy to read, of the tectonic processes that produced them.

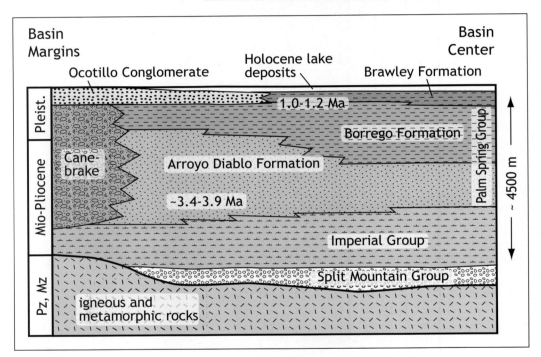

Figure 5.4
Stratigraphy of the Northwestern Salton Trough.
Age of lower Palm Spring Group is based on micropaleontologic study of Quinn and Cronin (1984); age of the base of the Ocotillo Conglomerate is based on paleomagnetic studies by Brown et al. (1991) and Remeika and Beske-Diehl (1996). Modified from Abbott (1969), Dibblee (1954) and Sharp (1982).

The Neogene (Miocene to Pleistocene) stratigraphy of the western Salton Trough and Imperial Valley is illustrated in Figures 5.3 and 5.4. Ages of the deposits have been determined from studies of micropaleontology, vertebrate paleontology, geochronology, and paleomagnetism (see references in figure captions; see also related discussions in this volume: Remeika *Dating, Ashes and Magnetics*, Cassiliano *Mammalian Biostratigraphy in the Vallecito Creek and Fish Creek Basins*, McDonald *Anza-Borrego and the Great American Biotic Interchange*, and Sussman et al. *Paleoclimates and Environmental Change*). Figure 5.3 organizes the stratigraphy of the well-studied Vallecito Creek-Fish Creek area into subdivisions that reflect evolving ideas about the architecture and organization of this complex succession of strata (e.g. Winker, 1987; Kerr and Kidwell, 1991; Winker and Kidwell, 1996). Appendix Table 4 "Correlation of Stratigraphic Names" provides an overview of stratigraphic nomenclature in the Vallecito Creek-Fish Creek section (the far right column is the convention used in this volume).

Neogene deposits in the northwestern Salton Trough (Figure 5.4) show similarities and differences with strata in the Vallecito Creek-Fish Creek section. Upper Miocene sedimentary rocks in the northwestern Trough are sporadically exposed around the margins of the basin and typically are thinner and less complete than in Split Mountain Gorge, though they are locally abundant in the southern Santa Rosa Mountains (Hoover, 1965; Cox et al., 2002; Matti et al., 2002). The Imperial and Palm Spring Groups in the San Felipe Hills are similar to the same units in the Vallecito Creek-Fish Creek stratigraphic section, but the lacustrine (lake) Borrego Formation is much thicker than the Tapiado Formation, and the Ocotillo Conglomerate is a widespread coarse alluvial unit in the northwestern Trough that has only a limited extent in the Vallecito Creek-Fish Creek area (Figures 5.3, 5.4; Dibblee, 1954, 1984, 1996a, 1996b; Bartholomew, 1968). These similarities and differences suggest that the two areas may have occupied a single integrated basin during deposition of the Mio-Pliocene Imperial and early Palm Spring Groups, but became segregated into separate sub-basins in late Pliocene time (Dorsey et al., 2004).

Axen and Fletcher (1998) showed that the Imperial and Palm Spring Groups, and possibly the upper part of the Split Mountain Group, accumulated in a large sedimentary basin that was bounded on its western margin by the west Salton detachment fault system (tectonic stage 2, above) from late Miocene to early Pleistocene time. Detachment faults are low-angle normal faults that form in areas of strong regional extension; they are sometimes associated with high crustal heat flow and commonly produce large sedimentary basins in their upper plates (Figure 5.5; e.g. Wernicke, 1985; Friedmann and Burbank, 1995; Miller and John, 1999). The west Salton detachment fault system was recognized in prior studies and was widely believed to be early or middle Miocene age (e.g. Stinson and Gastil, 1996; Frost et al., 1996a, 1996b). The synthesis by Axen and Fletcher (1998) presented evidence that slip on the detachment system probably began in late Miocene time and continued through Pliocene into early

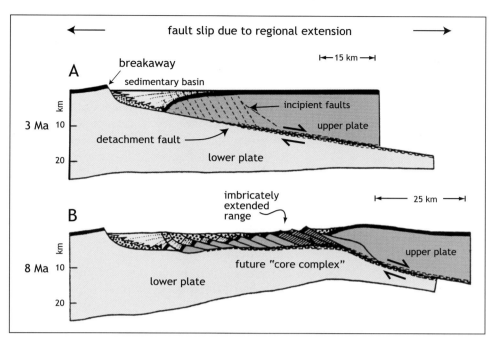

Figure 5.5
Conceptual Model for Partial Evolution of a Detachment Fault (low-angle normal fault) and Upper-plate (supradetachment) Sedimentary Basin Created by Regional Extension.
A. Early slip on the fault occurs by brittle shearing in the shallow crust and ductile deformation in the middle to lower crust. The curved, listric geometry of the breakaway produces a rollover monocline in the upper plate, which in turn produces a sedimentary basin that accumulates a thick section of syn-extensional deposits.
B. After about 8 million years of fault slip, the lower plate domes upward and the upper plate breaks apart along a series of closely spaced normal faults that disrupt sedimentation in the basin. Note that the upper plate of the west Salton detachment fault system *did not* experience break-up as shown in B, possibly because slip on the detachment was terminated by initiation of strike-slip faulting in late Pliocene or early Pleistocene time.
Modified from Wernicke (1985).

Pleistocene time. Detachment faulting resulted in widespread crustal subsidence and accumulation of thick sedimentary deposits that are now exposed in the western Salton Trough (Figures 5.2, 5.3). In contrast to many well-known detachment fault systems, the upper-plate basin of the west Salton detachment fault was not significantly broken apart by normal faults, perhaps because slip on the detachment was terminated by initiation of strike-slip faulting (tectonic stage 3). Moreover, it experienced an oblique, partially strike-slip component of movement that is unlike orthogonal detachment faults (Steely et al., 2004a, 2004b; Axen et al., 2004). These and other aspects of faulting and basin evolution in the Salton Trough are the subject of ongoing study by the author and her colleagues.

The three stages of tectonic evolution and basin development are briefly summarized below. The events are described from earlier to later, moving from lower to higher in the stratigraphic column. Stage 1 (Miocene) is best recorded in rocks exposed in and around Split Mountain Gorge. Stage 2 (Pliocene to early Pleistocene) is recorded in widespread deposits of the Imperial and Palm Spring Groups that are exposed extensively around the western Salton Trough region (Figure 5.2). Geomorphic, structural, and geophysical evidence for stage 3 (Pleistocene to modern) is ubiquitous in the landscape and is reflected in present-day mountain ranges and ridges, active fault scarps, alluvial fans, eroding badlands, and playa lakes.

1. Early to Late Miocene

Significant accumulation of Neogene strata began with deposition of lower Miocene continental sandstone and conglomerate of the Red Rock Formation, which occupies the lower part of the Split Mountain Group (Kerr and Kidwell, 1991; Winker and Kidwell, 1996). These deposits accumulated in rivers and eolian (wind-borne) sand dunes that filled in rugged paleotopography formed by earlier erosion of granitic and metamorphic rocks of the Peninsular Ranges batholith. In some places, they are conformably overlain by volcanic basalts, breccias, and interbedded basalt-clast conglomerates of the middle Miocene Alverson volcanics, which have been dated at approximately 22 to 14 Ma (Figure 5.3; Gjerde, 1982; Ruisaard, 1979; Kerr, 1982). Based on relationships between faulted volcanic and sedimentary rocks, Winker and Kidwell (2002) inferred that weak regional extension and slip on high-angle normal faults began during emplacement of the Alverson volcanics, prior to the late Miocene phase of strong extension and rift-basin development.

The Split Mountain Formation of earlier workers (e.g. Woodard, 1974) includes conglomerate, breccia, and sandy marine turbidites exposed in Split Mountain Gorge. Based on the conformable transition to Imperial marine strata, the upper, marine part of the Split Mountain Formation was reassigned to the lower Imperial Formation (Kerr and Kidwell, 1991), and later named the Latrania Formation of the Imperial Group (Winker and Kidwell, 1996; Remeika, 1998). The Anza, Alverson, and lower Split Mountain Formations were assigned to the Split Mountain Group in this revision (Figure 5.3; Winker and Kidwell,

1996). In the Vallecito Creek-Fish Creek basin, sedimentologically variable marine deposits of the Latrania Formation are conformably overlain by regionally extensive fine-grained marine deposits of the Deguynos Formation (Winker and Kidwell, 1996; Remeika, 1998).

The stratigraphy of the Split Mountain and lower Imperial Groups is very complex, exhibiting abrupt lateral changes in facies (texture, sedimentary structures, grain size, and composition) and local thickening into the Split Mountain Gorge area (e.g. Winker, 1987). These units include two megabreccias with huge clasts (boulder to house-size rock fragments) that were emplaced catastrophically by large rock avalanches, or "sturzstroms" (Figure 5.3; Kerr and Abbott, 1996; Rightmer and Abbott, 1996; Shaller and Shaller, 1996). The Elephant Trees Conglomerate (formerly Split Mountain Formation) is an impressive unit of coarse-grained debris flow and sheet flood deposits that are superbly exposed in the walls of Split Mountain Gorge (Figure 5.6). The age of the Elephant Trees is uncertain (e.g. Kerr and Kidwell, 1991; Winker and Kidwell, 1996). Pronounced lateral thickening of the conglomerate, its conformable association with sandstone of the underlying Red Rock Formation, and the presence of normal faults overlapped by sedimentary deposits, provide evidence that this area experienced sedimentation in steep alluvial fans and flanking braided streams in an active rift basin during late Miocene extension on high-angle normal faults (Kerr, 1982, 1984; Winker, 1987; Winker and Kidwell, 1996).

The Fish Creek Gypsum is a thick, discontinuous deposit that occupies the transition from nonmarine deposits of the Split Mountain Group to marine turbidites of the lower Imperial Group (Figure 5.3; Dean, 1988; 1996; Winker and Kidwell, 1996). Neither its age nor its origins are agreed upon by geologists at this time. Index species (taxa exclusively associated with a particular time interval) of calcareous nannoplankton indicate an age of 3.4-6.3 Ma (million years) for the gypsum (Dean, 1996). This age is refined by tentative placement of the Miocene-Pliocene boundary (5.3 Ma) in the overlying Latrania Formation by Winker and Kidwell (1996), suggesting that its age is approximately 6.3 to 5.5 Ma. The environment of formation for the Fish Creek Gypsum has been variably interpreted as a marginal-marine evaporite setting (Winker, 1987), a restricted shallow-marine basin (Dean, 1988; 1996), or a marine basin with precipitation of gypsum from a hydrothermal vent system (Jefferson and Peterson,

1998). These diverse interpretations highlight the existing uncertainty about the origin of the gypsum and its relation to tectonic evolution at Split Mountain Gorge.

The Fish Creek Gypsum and laterally equivalent lower Latrania Formation record a rapid transgression of marine waters that apparently was controlled by a change in the regional tectonic regime. Onset of late Miocene rifting and high-angle normal faulting recorded in the Split Mountain Group may have been related to early movement on the west Salton detachment fault system (Axen and Fletcher, 1998), or it may represent a distinct earlier phase of extension that pre-dates detachment faulting (Dorsey and Janecke, 2002; Winker and Kidwell, 2002). In either interpretation, it appears that a major tectonic change at about 6-7 Ma produced nearly synchronous marine incursion throughout the northern Gulf of California and Salton Trough region. This incursion flooded an area at least 400 km (250 miles) long, from San Felipe, Mexico, in the south, to San Gorgonio Pass in the north (Figure 5.1; Oskin and Stock, 2003) (see also Deméré, this volume, *The Imperial Sea*).

Recent studies on Isla Tiburon, Mexico, have shown that marine deposits there are younger than about 6.2 Ma, contrary to previous interpretations, and that tectonic opening of the northern Gulf of California was initiated when dextral (right lateral strike-slip) plate motion stepped into the Gulf at about 6.0-6.2 Ma (Gastil et al., 1999; Oskin et al., 2001; Oskin and Stock, 2003). Oskin and Stock (2003) noted that the age of the oldest marine deposits is remarkably similar throughout the northern Gulf and Salton Trough region. Micropaleontology of diatomite near San Felipe indicates that the oldest marine deposits there accumulated between about 5.5 and 6.0 Ma (Boehm, 1984). At San Gorgonio Pass, the Imperial Group is about 6.5 to 6.3 Ma based on micropaleontologic and geochronologic data (McDougall et al., 1999; and papers cited therein). An age of about 6.3 to 5.5 Ma for the Fish Creek Gypsum in the Split Mt. area (Dean, 1988, 1996) is consistent with the timing of marine incursion in other locations around the northern Gulf and Salton Trough region. Rapid marine flooding during this short time interval probably resulted from accelerated basin subsidence and crustal thinning related to initiation of the active plate boundary in the Salton Trough at about 6 Ma (Oskin and Stock, 2003). In addition, a rapid rise in global sea level in latest Miocene time (e.g. Haq et al., 1987) may have caused flooding of an area even larger than would have resulted from tectonic forces alone.

2. *Pliocene to Early Pleistocene*

The Pliocene was a time of deep basin subsidence and accumulation of thick marine and nonmarine sedimentary rocks of the Imperial and Palm Spring Groups throughout the western Salton Trough region (Figures 5.2, 5.3, 5.4). Widespread, fine-grained marine deposits of the Deguynos Formation rest on coarse-grained facies of the Split Mountain Group and Latrania Formation, and represent the culmination of the latest Miocene marine incursion (e.g. Winker and Kidwell, 1996). The tectonic setting was dominated by slip on the west Salton detachment fault system, whose bedrock and basin remains are exposed today around the western fringes of the Salton Trough (Figures 5.2, 5.3; Axen and Fletcher, 1998). The Miocene-Pliocene boundary at Split Mountain has tentatively been placed in the upper part of the Latrania Formation, above the upper megabreccia and at the base of the oldest recorded Colorado River-derived sandstones (Figure 5.3; K. McDougall, personal communication, as cited in Winker and Kidwell, 1996; Gastil et al., 1996). This stratigraphic transition is generally not exposed in the northwestern Salton Trough, but was penetrated by deep exploratory wells in the San Felipe Hills (Dibblee, 1984). The change from locally variable, coarse-grained Latrania deposits of upper Miocene age to regionally extensive fine-grained marine deposits of the lower Pliocene Deguynos Formation may be related to rapid subsidence rates (~5 mm/yr [1/5 in.]; Johnson et al., 1983) that overwhelmed the sediment supply and submerged the Salton Trough basin during early Pliocene time. This period of rapid subsidence probably was driven by the same tectonic forces that produced the latest Miocene marine incursion: initiation or acceleration of relative plate motion in the northern Gulf-Salton Trough region and initiation or integration of the detachment fault system.

Through a combination of tectonic controls, the Salton Trough and northern Gulf of California became a large elongate seaway in early Pliocene time that accumulated a thick succession of marine fossiliferous claystone, siltstone, sandstone, and minor limestones of the Imperial Group (Figures 5.3, 5.4, 5.7). During this time, southern California was located about 200 km (125 miles) southeast of its present location relative to North America, and the Salton Trough was part of a long marine embayment that extended a large distance to the north (Figure 5.9A; Winker, 1987; Winker and Kidwell, 1986). Shortly after the marine transgression that produced the Imperial seaway, this region became the site of a distal prodelta (outermost delta) where only very fine-grained clay and silt derived from the ancestral Colorado River were deposited by suspension settling from the marine water column. This is recorded in mudstone and silty rhythmites of the Deguynos Formation (Figure 5.3; Winker and Kidwell, 1996; 2002) and by similar deposits of the Imperial Group in the San Felipe Hills (Figure 5.4; Dibblee, 1954, 1984; Quinn and Cronin, 1984). Later, as the Pacific plate moved northwest relative to North America, fine-grained sand from the ancestral Colorado

Figure 5.7
Upper Imperial Group.
Tan fine-grained marine mudstone capped by oyster beds in the upper Imperial Group, Vallecito Creek-Fish Creek basin; white patch in distance is Fish Creek Gypsum in the western Fish Creek Mts. (Photograph by Rebecca Dorsey)

River advanced into the basin via dilute turbidity currents. This produced a coarsening-up trend in sediments of the upper Imperial Group that reflects gradual shallowing of the basin as it filled with Colorado River-derived sediments (Figures 5.3, 5.4). The youngest deposits of the Imperial Group include fossiliferous claystone disturbed by burrowing marine animals, wavy-bedded sandstone, and foraminifers (shelled protozoans) that indicate intertidal brackish water conditions; this suggests deposition in a low-energy intertidal environment similar to the broad modern tidal flats that occupy a large area of the present-day lower Colorado delta at the north end of the Gulf of California (Figure 5.1; Woodard, 1974; Quinn and Cronin, 1984; Winker, 1987; Winker and Kidwell, 1996). Marine deposits of the Imperial Group can be viewed along the sides of Fish Creek Wash, south of Split Mountain Gorge.

Shallow marine units of the upper Imperial Group are gradationally overlain by the Arroyo Diablo Formation, a thick unit of sandstone and mudstone that is exposed over much of the Salton Trough region (Figures 5.2, 5.3, 5.4). Quartzose sand of the Arroyo Diablo and Olla Formations (Figure 5.8) was eroded from the Colorado Plateau and deposited in the ancestral Colorado River delta which, at about 3.0 Ma, was located approximately 60-70 km (40 miles) southwest of the modern point of entry of the Colorado River into the Salton

Figure 5.8
Lower Palm Spring Group.
Channelized sandstone (tan color) and red mudstone of the Olla Formation (Palm Spring Group) that was deposited in the ancestral Colorado River delta plain. (Photograph by Rebecca Dorsey)

Trough (Figure 5.9B; Girty and Armitage, 1989; Guthrie, 1990; Winker and Kidwell, 1986). Deposition took place in a subaerial delta-plain setting that was characterized by laterally shifting distributary channels and interchannel swamps and marshes, with overall transport toward the southeast (Winker and Kidwell, 1986). The presence of fossil wood varieties including walnut, ash, and cottonwood suggests that the Pliocene climate was wetter and cooler than today (Remeika et al., 1988; Remeika and Fleming, 1995; see Remeika, this volume, *Ancestral Woodlands of the Colorado River Delta Plain*). The Canebrake Conglomerate, a coarse-grained lateral equivalent of the Arroyo Diablo Formation and other younger units, accumulated in alluvial fans and braided streams on the flanks of steep mountains around the margins of the delta plain (e.g. Dibblee, 1954, 1984; Hoover, 1965; Winker, 1987).

The base of the lacustrine (lake) Tapiado and fluvial Hueso Formations (Figure 5.3) marks an abrupt end of Colorado river input in the Fish Creek area (Winker, 1987; Winker and Kidwell, 1986, 1996). This transition coincides

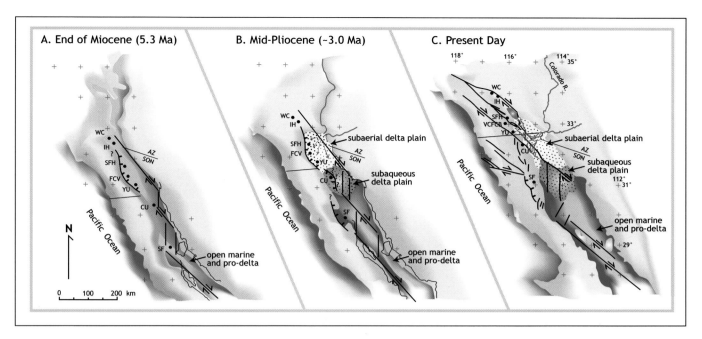

A. End of Miocene (5.3 Ma)

B. Mid-Pliocene (~3.0 Ma)

C. Present Day

approximately with the end of stratigraphic similarities between sediments in the Fish Creek area and the northwestern Salton Trough, and may have resulted from reorganization of the basin (Figures 5.3, 5.4; Dorsey et al., 2004). The Borrego Formation is a very thick succession of lake deposits exposed in the Borrego Badlands and San Felipe Hills that may be partially equivalent to the Tapiado Formation, but its age and stratigraphic architecture are not well known. The Borrego Formation contains abundant claystone and siltstone and rare sandstone beds with both Colorado River- and locally-derived sandstones. Our knowledge of the Borrego Formation and its paleontology, paleogeography, and Pliocene evolution in the northwestern Salton Trough is based largely on previous studies by Tarbet and Holman (1944), Morley (1963), Dibblee (1954, 1984), Hoover (1965), Merriam and Bandy (1965), Bartholomew (1968); Dronyk (1977), Feragen (1986), Wells (1987), as well as recent studies by Kirby et al., (2004a, 2004b), Steely et al. (2004a, 2004b), and Dorsey et al. (2004). Ostracodes (small crustaceans, mussel shrimp) and benthic foraminifers reflect deposition in fresh water to brackish and alkaline conditions (see Appendix, Table 2). The Borrego Formation represents a large perennial lake basin that became isolated from the

Figure 5.9
Paleogeographic Reconstructions of Sedimentary Basins and Faults in the Salton Trough and Northern Gulf of California Since the End of Miocene Time.
Southern California and northern Baja California have been moving to the northwest relative to stable North America since localization of the plate boundary in the Gulf of California at ~6.5-6.0 Ma (Oskin and Stock, 2003) or possibly earlier.
A. End of Miocene time, shortly after widespread marine incursion in the Salton Trough and northern Gulf of California.
B. Deposition of Palm Spring Group in the ancestral Colorado River delta.
C. Present-day geography, active faults, and environments. Explanation: CU, Sierra Cucapas; VCFCB, Vallecito Creek-Fish Creek basin; IH, Indio Hills; SF, San Felipe; SFH, San Felipe Hills; and WC, Whitewater Canyon. Thick lines with arrows are strike-slip faults showing relative movement; thick lines with tic marks are normal faults (low-angle detachment faults in A and B). Red arrows in B indicate inferred sediment transport directions (based on Winker and Kidwell, 1986) (see text for explanation). Redrafted from Winker (1987) with modifications from Axen (1995), Axen and Fletcher (1998), and Oskin et al. (2001).

Gulf of California as it moved tectonically to the northwest past the Colorado River delta into its present position (Figure 5.1; Kirby et al., 2004a, 2004b, Dorsey et al., 2004).

The youngest deposits of the Palm Spring Group are early Pleistocene, locally derived sandstone and conglomerate of the Hueso Formation in the Vallecito Creek-Fish Creek area (Figure 5.3; Winker and Kidwell, 1996; Cassiliano, 2002) and Ocotillo Conglomerate in the Borrego and Ocotillo Badlands (Figure 5.4; Dibblee, 1954; 1984; Brown et al., 1991; Bartholomew, 1968, 1970; Lutz, 2005). These deposits accumulated in alluvial fans and ephemeral streams that drained nearby fault-bounded mountain ranges. Deposition took place in sedimentary basins that were shaped by slip on early strands of the San Jacinto fault zone (Bartholomew, 1970; Pettinga, 1991; Lutz and Dorsey, 2003; Kirby et al., 2004a, 2004b; Kirby, 2005; Lutz, 2005). The Hueso Formation is exposed in View of the Badlands Wash (Vallecito Creek-Fish Creek Badlands), and the Ocotillo Conglomerate can be seen in the cliffs directly beneath Fonts Point in the Borrego Badlands.

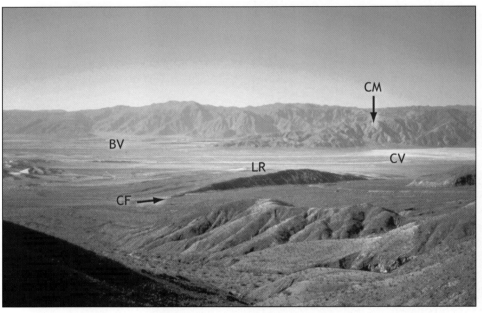

3. Early Pleistocene to Present

Initiation of the San Jacinto and Elsinore fault zones marks the onset of complex dextral strike-slip faulting and north-south compression in the western Salton Trough, which continues today (Figure 5.10, 5.9C). Knowledge of this stage is based on studies by Sharp (1967), Wesnousky (1986), Hudnut and Sieh (1989), Hudnut et al. (1989), Rockwell et al. (1990), Brown et al. (1991), Petersen et al. (1991), Sanders and Magistrale (1997), Heitmann (2002), Dorsey (2002), Ryter (2002), Janecke et al. (2003, 2004), Kirby et al. (2004a, 2004b), Lutz and Dorsey (2003), Lutz et al. (2004), Lutz (2005), and others. In spite of its young age, the timing and nature of the transition from transtensional detachment faulting to transpressive strike-slip faulting is poorly understood. Strike-slip faulting may have overlapped in time with movement on the detachment fault system, and the San Jacinto and Elsinore faults could have started at similar or different times. Dorsey (2002) suggested that progradation (spreading into the basin and over the top of the Borrego Formation lacustrine deposits) of the Ocotillo Conglomerate in the Borrego Badlands may have resulted from initiation of the San Jacinto fault at approximately 1.5 Ma, a date consistent with some prior estimates (e.g. Bartholomew, 1970; Morton and

Figure 5.10
Geomorphology Along the San Jacinto Fault Zone.
View looking west across Coyote Mountain (CM), Clark Valley (CV), Borrego Valley (BV), and Lute Ridge (LR). Lute Ridge is a deposit of Pleistocene coarse alluvial gravels that have been displaced and translated by right-lateral slip on the Clark fault (CF). (Photograph by Rebecca Dorsey)

Matti, 1993). Other studies have inferred an earlier, Pliocene age for the San Jacinto fault zone based on total fault offset and late Pleistocene slip rates (e.g. Rockwell et al., 1990). The Fonts Point Sandstone is a thin Pleistocene fluvial deposit with a well developed calcic paleosol (carbonate cemented ancient soil) layer. This paleosol records the end of sediment accumulation in the Borrego Badlands during slip on the Coyote Creek fault (Ryter, 2002; Lutz et al., 2004; Lutz, 2005). Based on the age of deformed sediments in the Vallecito Creek area (Johnson et al., 1983), Ocotillo Badlands (Brown et al., 1991), San Felipe Hills (Kirby et al., 2004a, 2004b), and Borrego Badlands (Remeika and Beske-Diehl, 1996; Lutz, 2005), combined with known structural relationships in the region (e.g. Dibblee, 1984; Brown et al., 1991; Janecke et al., 2003, 2004; Kirby, 2005), it is likely that the San Jacinto and Elsinore fault systems were initiated in late Pliocene or early Pleistocene time.

Early to middle Pleistocene age conglomerate, sandstone and mudstone, exposed along the northwestern San Jacinto fault zone, was informally named "Bautista beds" by Frick (1921) and later mapped and studied by Sharp (1967) and Dorsey (2002). Sharp (1967) expanded the name "Bautista beds" to include Pleistocene sedimentary rocks exposed around Clark Lake and the northern Borrego Badlands, but these deposits had already been named "Ocotillo Conglomerate" by Dibblee (1954). Recent study by Dorsey and Roering (in press) shows that the Bautista beds were deposited by west- to northwest-flowing streams on the high west flank of the Peninsular Ranges during an early phase of slip in the San Jacinto fault zone. These low-gradient streams were later captured by headward erosion in steep streams flowing southeast along the modern fault zone. The Ocotillo Conglomerate in the Borrego Badlands ("Ocotillo Formation" of Remeika and Beske-Dehl, 1998; Lutz, 2005) was deposited in a low-lying basin (depocenter) at the western margin of the Salton Trough, in a physiographic setting quite different than that of the Bautista beds.

The modern phase of active faulting and seismicity has created a rugged landscape characterized by northwest-trending ridges and fault-controlled features such as Coyote Mountain, Clark Valley, and Lute Ridge (Figure 5.10). Active faults and related uplift have produced young landforms in areas such as the Borrego Badlands, Superstition Mountain, and Superstition Hills, causing older basin deposits to be eroded and reworked into young terrace deposits and modern washes (e.g. Dibblee, 1954, 1984; Ryter, 2002). The Salton Sea is a large topographic depression that exists because of ongoing oblique extension and subsidence within a releasing step-over between the Imperial and San Andreas faults, which has produced an oblique spreading center (the Brawley seismic zone) (Figure 5.1; Elders et al., 1972; Fuis et al., 1982; Fuis and Kohler, 1984; Elders and Sass, 1988). This region has repeatedly dried out and filled with waters of ancient Lake Cahuilla, a Pleistocene to Holocene lake that previously lapped against the flanks of the San Felipe Hills and Santa Rosa Mountains (Waters, 1983). These lake-level highstands created distinctive calcareous algae-derived tufa deposits that are encrusted on granitic bedrock northwest of Salton City and in the Fish Creek Mountains along the U.S.

Gypsum mine railroad. Lake Cahuilla represents the most recent expression of a large ephemeral lake system that was repeatedly flooded and dried out during deposition of the Pleistocene Brawley Formation (Kirby et al., 2004a, 2004b; Kirby, 2005), but with a more restricted distribution that reflects active faulting controls on the modern depocenter.

Conclusions

The above summary provides a brief overview of the tectonic, basinal, and sedimentary history of the western Salton Trough region. We have seen that Anza-Borrego Desert State Park lies within an active plate-boundary zone – the San Andreas fault system – which has been absorbing relative movement of the Pacific and North American plates since about 30 Ma. During Pliocene to early Pleistocene time, a large sedimentary basin associated with slip on a regional detachment fault system accumulated a thick section of marine and nonmarine sediments, recording a wide range of environments that supported the evolution and preservation of ancient plants and animals. Climate also appears to have changed during this time, shifting from a wetter and cooler climate in late Miocene time to the hyper-arid desert setting of today. The modern phase of strike-slip faulting has resulted in uplift and erosion of older sediments, creating a rich natural archive ideal for studying ancient life forms and the environments in which they lived.

Dating, Ashes, and Magnetics: New Times for Old Bones

Theories that have the earth eroded
may all with safety be exploded
For of the Deluge we have data,
Shells in plenty mark the strata
And though we know not yet awhile
What made them range,
What made them pile,
Yet this one thing full well we know –
How to find them ordered so.

William "Strata" Smith,
1829 in *"The Abyss of Time"*

Paul Remeika

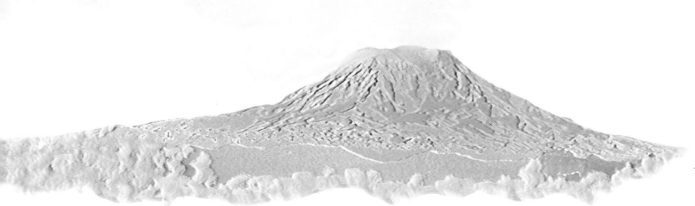

Dating, Ashes, and Magnetics: New Times for Old Bones

Volcanic Rocks South of Fish Creek. (Photograph by Paul Remeika)

Introduction

During the Pliocene and Pleistocene Epochs of geologic time, when ground sloths, camels, mammoths, and other large, bizarre, and now-extinct animals roamed Anza-Borrego, the environment was very different from the desert setting we see today. Climatic and geologic conditions prevailing then sometimes permitted the remains of these animals to be deposited where they lay, relatively protected from the elements, under mud or sand, in places that allowed them to be partially transformed into fossils over the millennia. All the protein and other soluble organic material contained in living bone decayed and dissolved away, leaving only its mineral matrix, a substance quite brittle all by itself. The fragile bits of vertebrae, skulls, teeth, limb bones, and jawbones found in Anza-Borrego are generally more delicate than the rock strata in which they are embedded; they usually have not been hardened, as have fossils from many other settings, by exposure to minerals dissolved in the groundwater.

Although nearly all Anza-Borrego vertebrate fossils are shattered fragments, chipped and cracked, they are an unparalleled collection of past life. Recovered as they weather out of the hills and washes, identified, catalogued and held in labeled trays, steel cabinets, and display cases, *these fossils in their thousands tell a more complete and continuous story of vertebrate life than those from any other site of comparable age in North America.* They have always captivated Park visitors, many of whom ask, "How old are the bones?" This question is easy to phrase but difficult to answer. After all, how do we know what we know?

How do we know what we know?

Dating, or determining the age of events and objects that are older than the historical record, has always been based on prevailing theories of the formation and development of the world. Before the birth of modern geology in the late eighteenth century, the seventeenth-century Irish archbishop James Ussher added up the life times of people recorded in the Bible and calculated the age of the earth to be about six thousand years with a specific creation date of 23 October 4004 BC. More modern theories of planetary formation led the French encyclopedist G.L.L. de Buffon, in 1749, to come up with an age of 75,000 years, based on the cooling time for an earth-sized mass of molten rock ejected from the Sun. Both of these estimates were too conservative by over six orders of magnitude. Not until the geologists of the nineteenth century began to correlate the fossils seen in sedimentary strata with other features of the rock they were exploring, could a realistic geologic time scale begin to be imagined. Even before Charles Darwin published his observations in the mid-nineteenth century, it was dawning on geologists that the relationships among the fossils they were finding in the strata could be used to pin down the relative ages of the rocks themselves.

Figure 6.1
The Bishop Tuff.
In Ash Wash this over 1 meter (3 ft) thick tephra provides a marker bed for correlation of the Coyote Canyon Badlands strata with the Borrego Badlands Stratigraphic section. Rock hammer for scale. (Photograph by George J. Miller)

Relative and Absolute Dating — "Once upon a time . . ."

Prior to the advent of modern dating techniques in the twentieth century, our two forms of evidence – the rock record and the fossil record – could provide only sequential or relative age information. The principle of superposition (Steno, 1669) tells us that, because sediments are deposited by wind or water on top of older deposits, in an otherwise undisturbed section of sedimentary strata, the layers at the bottom are older than the layers higher in the stratigraphic section. The biological characteristics and the relative positions of fossils within this sequence of layers allowed paleontologists to characterize and organize assemblages of species into associations that existed during a particular time interval at a particular locality – a local fauna (see Cassiliano, this volume, *Mammalian Biostratigraphy in the Vallecito Creek-Fish Creek Basin*). A kind of "fossil calendar," based on the collection and correlation of thousands of observations of fossils within strata, was compiled. These techniques, developed in Europe, were extended to western North American in the mid-nineteenth century, when enormous amounts of geologic and fossil data were obtained in the course of mining, quarrying, and railroad-building activities.

But a fossil calendar, developed as described above, can only tell us about the order in which the various species of an assemblage came to exist, to develop and interact with each other, and eventually, perhaps, to die out or evolve into other species. Similarly, purely geologic events other than sedimentation, such as volcanic eruptions, landslides, movement along faults, and the rising or sinking of landforms, can be read in the rock column to tell a local story of what came before, and what else was happening around the same time in nearby areas. Unless some events preserved in these sequences of strata and fossils can be pinned down to a definite time interval, we cannot really know whether our paleontological stories occurred fifty thousand or five hundred million years ago. "Once upon a time" will be about all we can say, and our understanding of the world and its processes will be much less complete.

The development of radiometric, or isotopic, methods of dating in the twentieth century marked a qualitative change in our ability to determine absolute rather than relative ages for rocks and fossils. Isotopes (atoms of the same element with different atomic numbers because of different numbers of neutrons in the nucleus) of many elements undergo natural processes of radioactive decay, which transforms them into other isotopes at fixed rates over time spans ranging from microseconds to millions of years. Based on measurements of radioactive levels, or determinations of the relative proportions of original and decay isotopes of specific elements, the absolute age of a rock can be determined as a function of the known rate of decay for those elements. Because different isotopes have different decay rates, certain isotopes are used for older or younger time periods, and sometimes for different types of rocks. This ability added well-constrained dates to the vast and complex accumulation of information about fossil assemblages and rock formations. Thus, the temporal ranges of biochronologic units (relative age dating based on fossils) like North

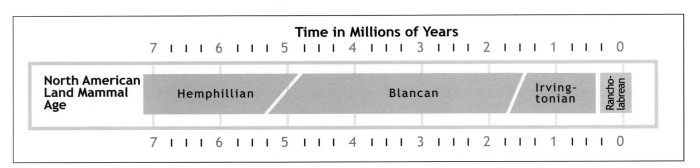

Time in Millions of Years

North American Land Mammal Age

Hemphillian Blancan Irvingtonian Rancholabrean

Figure 6.2
NALMA.
Paleontologists have subdivided later geologic time into a non-overlapping succession of land mammal ages that represent widespread mammalian faunas sychroneus throughout North America.

American Land Mammal Ages (NALMA) can be further refined (see Cassiliano, this volume).

The over 7 million years of prehistory preserved in Anza-Borrego's thick, extensive, continuous sedimentary deposits, revealed because of ongoing erosion and tectonic activity in this geologically lively area, provide a good look at a long story, especially with the combined use of relative and absolute age dating methods. Local fossil faunas could be positioned in the NALMAs, such as the Hemphillian (late Miocene, 8.9-4.9 Ma), the Blancan (latest Miocene through Pliocene, 4.9 to 1.4-1.7 Ma) and Irvingtonian (early to late middle Pleistocene, 1.4-1.7 to 0.5-0.2 Ma), each named for a specific site in North America (Kurtén and Anderson, 1980; Lindsay et al., 1987; Lundelius et al., 1987; Bell et al., 2004) (Ma = mega-annum, one million years). This local paleofauna record, serves as a natural archive. It portrays evolutionary past life based upon the recognition of species diversity, geographic distribution, progressive optimization of habitat, degree of modernity, and stratigraphic range (Downs and White, 1968; Remeika, 1992; Remeika and Jefferson, 1993).

Tephrochronology and paleomagnetism – recent advances in the absolute dating of strata – now offer the ability to assign specific dates to the Anza-Borrego Desert's orderly succession of long sedimentary sequences. Not only can we establish the age of a geological event or a fossil, but we can place it in a comprehensive time-stratigraphic framework using physical properties of the rocks, such as their chemical composition, microscopic appearance, and magnetization. The two techniques are independent, but especially powerful when they can be used to complement each other. The following section describes these dating methods and applies them to the western Borrego Badlands.

Tephrochronology of Anza-Borrego

Figure 6.3
Mt. St. Helen's.
In 1980, Mt. St. Helen's erupts large volumes of volcanic ash into the upper atmosphere. (Photograph courtesy of the U.S. Geological Survey)

Sigurdur Thórarinsson, a volcanologist, coined the term tephrochronology in 1970 to mean the dating of geologic events by reference to their position in a sequence of ash, or tephra deposits (Thórarinsson, 1970.) The word tephra, from the Greek for "volcanic ash", describes the debris, an assortment of fragments ranging in size from large blocks to fine dust that is ejected into the air during a volcanic eruption. Tephras from large-magnitude eruptions are blasted high into the atmosphere, dispersed by wind currents, and are deposited on a continental scale as ash layers across basin-wide landscapes.

On a geologic time scale these layers form time horizons (chronohorizons), each deposited almost instantaneously. Tephra layers may be characterized and identified by several methods: 1. chemical analyses determine its diagnostic "signature" of major, trace, and rare Earth elements; 2. microscopic examination of the shape and nature of the mineral inclusions it contains; and 3. special techniques such as energy-dispersive X-ray fluorescence and electron microprobe analysis of specific fractions of the tephra, such as its volcanic glass shards. Some tephra layers also have been isotopically dated, providing absolute ages for the eruptions that produced them (Izett, 1981) When ash layers are deposited within sedimentary sequences that contain fossils, the precision of dating these fossils is increased enormously.

In the western Borrego Badlands, for example, a sequence of closely spaced tephra layers has been identified in the Pleistocene (Irvingtonian NALMA) fluvial-floodplain and lacustrine (lakebed) sediments of the Bautista beds and Ocotillo Conglomerate. The glass and mineral chemistry of the tephra indicates that they originated in the Long Valley magma chamber, hundreds of kilometers to the north, near Mammoth Lakes in east central California. This huge body of granitic magma, which still exists several kilometers within the earth, has been the source of a variety of eruptions over the past three million years, most dramatically the Long Valley eruption of 758,000 years ago.

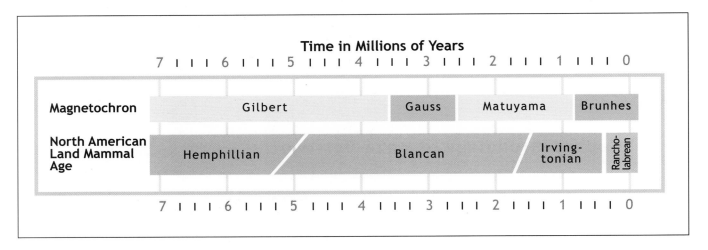

Figure 6.4
Land Mammal Ages Plotted Against the Geomagnetic Polarity Time Scale.
The magnetic polarity sequence is calibrated into time of dominantly normal polarity (brown) and times of reversed polarity (yellow). Time scale follows Lindsay et al. (1987), Berggren et al. (1995), Cande and Kent (1995), and Woodburne and Swisher (1995).

This 0.758 Ma event ejected about 225 km³ (150 cubic miles) of ash into the air. Long Valley itself was formed when the ground collapsed into the void left by the enormous eruption (Izett and Naeser, 1976). Traces of the expelled ash, known as the Bishop Tuff, have been identified from the Pacific Ocean west of Los Angeles to western Nebraska. Temperatures were so high (over 700 degrees Centigrade) that in some places the ash bed was welded into solid rock before cooling. The age of the Bishop Tuff north of Owens Valley has been determined by laser-fusion argon 40Ar/39Ar isotopic dating to be 0.758 +/- 0.002 Ma (Sarna-Wojcicki and Pringle, 1992).

The Bishop Tuff tephra was discovered in the western Borrego Badlands in 1992 by the author and was positively identified based on the similarity of its major and trace element chemistry (Remeika, 1995, 1997). Stratigraphically, the Bishop Tuff is located 67 m (220 ft) above the base of the Ocotillo/Bautista formational contact. It has a normal paleomagnetic declination that is consistent with its chronostratigraphic position directly above the Matuyama/Brunhes geomagnetic boundary dated at 0.780 Ma (Brown et al., 1994; Hillhouse and Cox, 1976; Singer and Pringle, 1997; Tauxe et al., 1996; Valet et al., 1988).

The Bishop Tuff is a master tephra marker bed of the middle Pleistocene Epoch (Figures 6.1, 6.5). Three reasons for this include: 1. it is particularly widespread; 2. it represents a time plane of known absolute age; and 3. it provides independent control on the age of the western Borrego Badlands which otherwise was poorly constrained prior to the magnetostratigraphic study of Remeika and Beske-Diehl (1996). Exposed in floodplain deposits, it is the most reliably correlated tephra layer in the western Borrego Badlands. It may correlate with a stratigraphically high tephra in the Vallecito-Fish Creek Badlands (Remeika, 2001). Regionally, it is reported in badlands exposures near Anza (Sharp, 1965) and Murrieta (Mann, 1955) in Riverside County, and bears a strong resemblance to exposures on the east side of the Salton Trough in the Indio Hills (Rymer, 1991), Mecca Hills (Merriam and Bishcoff, 1975; Babcock, 1974; Rymer, 1991), and Durmid Hills (Babcock, 1974).

Figure 6.5
Bishop Tuff.
A white layer of powdery Bishop Tuff, composed of fine-grained volcanic glass particles, caps this outcrop in the western Borrego Badlands. Rock hammer for scale. (Photograph by Lowell Lindsay)

Other local tephra layers also are associated with both earlier and later eruptions from the Long Valley magma chamber. Several closely spaced discrete white ash layers, chemically and morphologically indistinguishable from each other, are earlier and lie closely spaced below the Bishop Tuff. They are thought to have originated during the Upper Glass Mountain volcanic events of the early to middle Pleistocene, around 900,000 years ago. The lowest two layers have been found in fluvial-floodplain sandstones of the Ocotillo Conglomerate. Some of these ash layers are in direct association with diagnostic early Pleistocene vertebrates, namely *Mammuthus meridionalis* (see McDaniel, this volume, *Mammoths and Their Relatives*).

Above these layers, in the western Borrego Badlands, two more layers occur approximately 30 m (97.5 ft) and 10 m (32.5 ft) respectively below the Bishop Tuff. Both are compositionally indistinguishable from each other. One layer is noteworthy for containing an assemblage of age-diagnostic freshwater ostracodes of the genus *Limnocythere*, and the other layer contains vertebrate footprint impressions of *Lamaichnum borregoensis* (see Remeika, this volume, *Fossil Footprints of Anza-Borrego*).

In total, the Upper Glass Mountain family of tephra beds represent products of several volcanic eruptions from the same eruptive source area that are spaced closely in time.

The stratigraphically youngest tephra layer in the western Borrego Badlands overlies the Bishop Tuff by 40 meters (130 ft), and has been identified as the middle Pleistocene ash of Thermal Canyon based on electron microprobe analysis. It has not been directly dated. However, it closely matches the Bishop

Tuff in its field characteristics, and has a major-element composition that chemically matches correlative tephra that are assigned an age of around 0.72 Ma (Sarna-Wojcicki et al., 1997). This ash is named after Thermal Canyon, located on the east side of the Salton Sea, and came from the resurgent Long Valley caldera, which erupted around 18,000 years after the caldera-forming eruption that produced the Bishop Tuff and associated tephras.

Paleomagnetism

It has long been known that the Earth's rotation generates a geomagnetic force field which blankets the entire planet in much the same way that a magnetic force field surrounds a simple bar magnet. The magnetic north vector was defined with the invention of the first compass. The present polarity, or orientation of the Earth's magnetic field, is such that the north-seeking end of a free compass needle will point in the general direction of the geographic North Pole, and not toward Antarctica. Interestingly, in China, the South Pole traditionally was used as the primary reference compass direction. This North Pole orientation is something we are very comfortable with, but it has not always been so. In the geologic past, as revealed by studying the magnetic properties of iron-containing rock, there have been unpredictable flips in the polarity of the field a few times every million years. For reasons that remain unclear, magnetic north becomes geographic south and magnetic south shifts to geographic north, or vice versa. Periods like the present are called *normal polarity intervals* and periods when magnetic north is toward the geographic south are called *reversed polarity intervals.*

A polarity flip may take thousands of years to spread over the planet from initial foci of change. The very first event in this process is a drop in magnetic field strength, followed by full reversal of the local magnetic field orientation. At the present time, the southwest Pacific Ocean is beginning to reverse, although most of the Earth still retains a normally oriented magnetic field. Compared with the tremendous slowness of most geologic processes, however, even a polarity reversal requiring thousands of years will seem almost instantaneous in the evidence it leaves in the iron-containing minerals of the Earth's crust. Since ferromagnetic (iron-containing magnetic) minerals line up with their poles parallel to the prevailing magnetic field in which they are formed (crystallized or deposited), a permanent record of ancient polarity reversals is available in both igneous and sedimentary rocks.

After bursting through the Earth's crust from fiery depths below, the ferromagnetic minerals that crystallize from a volcanic magma record the orientation of the magnetic field at the time the molten mass cools. Since the Earth's magnetic field vector varies in orientation from the poles to the equator, this rocky "tape recorder" can also be used to determine the latitude where the magma cooled – a capability that has been helpful in tracking the travels of continental landmasses as they ride the crustal plates over the Earth's mantle.

More important in Anza-Borrego, where the continuous sequence of sediments is nearly three miles thick (triple the depth of the Grand Canyon, Figure 6.6), ferromagnetic mineral grains preserved a record of the magnetic polarity as they settled out of suspension. *There have been at least 11 polarity reversals in the past 6 million years, and about 170 reversals over the past 100 million years* (Lowrie and Alvarez, 1981). The durations of these intervals were relatively short, varying from a few tens of thousands of years to hundreds of thousands of years. However, they form a clearly identifiable pattern of reversed and normal magnetic polarity intervals. All over the world, in many different successions of rocks, parts of this magnetic barcode pattern have been identified and correlated. Combined with radiometric dating methods, such as potassium-argon, such correlations have led to the development of a global magnetic polarity time scale (Lindsay et al., 1987; Berggren et al., 1995; Cande and Kent, 1995; and Woodburne and Swisher, 1995). The detailed record of paleomagnetism opens an absolute time window into sedimentary sequences that is especially useful to paleontologists. Fossils found between magnetic signatures in the sediments can be closely constrained, and a minimum and maximum numerical age can be assigned, based on the magnetic polarity time scale.

The basic units for this universally recognized time scale are the long-lived magnetochrons and the shorter-lived subchrons (see Table 6.1). The three most recent magnetochrons are named for scientists who made great contributions to studies of paleomagnetism. For example, the Gauss normal-polarity magnetochron (3.59-2.58 Ma) is named after Carl Gauss, a German physicist who invented the magnetometer in 1832. The Matuyama reversed-polarity magnetochron (2.58-0.78 Ma) is named for

Figure 6.6
The Grand Canyon.
The Grand Canyon sequence of sediments is 1.6 km (1 mile) thick. This is only 1/3 the thickness of the Anza-Borrego strata. (Photograph by Paul Remeika)

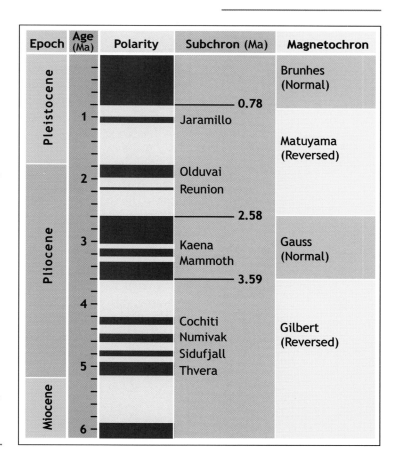

Table 6.1 (At right)
Geomagnetic Polarity Time Scale,
Late Miocene Through Pleistocene.
Black bands represent time intervals of normal polarity; white intervals are reversed polarity. Ages are in million years (Mega-annum, Ma). (Adapted by Lowell Lindsay from Cande and Kent, 1995)

Montori Matuyama, who documented polarity reversals in volcanic rocks in Japan and Korea in 1929 with magnetic polarity stratigraphy. He conjectured that the geomagnetic polarity of the Earth changed several times during the Quaternary Period. The Brunhes normal-polarity magnetochron (0.780 Ma-now), which we are currently experiencing, is named after Bernard Brunhes, a French physician who first determined in 1906 that the Earth's magnetic field was not fixed. He discovered lava flows in the Massif Central Mountains of France with a magnetic polarity that is exactly opposite to the current field. Subchrons, intervals of shorter duration on the geomagnetic time scale, are named for the locations where they were discovered, such as the Olduvai normal-polarity subchron (1.95-1.77 Ma), named for Olduvai Gorge in Kenya, Africa, and the Jaramillo normal-polarity subchron (1.07-0.99 Ma), named for Jaramillo Creek in New Mexico.

Paleomagnetic Sampling Procedures

Paleomagnetic investigation begins with meticulous and demanding fieldwork. Because small mineral grains align better with the ambient magnetic field as they are deposited than do large particles, lacustrine (lake deposited) horizons of clays and silts are best for obtaining samples with reliable paleomagnetic properties. These are collected in stratigraphic order, either by hand or by boring with a portable core drill, after removing up to 0.5 m (1.6 ft) of the exposed surface to reach unweathered sediment. It is a tedious process. The orientation of each sample with respect to the ambient magnetic field and to the azimuth is measured *in situ* with a Brunton compass prior to removal, a necessary preliminary step to determining the parameters of the past magnetic field. Because individually oriented samples are not identical, a minimum of three and usually six samples per site are generally necessary for statistical purposes.

In the laboratory, each sample is trimmed to a one-inch (25 mm) cube. This is placed in a highly sensitive magnetometer to determine its magnetic signature. The inclination and declination of the natural remanent magnetization is calculated from magnetometer readings. Often, this natural remanent magnetization, developed when the sediment was laid down, has been overprinted to some extent by the present magnetic field. This overprinting is removed by a process of stepwise heating or electrical demagnetization techniques, with the sample being re-measured in a spinner magnetometer after each step until a stable, unambiguous reading is obtained. Reliable readings are plotted against stratigraphic levels to construct a composite magnetic polarity stratigraphy for each measured stratigraphic section.

Magnetic Polarity Stratigraphy of the Western Borrego Badlands

Based on various measured sections of the western Borrego Badlands (Scheuing, 1989; Remeika and Beske-Diehl, 1996; Remeika and Liddicoat, in prep.), the continuous fluvio-lacustrine (floodplain-lakebed) basin fill sequence allows for the identification of the magnetic polarity stratigraphy. From top to bottom, the Pliocene and Pleistocene sediments are crossed by several important magnetic zones of both normal and reversed polarity. These are stratigraphically distinct, and can be directly correlated to the geomagnetic polarity time scale.

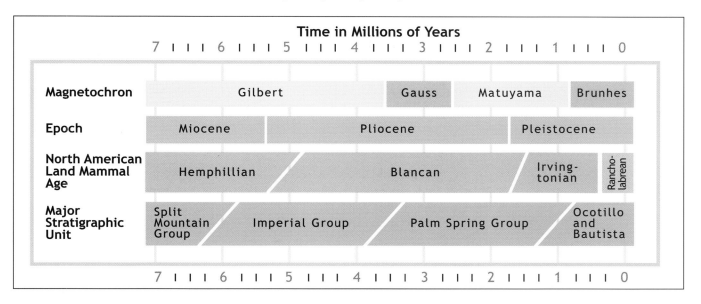

The sediments of this basin are divided into two depositional sequences: 1. locally-distributed floodplain alluvium of the Plio-Pleistocene Ocotillo Conglomerate (Remeika, 1992) and Pleistocene Bautista beds (Frick, 1921), and 2. the ancestral Colorado River-derived lakebeds of the Plio-Pleistocene Borrego Formation (Tarbet and Holman, 1944) and Pleistocene Brawley Formation (Dibblee, 1954).

The Borrego and Brawley Formations are compositionally distinct, and consist of fine-grained gray and red claystones, siltstones, and subordinate sandstones that grade up section into the Ocotillo Conglomerate and Bautista beds, respectively. The Ocotillo Conglomerate is a fluvial-floodplain succession of gravelly sandstones, siltstones, and clays. The Bautista beds are composed of marginal fluvio-floodplain and lacustrine (lakebed) claystones, siltstones, and gravelly sandstones. Complex intertonguing (overlap and interfingering) between these various lithofacies (gradational sedimentary rock types) is common along the basin margins. Here an overall upward coarsening, and increase in heterogeneity (mixed types) of clastic (rock fragments and mineral grains) sedimentation takes place.

Figure 6.7
Generalized Correlation of Anza-Borrego's Magnetic Stratigraphy to the Geomagnetic Polarity Time Scale.

The age of strata depends on its stratigraphic position, fossils, and interpretation of reversal intervals (yellow) in magnetic polarity. Neogene/quaternary time scale follows Berggren et al. (1987) and Cowie and Bassett (1989).

The uppermost 123 m (400 ft) of the composite measured section contains age-diagnostic ostracode fossils, small freshwater crustaceans (*Limnocythere* sp.), which are considered indicator fossils of middle Pleistocene age. This age is in excellent agreement with the paleomagnetic data. The normal polarity of the samples suggests that sediments were deposited during the Brunhes normal-polarity magnetochron. Furthermore, the presence of the Bishop Tuff and the ash of Thermal Canyon within this interval support the biostratigraphic, sedimentary, and paleomagnetic data, and serve as excellent stratigraphic markers for age control.

Below 123 m, the first polarity reversal is identified, from dominantly normal to dominantly reversed. It represents the Brunhes/Matuyama geomagnetic boundary at 0.78 Ma. Age-diagnostic fossils in these strata of the upper Matuyama magnetochron are consistent with an early Pleistocene age or Irvingtonian NALMA. These include the ostracode *Elkocythereis bramlettei* in the Brawley Formation, and presence of a vertebrate assemblage including *Pitymys meadensis*, *Mammuthus imperator, M. meridionalis, Equus scotti,* and *Camelops huerfanensis* or *C. hesternus* (Remeika and Beske-Diehl, 1996; Remeika, 1998; McDaniel and Jefferson, 2003a, 2003b) in the Ocotillo Conglomerate.

Samples obtained between 271 and 304 m (880 to 988 ft), show normal polarity, indicating that the Jaramillo subchron was recorded in the Ocotillo deposits. Below this depth, for 119 m (387 ft), the dominantly reversed polarity continues to correlate well with the lower Matuyama Magnetochron.

The next distinctive polarity shift, from reversed to normal and back, occurs in the Borrego Formation between 423 and 545 m (1347 to 1771 ft), and correlates to the Olduvai subchron. This short interval is notable because it encompasses the internationally accepted Plio-Pleistocene boundary at 1.8 Ma (Pasini and Colalingo, 1997). Below 545 m (1770 ft), the dominantly reversed polarity correlates well with the lowermost portion of the Matuyama. At 758 m (2462 ft), a polarity shift to normal, clearly identified in mudstones and silts in the southwestern Borrego Badlands, signifies the boundary of the predominantly normal polarity period represented by the Gauss magnetochron of the Pliocene Epoch.

Summary

To improve dating and correlation of basin-wide sedimentary sequences, it is necessary to combine lithostratigraphic (see Dorsey, this volume, *Stratigraphy, Tectonics and Basin Evolution in the Anza-Borrego Desert Region*, and biostratigraphic, see Cassiliano, this volume) data with tephrochronologic and paleomagnetic observations. This establishes a comprehensive and reliable time-stratigraphic framework (Sarna-Wojcicki et al., 1980).

The presence of the Bishop Tuff (0.758 Ma) and the slightly younger ash of Thermal Canyon within the western Borrego Badlands extends the known dispersal range of these tephra into Anza-Borrego. Regionally, both tephras reside as a well-documented stratigraphic pair in middle Pleistocene (Irvingtonian NALMA) age deposits of Bautista beds. Their occurrence is in excellent agreement with magnetic polarity stratigraphy, providing a strong correlation of chronologies in the composite measured stratigraphic section for the Borrego Badlands. The paleomagnetic data clearly indicate that the normal polarity chron in the upper stratigraphic levels of the section is the Brunhes magnetochron, with the Matuyama reversed polarity magnetochron a short distance stratigraphically below.

The boundary of the Brunhes and Matuyama magnetochrons is evidenced by a paleomagnetic signature preserved in a prominent pebbly sandstone layer of Bautista beds. Furthermore, the data give us added confidence that the normal polarity interval below this boundary is the Jaramillo subchron in the Ocotillo Conglomerate. The next older normal polarity interval, documented in the Borrego Formation is the Olduvai subchron, marking the Plio-Pleistocene boundary.

It is our good fortune that here, in the Anza-Borrego Desert, exposures of these thick sedimentary sequences are marked by ash layers from the cataclysmic, well-studied, and radiometrically-dated eruption sequence at Long Valley. These ash layers bear in their iron-containing sediments the traces of ancient magnetic polarity reversals of known ages. Because of these ash layers and magnetic signatures in the strata, our dating of the time-characteristic fossils found in the western Borrego Badlands is further supported and confirmed. *How do we know how old the bones are? It is written in stone, in several dialects of the language of science.*

7 Mammalian Biostratigraphy in the Vallecito Creek-Fish Creek Basin

Discovery consists of seeing what everybody has seen and thinking what nobody has thought.

Albert von Szent-Gyorgi (1893-1986)

Michael Cassiliano

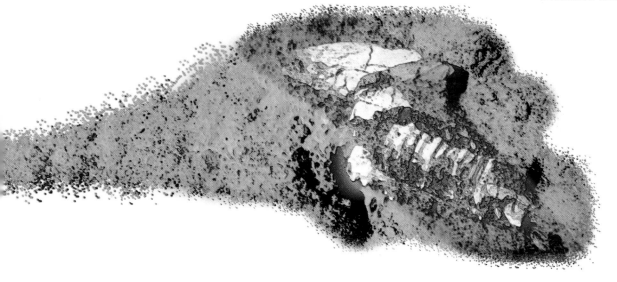

Mammalian Biostratigraphy in the Vallecito Creek-Fish Creek Basin

Deltaic Sediments Above Fish Creek Basin. (Photograph by Paul Remeika)

Introduction

Biostratigraphy is the study of the stratigraphic distribution of fossils. It addresses the questions: 1) in a given thickness of sedimentary rock, what are the fossil taxa (the different types of organisms), and 2) where do they first appear and then disappear? According to the North American Stratigraphic Code (1983, page 862): "A biostratigraphic unit is a body of rock defined or characterized by its fossil content. The basic unit in biostratigraphic classification is the biozone, of which there are several kinds." Biostratigraphy combines the disciplines of paleontology, the study of ancient life, and stratigraphy, the study of the layers of sedimentary rocks.

Biostratigraphy is crucial to understanding the paleontologic and geologic history of Anza-Borrego Desert State Park (ABDSP) because the sequence of appearances and disappearances of animals forms the chronologic background for that history. This chapter deals with the mammalian biostratigraphic record in the Vallecito Creek-Fish Creek Basin in the southern part of ABDSP.

Biostratigraphy

Biostratigraphy is founded on two basic observations about rocks and fossils. The first is that *in a column of sedimentary rock, the oldest rocks are at the bottom of the column and the youngest rocks are at the top* (Figure 7.1). This is the Principle of Superposition, first proposed by the Danish geologist Nikolas Steno in 1669.

The second observation is the Principle of Faunal and Floral Succession (Figure 7.2). Paleontologists saw that particular types of fossils always occurred together as assemblages no matter where they looked. They also noted that the fossils in one assemblage were distinct from those in others. Finally, they observed that each distinctive fossil assemblage was always in the same stratigraphic position (above or below) relative to other assemblages.

For example, fossil assemblage A always contains the same types of animals and plants. Fossil assemblage B also contains a unique set of animals and plants, different from that in assemblage A, as does fossil assemblage C. Furthermore, in a section of sedimentary rock, fossil assemblage A always occurs below the rocks containing fossil assemblage B, and the rocks containing fossil assemblage B are always below the rocks containing fossil assemblage C. If we combine this with the Principle of Superposition, we can say that fossil assemblage A is older than assemblage B, and assemblage B is older than assemblage C. Thus, the Principle of Faunal and Floral Succession was derived: *assemblages of fossil taxa are unique and distinctive from each other, and assemblages succeed each other in an orderly and predictable way.*

The uniqueness of each fossil assemblage is due to evolution and extinction. During its existence, each taxon evolves a distinctive suite of characteristics that sets it apart from all other taxa. These distinct taxa combine to form a unique assemblage of organisms. When a taxon becomes extinct, its genes are lost forever and it does not re-evolve. This irreversibility of evolution is what makes fossil taxa and fossil assemblages unique.

The succession of fossil assemblages produces a relative scale of geologic time. Based on fossils alone, no numerical dates can be assigned to the time in which the fossil assemblage lived. All the fossils can tell us is that a particular layer of rock is older or younger than another layer.

North American Land Mammal "Ages" and Biostratigraphy

Paleontologists have long known that particular sequences of terrestrial sedimentary rock are characterized by the presence of distinctive groups of mammalian fossils. These groups have been used by vertebrate paleontologists to subdivide the Cenozoic Era (the past 65 million years) into smaller time divisions. In North America, these mammal-based time divisions are called North American Land Mammal "Ages" (NALMA) (Figure 7.3). Each NALMA is characterized by the first and last appearances of different taxa of mammals, and by unique associations of mammalian taxa within it.

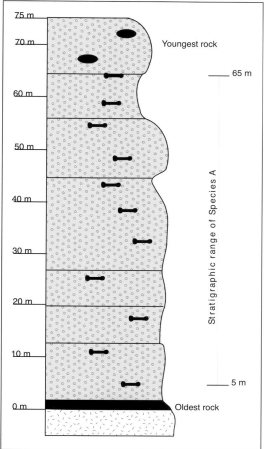

Figure 7.1
Example of Practical Biostratigraphic Practice.
The columnar section consists of several layers of sedimentary rock containing fossils. Specimens of species A are shown by the bone-shaped symbols, specimens of species B by the ovals. The columnar section is divided into 10 meter (32.5 ft) increments measured from the top of the underlying crystalline rock (confetti pattern). The stratigraphic range of species A is measured from the lowest fossil specimen (lowest stratigraphic datum, LSD) to the highest specimen (highest stratigraphic datum, HSD).

A NALMA is a special type of time division called a biochron, a span of time defined by the animals and plants that lived during it. Although based in part on the stratigraphic ranges of the taxa, biochrons are not strictly tied to columns of sedimentary rock, and do not depend on boundaries between fossil-based units defined on the first and/or last appearances of taxa. Therefore, they are not recognized as biostratigraphic units in terms of the codes used by paleontologists and geologists to establish stratigraphic and biostratigraphic units.

NALMAs Originally Called "North American Provincial Ages"

NALMAs are the result of the Wood Committee's publication of 1941 and were originally called "North American Provincial Ages." These were established only for the Tertiary Period of the Cenozoic Era; the earlier Cretaceous Period and later Pleistocene Epoch were not included. Savage (1951) later proposed provincial age names for the Pleistocene. The goal of the committee was to establish a series of purely temporal units for terrestrial rocks containing mammal fossils. This time scale was to be independent of the global geologic time scale (based on marine invertebrates in Europe) and was not defined in relation to that classic time scale.

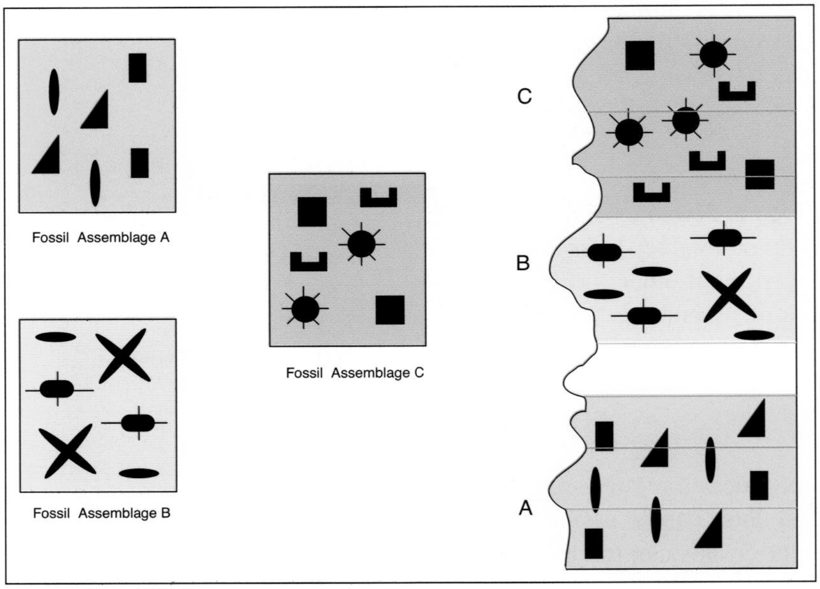

The North American Provincial Ages were based on diagnostic assemblages of fossil mammals thought to indicate specific intervals of geologic time. But insufficient attention was paid to stratigraphic details, with the result that the North American Provincial Ages came to be based on biologic rather than lithologic criteria, on fossil assemblages instead of sedimentary strata. Thus, they are not true biozones, but biochrons. The results achieved by the Wood Committee proved to be less than satisfactory. The North American Provincial Ages, although characterized by biologic criteria, were in fact bounded by the physical boundaries of the rock units lending their names to the age. For example, the Barstovian North American Provincial Age (Figure 7.3) took its name from the Barstow Formation in southern California and the temporal duration of the Barstovian was considered coeval (same age span) with that of the Barstow Formation.

NALMA Replaces "North American Provincial Ages"

Savage (1962) proposed that the term North American Provincial Age be replaced by North American Land Mammal "Age." The word "age" is usually placed within quotation marks to show that the NALMAs are not true

Figure 7.2
Faunal and Floral Succession.
Three fossil assemblages, A, B, and C consist of different taxa as shown by the different symbols. In the columnar section, these assemblages are placed in their superpositional order. Note that each assemblage consists of a unique group of taxa that can be used to infer distinct spans of geologic time.

geochronologic ages, since they are not linked by documented first and/or last appearances to a specific unit of rock formed during a defined span of geologic time.

Despite their drawbacks, the NALMAs have been useful in providing a mammal-based relative time scale for terrestrial sedimentary rocks in North America. They aid communication among vertebrate paleontologists by allowing those studying fossil mammals to establish a relative age for their specimens and to order the specimens in relation to older and younger assemblages.

Only recently (using the techniques of radiometric dating and magnetostratigraphy; see Remeika, this volume, *Dating, Ashes and Magnetics*), in the setting of mammalian assemblages discovered in terrestrial rocks that interfinger with marine rocks, has it been possible to begin correlating the NALMAs with the marine-based Epochs of the global geologic time scale established in Europe. As a result, many of the traditional NALMA-Epoch correlations have been revised, and it has been determined that NALMA boundaries do not necessarily coincide with Epoch boundaries. For example, the Chadronian NALMA (Figure 7.3), once placed in the Oligocene Epoch, is now placed in the Eocene Epoch. The Oligocene-Miocene Epoch boundary is now located within the Arikareean NALMA.

Bell et al. (2004) make the convincing argument that extreme faunal provinciality in North America from the Blancan, about 5 million years ago, to the present, implies that faunas from high northern latitudes (> 55°N latitude) are very different from those that lived at the same time in more southerly latitudes. The NALMAs in common use are defined and characterized by lower latitude faunas, and thus do not apply to the fauna of higher latitudes. They recommend that the NALMAs be restricted in their use to faunas below 55°N latitude.

There has been a growing movement among North American vertebrate paleontologists to define the NALMAs according to the procedures set forth in the formal stratigraphic codes such as the North American Stratigraphic Code (1983). This would give the NALMAs a firm stratigraphic basis to be formally recognized as chronostratigraphic units (bodies of rock representing all the geologic time between their boundaries).

Era	Period	Epoch	NALMA	Ma
Cenozoic	Quaternary	Holocene	No NALMA	10,000 years ago
		Pleistocene	Rancholabrean	72,000 to 210,000 years ago
			Irvingtonian	
		1.77		1.35
	Tertiary	Pliocene	Blancan	4.6 to 5.2
		5.335		
		Miocene	Hemphillian	8.9
			Clarendonian	12.5
			Barstovian	15.9
			Hemingfordian	21
		23.5	Arikareean	
		Oligocene	Whitneyan	30 / 32
		33.5	Orellan	33.5
		Eocene	Chadronian	37
			Duchesnean	40
			Uintan	46.5
			Bridgerian	50.4
		55	Wasatchian	55.5
		Paleocene	Clarkforkian	56
			Tiffanian	61
			Torrejonian	63.5
		65	Puercan	

Figure 7.3
List of North American Land Mammal "Ages."
The NALMAs are correlated with the standard geologic time scale of Eras, Periods, and Epochs. The numerical dates show the time, in millions of years ago, that the Epoch or NALMA began (Prothero, 1998; Bell et al., 2004).

The NALMAs of the Pliocene and Pleistocene Epochs

The Pliocene is the youngest epoch of the Tertiary Period of the Cenozoic Era. It was established by the English geologist Charles Lyell in 1833. He originally used the term Older Pliocene, defined by the presence of between 40 and 70 percent of Recent marine molluscan species in a fossil assemblage. He also proposed a Newer Pliocene that contained more than 70 percent Recent marine molluscan species. In 1839, he changed the name Newer Pliocene to Pleistocene. Note that the Pliocene and Pleistocene were established in Europe on the basis of marine mollusks, not in North America and not on the basis of mammals. This means that dating any North American mammal fossils as Pliocene or Pleistocene must be done by correlating them with marine sedimentary rocks containing mollusks of the appropriate age. This is rarely possible in North American terrestrial fossil assemblages.

The Wood Committee established three North American Provincial Ages (now NALMAs) for the Pliocene; from oldest to youngest, they were the Clarendonian, Hemphillian, and Blancan. Recent work on the type section of the Pliocene in Europe, using radiometric dating and magnetostratigraphy, has shown that the Pliocene is much shorter than once thought. In terms of NALMAs, it consists only of the latest Hemphillian and Blancan (Figure 7.3). The Clarendonian and most of the Hemphillian are now assigned to the older Miocene Epoch. The Pliocene Epoch is thought to have lasted from about 5.4 million years (Ma) ago to 1.8 Ma ago (Pasini and Colalongo, 1997; Bell et al., 2004).

The latest Hemphillian contains a mix of taxa, some from the Hemphillian and others characteristic of the Blancan. This suggests that the latest Hemphillian was a time of transition between the typical older Hemphillian mammals and those of the younger Blancan. According to Tedford et al. (1987), the more important taxa in the latest Hemphillian are odocoiline deer (the whitetail and mule deer), *Felis* (mountain lion), *Megantereon* (sabertooth cat), *Ochotona* (the mountain-dwelling pika or conie, a relative of rabbits and hares), *Plesiogulo* (an extinct genus of wolverine), *Agriotherium* (an extinct genus of bear), and *Promimomys* (an extinct genus of vole). The Hemphillian is thought to have ended between 5.2 and 4.6 Ma ago, depending on which date is accepted for the beginning of the Blancan (Bell et al., 2004).

Most of the Pliocene Epoch in North America is correlated with the Blancan NALMA, which lasted from between 5.2 or 4.6 to about 1.4 Ma ago (Bell at al., 2004). The new (2004) younger date (1.4 Ma) for the end of the Blancan means that the very youngest Blancan fossil assemblages are now included in the earliest Pleistocene Epoch (Bell et al., 2004). Lundelius et al. (1987) listed the following genera as having first appeared at or very near to the Hemphillian-Blancan boundary: *Ursus* (bear), *Equus* (*Dolichohippus*) (zebra), *Odocoileus* (deer), *Chasmaporthetes* (an extinct hyena), *Geomys* (a burrowing gopher), *Trigonictis* (an extinct mustelid), *Sigmodon* (cotton rat), *Mimomys* (an extinct rodent), and *Pliophenacomys* (an extinct rodent). However, they did not choose a single genus to define the boundary between the Hemphillian and Blancan in the sense of Murphy (1977) and Woodburne (1977, 1987, 1996). Bell et al. (2004) suggested that the earliest appearance of *Mammuthus,* the arvicoline rodent *Mimomys*, or the *Mimomys*-like arvicolines *Ogmodontomys* and *Ophiomys,* be used to define the beginning of the Blancan below 55°N latitude. In ABDSP, we find all of the genera listed above, with the exceptions of *Mimomys, Ophiomys, Ogmodontomys,* and *Pliophenacomys.* Thus, at present, the Hemphillian-Blancan boundary can only be approximated in the Park's thick stratigraphic section.

The criteria and dates for the beginning of the Pleistocene Epoch have been debated for some time. The most recent data from the type section in Italy suggest that the Pleistocene began around 1.8 Ma (Pasini and Colalongo, 1997). The Pleistocene Epoch in North America is divided between two NALMAs, the older Irvingtonian and the younger Rancholabrean (Savage, 1951; see Figure 7.3). Cassiliano (1999) and Bell et al. (2004) discuss the problems associated with defining the Blancan-Irvingtonian boundary. Traditionally, the Irvingtonian is defined by the appearance of the mammoth, *Mammuthus,* an immigrant from Eurasia; Bell et al. (2004) proposed retaining this definition. According to these authors, the oldest reliable radiometric date for the first appearance of *Mammuthus* is between 1.35 and 1.4 Ma. This means that the beginning of the Pleistocene Epoch and the Irvingtonian NALMA do not coincide; the Blancan-Irvingtonian boundary at about 1.4 Ma does not match up with the slightly older Plio-Pleistocene boundary at 1.8 Ma. Cassiliano (1999) suggested that other, better-dated taxa might better define the boundary because of problems in dating early appearances of *Mammuthus*. Other important genera of mammals that are characteristic of the Irvingtonian include: *Smilodon* (sabertooth cat), *Euceratherium* (an extinct bovid), and *Lepus* (hare). Genera once thought to have first appeared in the Irvingtonian, such as the extinct pronghorn *Tetrameryx*, true horses *Equus (Equus),* and *Lepus* now fall within the late Blancan as a result of the redefinitions of Bell et al. (2004).

The Rancholabrean is characterized by the earliest appearance in North America of *Bison,* another immigrant from Eurasia. The radiometric age of the Irvingtonian-Rancholabrean boundary is still not settled; estimates range from 72,000 to 210,000 years ago (Bell et al., 2004). There is a significant problem with trying to determine if an assemblage of fossil mammals is Irvingtonian or Rancholabrean because the two NALMAs have many taxa in common. The presence of *Bison* is a sure way to distinguish a Rancholabrean assemblage from an Irvingtonian one, but *Bison* is not found in all Rancholabrean assemblages. The problem is compounded by the fact that *Mammuthus* survived from the Irvingtonian through the Rancholabrean. The presence of *Mammuthus* without *Bison* in an assemblage is no guarantee that the assemblage is either Rancholabrean *or* Irvingtonian. Taxa known only from the Rancholabrean have not been found in the Vallecito Creek-Fish Creek Basin area of the Park; consequently, rocks formed during that NALMA are not recognized here.

North American Land Mammal "Ages" and Anza-Borrego Desert State Park

One of the main areas where land mammal fossils are found in ABDSP is the Vallecito Creek-Fish Creek Basin (VCFCB). These fossils were collected primarily from sedimentary rocks of the Palm Spring Group (Cassiliano, 2002), a suite of mainly deltaic and fluvial rocks with some lacustrine rocks; that is, the Palm Spring Group represents a terrestrial environment. Below the Palm Spring Group are rocks called the Imperial Group, a marine unit. Below the Imperial Group is a mix of terrestrial and volcanic units (Winker and Kidwell, 1986).

For several reasons, the VCFCB area is one of the best places in North America to study the biostratigraphy of land mammals that lived during the Blancan and early Irvingtonian NALMAs. The Pliocene and Pleistocene sediments in the Anza-Borrego Desert are more than 4,300 meters thick (13,975 ft), which permits study of the stratigraphic distribution of fossils in a single geographic area without the need to correlate from one distant area to another. Elsewhere in North America, the exposures of sedimentary rock containing mammalian fossils are usually very limited in their vertical and lateral extent. Consequently, only portions of a fossil mammal's stratigraphic range can be studied in any one area; hence the need to correlate the rocks and fossils from one area to another. In the VCFC basin a collecting transect was established to permit the stratigraphic ordering of the fossils found there (see below). This transect is an invaluable tool for plotting the appearances and disappearances of fossil taxa through time. It allows us to study changes in the composition of Pliocene and Pleistocene faunas (assemblages of animals that lived in particular place at a particular time) in a single area and see how those faunas changed or remained stable as their environment changed.

Also, strata in the Anza-Borrego Desert are very well exposed because of erosion, and there is little or no soil and vegetation covering them. This makes finding fossils, describing the deposits, measuring stratigraphic sections, and placing fossil locations in the section much easier.

Finally, VCFCB strata have been studied with respect to their magnetostratigraphy, the pattern of polarity reversals of the earth's magnetic field. This, combined with radiometric and ash bed dating, allows the rocks and fossils to be correlated with the global geologic time scale (Opdyke et al., 1977; Johnson et al., 1983; Cassiliano, 1999; see Remeika, this volume).

Biostratigraphic Practice in ABDSP

Mammal fossils from the sedimentary rocks of ABDSP have true biostratigraphic meaning because the fossils are linked to documented stratigraphic levels of the sedimentary formations in the Park. The late Dr. Theodore Downs of the Natural History Museum of Los Angeles County devised a method for doing this, using air photographs (provided by the Soil Conservation Service of the Department of Agriculture). These showed the ground surface in great detail and allowed accurate plotting of fossil localities in the rugged terrain. Fossil localities were initially plotted on the air photographs by marking the sites with insect-mounting pins. All of these points are now on geographic information system, GIS, layers; all new sites are located on digital ortho quarter quadrangle ground imagery with a global positioning system or GPS.

The stratigraphic ordering of fossil localities was achieved by using a collecting transect drawn on the air photographs. As used in the VCFCB area, this is a line that divides the stratigraphic section into a series of sampling horizons (positions in a stratigraphic column that associate the stratum with a particular geologic time) called collecting units (CUs). Each CU is the equivalent of about 61 m (200 ft) of rock thickness. The system allows fossil localities to be placed to within about six meters (20 ft) of each other. The 66 numbered CUs are parallel to the strike (the linear trend of the stratum measured as a compass direction) of the sediments. The baseline for the transect is the base of CU 56, near the top of the paleomagnetic Olduvai Subchron at the Plio-Pleistocene boundary (Appendix, Table 5).

Tables 5A to 5C (Appendix, Table 5) show the vertical ranges of the different mammalian taxa in the VCFCB stratigraphic section (these figures should be referred to for the following sections of this chapter). These ranges are plotted against several metrics: thickness of the rock column, CUs, magnetostratigraphy, geologic formation, NALMA, and Epoch. By structuring the data in this manner, one can see the stratigraphic and temporal range of any mammalian taxon in purely local terms (position in the local rock column) or in terms that allow correlation with other Blancan and Irvingtonian paleofaunas. These figures plot the LSD (lowest stratigraphic datum or level) and HSD (highest stratigraphic datum or level) for each genus and species found in the VCFCB area. These are not necessarily the first and last geologic appearances of these animals, which must be determined by correlation with all occurrences on a continent-wide basis. The LSDs and HSDs show when each genus and species made its first and last appearances in the VCFC basin only, and are always subject to change with new fossil discoveries or changes in the identification of a specimen used to establish the LSD or HSD. The line connecting each animal's LSD and HSD does not imply that there is a continuous series of fossils between the LSD and HSD. Gaps in the fossil record are the rule rather than the exception. However, the gaps likely are caused by non-preservation (Cassiliano, 1997) rather than by repeated emigrations and immigrations. Thus, a reasonable assumption can be made that each taxon was present in the VCFCB area continuously between its LSD and HSD.

Hemphillian and Older NALMA in ABDSP

Hemphillian age mammals are very rare from ABDSP. Outside the VCFC basin in the Coyote Mountains south of the Park, sediments associated with the volcanic Alverson Formation yield fossils of the rodent *Cupidinimus*. The specimen has been correlated by radioisotope dating of the volcanics with the Hemingfordian and/or Barstovian NALMAs (S. Walsh, personal communication, 2002; see Figure 7.3). Also, fossils from the Hawk Canyon locality southeast of Borrego Springs may demonstrate the existence of Clarendonian or Hemphillian age mammals in the Park (Jefferson, 1999). However, these are fragmentary and cannot be positively identified, and there are no reliable dates from this deposit. Rare specimens of the camelid *Hemiauchenia vera* recovered from the marine Latrania and Deguynos Formations of the Imperial Group indicate that Hemphillian-aged strata do occur in the Park (Deméré, 1993; Webb et al., this volume, *Extinct Camels and Llamas of Anza-Borrego*). The search for land mammals of the older NALMAs is potentially a fruitful area for research in the Anza-Borrego Desert.

Blancan NALMA in ABDSP

Of the Blancan genera found in the Park, all except for *Sigmodon* (cotton rat) occur very high stratigraphically, above any possible placement for the boundary with the Hemphillian. At present, the best approximation for the Hemphillian-Blancan boundary is the LSD of *Sigmodon* because the suggested defining taxa (*Mimomys, Ophiomys,* and *Ogmodontomys* (Bell et al., 2004) are not found here. The *Sigmodon* LSD occurs in CU 6.2 (approximately 3,980 m below the top of the section) and has an estimated date of 4.2 Ma. This is younger than proposed ages of between 5.2 and 4.6 Ma for the beginning of the Blancan (Lundelius et al., 1987; Bell et al., 2004). Other mammalian genera, such as *Dinohippus* (extinct horse) and *Perognathus* (rodent), occur in rocks below the *Sigmodon* LSD, but these are genera that first appeared in the Hemphillian. Their presence suggests the presence of Hemphillian-aged rocks below the *Sigmodon* LSD. However, both *Dinohippus* and *Perognathus* survive into the Blancan, so they do not automatically mean that the rocks containing their fossils are Hemphillian in age. Until better data, in the form of stratigraphically documented earliest Blancan and latest Hemphillian mammal fossils are found, the *Sigmodon* LSD remains our best indicator of the Hemphillian-Blancan boundary in the Park. Most importantly, the fossil data from VCFCB suggest that the entire Blancan is represented by a single body of rock in a single geographic area, a critically important point if the Blancan NALMA is to be converted to a true biostratigraphic unit rather than remaining a biochron.

During the Blancan, the Isthmus of Panama formed, connecting North America with South America. This major plate tectonic event, which occurred about 3.0 million years ago, allowed animals and plants to easily disperse between the two continents, an event called the Great American Biotic Interchange (Webb, 1976; see McDonald, this volume, *Anza-Borrego and the Great American Biotic Interchange*). Animals such as ground sloths, glyptodonts, and porcupines entered North America. Although a few sloth bones have been recovered from the Hemphillian-age Deguynos Formation, they are not identifiable to genus. Otherwise, the first South American immigrant to the VCFCB area was the porcupine, *Coendu stirtoni.* The stratigraphically lowest fossils of this arboreal rodent occur in CU 47.2, with an estimated age of 2.5 Ma. Surprisingly, the large armored glyptodonts, relatives of the living armadillo, apparently never entered the VCFCB area. Some barrier, such as the ancestral Colorado River, may have prevented their dispersal here (see McDonald, this volume).

The Blancan age rocks in VCFCB also record the survival of several mammals of Hemphillian and older NALMAs that were once thought to have become extinct much earlier, such as the horse *Dinohippus* (Table 7.1). Previous studies suggested that *Dinohippus* became extinct during the latest Hemphillian, about 4.89 Ma (Lindsay et al., 1984). The specimens from VCFCB show that this horse survived well into the Blancan, becoming extinct at about 3.0 Ma in CU 43.0 (Downs and Miller, 1994; Cassiliano, 1999; see Scott, this volume, *Extinct Horses and Their Relatives*). Another example is the proboscidean *Gomphotherium*, a hippopotamus-sized distant relative of elephants. Jefferson and McDaniel (2002) reported fossils of *Gomphotherium* from upper Deguynos Canyon in the VCFCB area. The approximate age of the site is 3.6 Ma (CU 26), which is well within the Blancan, and represents a temporal range extension for the taxon that previously was considered restricted to the Hemphillian.

TABLE 7.1

Hemphillian Mammalian Survivors Found in the Blancan-Aged Rocks of the Vallecito Creek-Fish Creek Basin, Anza-Borrego Desert State Park.

SCIENTIFIC NAME	INFORMAL OR COMMON NAME
Calomys (Bensonomys)	extinct genus of rodent
Dinohippus	extinct genus of horse
Gomphotherium	extinct genus of proboscidean
Hemiauchenia	extinct stilt-legged llamas
Perognathus	pocket mice

Additionally, the Blancan rocks in VCFCB show that mammals characteristic of the succeeding Irvingtonian NALMA began to appear earlier than previously thought. The Blancan occurrences of these Irvingtonian mammals demonstrate that the change from the Blancan to the Irvingtonian was a transition rather than an abrupt faunal turnover (Table 7.2).

TABLE 7.2

Mammals Once Thought to Have First Appeared in the Irvingtonian But Have Been Discovered in the Blancan of Vallecito Creek-Fish Creek Basin, Anza- Borrego Desert State Park.

SCIENTIFIC NAME	INFORMAL OR COMMON NAME
Arctodus simus	extinct short-faced bear
Camelops	extinct camel-like llamas
Canis priscolatrans	extinct coyote
Equus (Equus)	modern horse genus
Navahoceros	extinct mountain deer
Tetrameryx	extinct four-horned pronghorn
Tremarctos	spectacled bears

These refinements in our determination of the geologic range of these mammals are based on the fact that the deposits of the VCFCB preserve nearly all the time encompassed by the Blancan in a single stratigraphic section. Because so little of the Blancan rock record is missing, the fossil record here is more complete, and more likely to preserve biological and evolutionary events not preserved in the less complete rock and fossil records in other areas.

An interesting example of ecologic replacement also can be documented in the Blancan rocks of the VCFCB, in the Lagomorpha (rabbits, hares, and pika), with extinction of an entire lagomorph subfamily. One of the major groups of lagomorphs is the subfamily Archaeolaginae, represented in VCFCB by *Hypolagus* and *Pewelagus*. As shown by White (1984, 1987) (see White and others, this volume, *The Small Fossil Mammals*), *Hypolagus* is ecologically similar to the living cottontail, *Sylvilagus,* and *Pewelagus* is ecologically similar to the pika, *Ochotona.* That is, these are ecomorphs of each other (two independent taxa adapted to the same environment). The fossil record shows that the HSD of both *Hypolagus edensis* and *H. vetus* (CU 39.8, about 3.2 Ma) in VCFCB is most probably its last occurrence in North America. Shortly thereafter its ecomorph, *Sylvilagus,* appears in VCFCB; a probable case of ecologic replacement of one lagomorph by another. The HSD of *Pewelagus* (CU 55.5, about 1.9 Ma) probably represents the last occurrence of the Archaeolaginae in North America.

Equally important, paleontologically, is the LSD of the modern horse lineage, *Equus (Equus),* in VCFCB (CU 50.6, about 2.1 Ma). This LSD may represent the first occurrence of the modern horse in North America (see discussion by Scott, this volume).

Such major events in the evolutionary history of the Lagomorpha and Equidae could not have been adequately documented outside the stratigraphically complete section of VCFCB.

Irvingtonian NALMA in ABDSP

The Irvingtonian NALMA is marked by the immigration of large and small mammals from Eurasia, as well as by the appearance of mammals that evolved from Blancan North American ancestors. Among the most important immigrants found in the VCFCB are *Microtus californicus* (meadow vole), *Mammuthus meridionalis* (southern mammoth), and an extinct form of shrub-oxen (? *Euceratherium* sp.). The record of *Microtus californicus* (CU 57.8, about 1.5 Ma) may be the oldest in North America. However, the validity of this record recently has been questioned by Murray and others (personal communication 2005).

In the VCFCB, *Mammuthus* is known from molar fragments found at three localities (McDaniel and Jefferson, 1999) (see McDaniel, this volume, *Mammoths and Their Relatives*). The oldest of these specimens is approximately 1.4 Ma, which falls within the 1.35 to 1.4 Ma for the earliest appearance of

Mammuthus noted in Bell et al. (2004). For the present, the beginning of the Irvingtonian in ABDSP is best approximated by the stratigraphic position of the oldest (LSD) *Mammuthus* specimen.

TABLE 7.3
Blancan Mammalian Survivors Found in the Irvingtonian-Aged Rocks of the Vallecito Creek-Fish Creek Basin, Anza-Borrego Desert State Park.

SCIENTIFIC NAME	INFORMAL OR COMMON NAME
Calomys (*Bensonomys*)	extinct genus of rodent
Equus simplicidens	extinct zebra
Stegomastodon	extinct proboscidean

The change from the Blancan to the Irvingtonian was a gradual transition rather than an abrupt event, as evidenced by genera of Blancan mammals that survived into the Irvingtonian (Table 7.3) as well as by the early-appearing Irvingtonian mammals (Table 7.2). This clearly shows that the change from one NALMA to another is not always a case of extinction and rapid replacement by immigration and evolution. But it would not have been possible to demonstrate this in areas where the Blancan and Irvingtonian strata form thin layers with no superposition of fossil specimens. Under such conditions, the change from one NALMA to the next would appear to be abrupt or geologically rapid because much of the record is missing. However, in the VCFCB, the rock and fossil record is essentially continuous from the Blancan through the Irvingtonian. Here, we are able to delineate a more accurate picture of the events that occurred during the change from one NALMA to the next.

TABLE 7.4
Core-Fauna Blancan and Irvingtonian-Aged Mammals of the Vallecito Creek-Fish Creek Basin, Anza-Borrego Desert State Park.

SCIENTIFIC NAME	INFORMAL OR COMMON NAME
Camelops	extinct camel-like llamas
Dipodomys	kangaroo rats
Equus	horses, zebras, asses
Geomys	pocket gophers
Hemiauchenia	extinct stilt-legged llamas
Lynx	bobcat, lynx
Neotoma	wood rats
Perognathus	pocket mice
Peromyscus	deer mice
Sigmodon	cotton rats

One interesting aspect of the mammalian paleofauna in the VCFCB area is that it contains a number of genera (Table 7.4) that persisted throughout the Blancan and Irvingtonian despite other biotic changes; at about 2.58 (arrival of South American immigrants) and 1.8 Ma (arrival of some Irvingtonian immigrants), and changes in the physical environment. This observation suggests that the VCFCB region supported a stable and resilient "base mammalian fauna" that tolerated significant biological and environmental change. Nonetheless, the overall composition of the Irvingtonian fauna in VCFCB was profoundly different from the Blancan. By this time, the mammalian paleofauna had taken on a definitely modern aspect. Many of the species that lived during the Irvingtonian, especially small mammals such as rodents, still exist today.

The Importance of the VCFCB Area to Mammalian Biostratigraphy

The VCFCB is critical to our understanding of faunal change in North America during the last six million years of earth history. Here, the pattern of appearances and disappearances of mammalian taxa in superposed assemblages of the Hemphillian, Blancan, and early to middle Irvingtonian NALMAs occur in a single stratigraphic section. Nowhere else in North America are these events as well preserved for these latter two NALMAs. Elsewhere, the pattern must be pieced together by correlating stratigraphically thin, geographically discontinuous occurrences, using methods such as magnetostratigraphy, radiometric dating, and the evaluation of the stage of evolution of taxa. The procedure is commonly used, but prone to error. In the VCFCB, there is no need for correlation to establish the appearance/disappearance pattern of taxa. These data are preserved in one area, in a stratigraphic section that contains no appreciable gaps in the rock record and, therefore, no appreciable gaps in the time record. Because strata in the VCFCB contain the entire Blancan NALMA, the area has the potential to serve as the standard reference for evaluating the stage of evolution of characteristic Blancan mammals. This will allow better age determinations of Blancan mammals in other parts of North America.

Defining the boundaries between NALMAs is best accomplished in areas where characteristic fossil assemblages occur in superposition. Superposition allows paleontologists to determine most accurately when taxa from different paleofaunas appeared and disappeared relative to one another. They work out this pattern by means of biostratigraphy. It helps to know if there are gaps in the stratigraphic section due to erosion or non-deposition of sediments, and to have an understanding of taphonomic biases, such as inconsistencies in the preservation of fossil remains. Both factors allow paleontologists to determine if a taxon's range may have been locally truncated, or if apparent gaps in a taxon's range are the result of emigration and later immigration, or simply due to an imperfect rock record. In the VCFCB, these problems are so minimal as to be non-existent.

Physical methods, such as magnetostratigraphy and radioisotopic dating, provide an independent check of the biostratigraphy in the VCFCB. These permit the mammalian assemblages of each NALMA to be correlated with the standard geologic time scale in both a relative (Periods and Epochs) and absolute sense (numerical dates) (see Remeika, this volume).

For all of these reasons, the paleontological information revealed in the VCFCB has the potential to define the boundary between the Blancan and Irvingtonian NALMAs, and between the Hemphillian and Blancan NALMAs. "Define" means to recognize the time of change between one NALMA and another by the earliest appearance in the geologic record of a particular species of mammal. This is an arbitrary decision on the part of paleontologists, but it is a necessary convention, adopted to bring accuracy and precision to the NALMA system.

Of these two boundaries, the evidence is best for the Blancan-Irvingtonian (Cassiliano, 1999). The superposition of Blancan and Irvingtonian mammalian paleofaunas is found in only two places in North America: the VCFCB, and the San Pedro Valley, Arizona (Lindsay, 1984; Lundelius et al., 1987; Bell et al., 2004). Both areas also are characterized by magnetostratigraphic and radioisotopic data that locate the Plio-Pleistocene boundary in the local stratigraphic columns (Johnson et al., 1975; Opdyke et al., 1977; Johnson et al., 1983; Lindsay, 1984; Cassiliano, 1999; Bell et al., 2004). The San Pedro Valley localities, however, contain no taxa thought to be definitive of the Irvingtonian, so in that area the Blancan-Irvingtonian boundary can only be approximated, not defined biostratigraphically.

The situation is better, but not perfect, in VCFCB. A review of the occurrences of *Mammuthus* (the accepted defining mammalian genus for the Irvingtonian) in North America, that are associated with radioisotopic dates, suggests that there is considerable ambiguity and/or range in the timing of its appearance on the continent (Cassiliano, 1999; Bell et al., 2004). Specimens of *M. meridionalis* (southern mammoth) from VCFCB occur in strata that date between 1.4 and 1.1 Ma , and *M. columbi* (Columbian mammoth) between 1.1 and about 0.7 (McDaniel and Jefferson, 1999; McDaniel, this volume). Additionally, these specimens have the distinct advantage of occurring in superposition with definitive Blancan paleofaunas. Other known occurrences of *Mammuthus* do not occur in superposition with Blancan assemblages. This suggests that only in the VCFCB will it be possible to precisely locate the Blancan-Irvingtonian boundary biostratigraphically.

The Future of Biostratigraphic Research in ABDSP

The stratigraphic distribution of mammalian fossils in the VCFCB clearly shows that the boundaries between NALMAs are not impassable walls at which the mammals of the older NALMA must go extinct and those from the younger NALMA must first appear. Rather, the succession of rocks and fossils in VCFCB demonstrates that NALMA boundaries are actually transitions lasting thousands of years during which the older mammalian paleofauna was gradually replaced by a younger one. For this reason, the definition of a NALMA boundary is of necessity an arbitrary decision that stands or falls based on good biostratigraphic data and sound reasoning.

The best place in North America to obtain the biostratigraphic data to accurately resolve the terrestrial chronology of the Pliocene and Pleistocene Epochs in North America is in the VCFCB of ABDSP. This chronology depends on the accurate and precise definition of the boundaries between the Hemphillian and Blancan NALMAs, and between the Blancan and Irvingtonian NALMAs. Of all the areas in North America that could serve as the boundary stratotype for these NALMAs, the VCFCB shows the most promise. A boundary stratotype is the stratigraphic section that serves as the standard or physical referent for a NALMA boundary; its designation is necessary and critical if the NALMAs are to be brought into conformance with the formal stratigraphic codes. Additionally, the VCFCB has the potential to allow paleontologists to change the status of the Blancan from a biochron to a true biostratigraphic age. This means that all the time represented by the Blancan can be referred to a physical entity, the body of rock between the Hemphillian-Blancan boundary and the Blancan-Irvingtonian boundary. The main problems to be resolved in this regard are accurately and precisely locating the Hemphillian-Blancan boundary, and deciding what species of mammal to use to define the Blancan-Irvingtonian boundary.

Solutions to these problems will be the first steps towards fulfilling the decades-long desire of vertebrate paleontologists to convert the NALMAs from imprecise biochrons to true biostratigraphic ages that then can become part of the standard geologic time scale for North America.

The Fossil Lower Vertebrates: Fish, Amphibians, and Reptiles

Blancan and Irvingtonian deposits from the Palm Spring Formation in Anza-Borrego Desert State Park . . . have yielded the most diverse Tertiary lizard assemblage yet reported.

Mark A. Norell (1989)

First to describe the fossil lizards from Anza-Borrego, Mark is a hunter of Asian dinosaurs, and Curator of Paleontology at the American Museum of Natural History, New York.

Philip Gensler
George T. Jefferson
Mark A. Roeder

The Fossil Lower Vertebrates: Fish, Amphibians, and Reptiles

The Lower Colorado River at Picacho State Park, California. (Photograph by Lowell Lindsay)

Introduction

The record of fossil fish, amphibians, turtles, lizards, and snakes from Anza-Borrego Desert State Park is remarkable in several respects. The diversity of these animals is exceptional and exceeds that found in other Pliocene or early and middle Pleistocene locations in North America. Furthermore, the presence of tropically related forms, like iguanas and the giant tortoise *Hesperotestudo,* have significant implications about past climates and climate change.

Discussion

At least three species, represented by three genera and two families of freshwater fish, occur in the Pliocene and Pleistocene deposits of Anza-Borrego. Although their numbers have been greatly reduced by water reclamation dams and the introduction of exotic game fishes, all of these species can be found in the present-day Colorado River (Swift et al., 1993).

One of the more interesting fossil freshwater fish records is that of the razorback sucker (*Xyrauchen texanus*) in the family Catostomidae (suckers) (see Appendix, Table 3). The present-day razorback sucker is one of the largest members of the sucker family and reaches lengths up to almost 1 meter (3 ft). A nearly complete fossil skeleton of a razorback sucker (Figure 8.1) was found in the mid-1970s on the north side of the San Felipe Hills (Stewart and Roeder, 1993; Hoetker and Gobalet, 1999). The fossil fish was preserved in a sandstone concretion eroded from the Arroyo Diablo Formation of the Palm Spring Group.

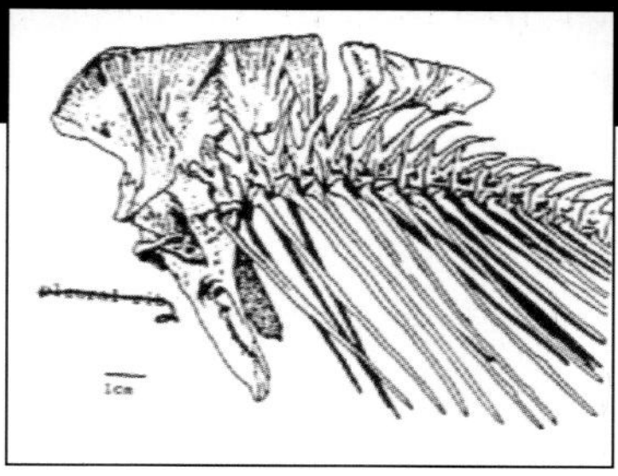

These rocks may be up to 4 million years in age and represent ancestral Colorado River delta sediments. According to Hoetker and Gobalet (1999), the San Felipe Hills *Xyrauchen* fossil is indistinguishable from the modern *X. texanus*. In addition to the San Felipe Hills specimen, *Xyrauchen* remains, which consist of isolated vertebrae and skull bones, are the most common fossil fish remains in the Anza-Borrego sediments and are found throughout the Hueso Formation and Ocotillo Conglomerate.

20 cm

Another fossil freshwater fish recorded from Anza-Borrego is the modern pike minnow (*Ptychocheilus lucius*). Pike minnows are one of the largest members of the family Cyprinidae (chubs), and may reach a length of up to 1.8 meters (6 ft). Unlike *Xyrauchen*, the fossils of *Ptychocheilus*, which consist of isolated vertebrae, are rare. Based on their stratigraphic occurrence, the age of the earliest fossil pike-minnow remains at Anza-Borrego is about one million years.

Based on isolated pharyngeal (throat) teeth and vertebrae, a second cyprinid fish was identified as *Gila* sp. (chub). Today, there are four species of *Gila* (*G. cypha, G. elegans, G. intermedius,* and *G. robusta*) that occur in the Colorado River drainage system. Also two other species, *Gila orcutti* and *G. bicolor* (= *Siphateles* of some researchers) occur to the west and north of Anza-Borrego respectively. Until more diagnostic elements of this animal are found, identification of the fossils will remain only at the generic level. The earliest occurrence of *Gila* remains in Anza-Borrego sediments is about one million years.

The only amphibian in the Anza-Borrego fossil record is a toad, identified as *Bufo* sp. Three species of toad can be found on the desert floor today: the California toad (*B. boreas*), the arroyo toad (*B. californicus*), and the red-spotted toad (*B. punctatus*). All of these species are restricted to isolated moist habitats within the generally dry desert environment.

Figure 8.1 (Above)
Xyrachen texanus.
Razorback sucker skeleton (ABDSP[SBCM]A169-1) preserved in a sandstone concreation from the Arroyo Diablo Formation (Paleolandscape 2). (Photograph by Barbara Marrs)

Figure 8.2 (Above, left)
Xyrachen texanus.
On the razorback sucker skeleton, the enlarged dorsal (upper) spines of the vertebrae, (called interneurals) immediately behind the skull, support the hump. (Picture from Hoetker and Gobalet, 1999, courtesy of Copeia.)

Figure 8.3
Hesperotestudo sp.
Foraging group of giant tortoises.
(Picture by John Francis)

At least five species, representing four genera and three families of tortoises and turtles, occur in the Pliocene and Pleistocene deposits of Anza-Borrego (Jolly, 2000) (Appendix, Table 3). These include, most notably, the giant tortoise *Hesperotestudo*. Living giant tortoises are now known only from isolated islands like the Galapagos. The fossils also include the desert tortoise (*Xerobates agassizii*), the Sonoran mud turtle (*Kinosternon sonoriense*), the western pond turtle (*Clemmys marmorata*), and an unidentified species in the family Emydidae, the sliders and their relatives.

There are apparently two extinct species of *Hesperotestudo* from Anza-Borrego, distinguished on the basis of their size. As with the other smaller fossil tortoise and turtle remains from the area, these giant tortoises are represented by numerous thick, plate-like bones from the shell. Skulls, limb bones, and dermal ossicles are less frequently found. A nearly complete shell of the larger species of *Hesperotestudo* (Figure 8.4A and 8.4B), measuring 1.15 m (3.74 ft.) long, 8.2 cm (2.7 ft.) wide and 5.7 cm (1.9 ft.) tall, was recovered from the 1.6 million year old deposits in the Vallecito Creek-Fish Creek Badlands.

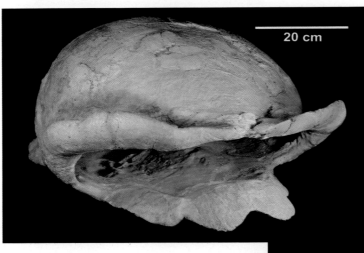

20 cm

These large animals were not able to burrow like their smaller relative *Xerobates*, and therefore could not tolerate temperatures below freezing. This implies that early and middle Pleistocene winters at lower elevations in the region were frost free. Living tropical giant tortoises frequently visit muddy pools to cool themselves. They are strictly vegetarian, and extinct *Hesperotestudo* may have been largely a grazer.

Figure 8.4 (Above and at right)
Hesperotestudo sp.
This 1.75 million year old complete giant tortoise shell (A. front view, B. left side view; ABDSP[LACM]1918/V16217) was recovered from the Hueso Formation. The large projections from the front, bottom of the shell identify this individual as a male. (Photographs by Barbara Marrs)

40 cm

Xerobates agassizii is locally extinct, but today occurs to the north in the Mojave Desert. Apparently, it is unable to cope with the extreme summer heat and aridity of the western Colorado Desert. The presence of fossil *Xerobates agassizii* in the middle Pleistocene of Anza-Borrego suggests that the region was less xeric (dry).

Remains of aquatic turtles are found as fossils far more frequently than those of terrestrial *Xerobates*. This is because they inhabit sites where fluvial deposition takes place, and their remains are more likely to become buried and fossilized. The western pond turtle (*Clemmys marmorata*) and the Sonoran mud turtle (*Kinosternon sonoriense*) (Figure 8.5), presently inhabit rivers, lakes, and ponds in coastal California, and southern Arizona and northern Sonora respectively. Neither animal is found in the Colorado Desert today, however the mud turtle lives along the Colorado River. These animals require moderately large, permanent bodies of fresh water, and feed on a variety of aquatic invertebrates (Appendix, Table 2), small fish and frogs, and water plants.

Twelve fossil lizard taxa have been recovered from Anza-Borrego. Four of the species are endemic, or known only from Anza-Borrego deposits (Appendix, Table 3). Of importance is the diversity of this assemblage. In comparison to other North America sites of similar age, over twice as many genera are known from Anza-Borrego than any other area.

The fossil remains of lizards are usually very rare in Pliocene and in all but the latest Pleistocene deposits. This scarcity of specimens, which are often incomplete, makes anatomical or osteological comparisons difficult and creates problems in identifying and naming possible new fossil lizard species. For example, there are two forms of spiny lizard present in the Anza-Borrego assemblage, a large and a small form which are also distinguished on the basis of dental characters. However, until more specimens are discovered form Anza-Borrego, these two lizards will have to remain informally named *Sceloporus* species A and *Sceloporus* species B.

Members of the Iguanidae (Family) are the best represented and include the desert iguana (*Dipsosaurus dorsalis*) (Figure 8.6), the extinct crowned leopard lizard (*Gambelia corona*), the extinct Anza horned lizard (*Phrynosoma anzaense*), another horned lizard identified only to the genus *Phrynosoma*, the extinct Novacek's small iguana (*Pumilia novaceki*) (Figure 8.7), the two spiny lizards, and a side-blotched lizard (*Uta* sp.). Also present is a member of the Family Teiidae, a ground lizard or whiptail (*Ameiva* or *Cnemidophorus*) (Figure 8.8).

The fossil assemblage was recovered from an over 2500 meter (1.54 mile) thick section of strata within the Palm Spring Group, spanning approximately 2 million years. However, not all taxa are distributed evenly through this period. The large herbivorous lizards, like the iguanids *Pumilia* and

Figure 8.5
Kinosternon sp.
This Arizona mud turtle suns itself on the bank of a Borrego Valley stream about 1 million years ago (see Paleolandscape 4). (Picture by John Francis)

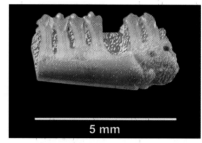

Figure 8.6
Dipsosaurus dorsalis.
Fossil desert iguana dentary (lower jaw) with teeth (ABDSP[LACM]6583/V78261). (Photograph by Barbara Marrs)

Figure 8.7
Pumilia novaceki.
About 1 million years ago, this Novacek's small iguana rests on a willow limb over an ancient Borrego Valley stream (see Paleolandscape 4). (Picture by John Francis)

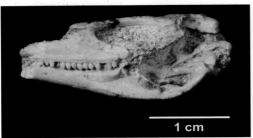

Figure 8.8
Ameiva or *Cnemidophorus* sp.
Articulated skull and mandible (lower jaws) (ABDSP[LACM]6831/V209929) of a fossil ground lizard or whiptail. This delicate specimen is over 3 million years old. (Photograph by Barbara Marrs)

Dipsosaurus, only occur in the older deposits. They were replaced about 2.5 million years ago by forms like the spiny lizards (*Sceloporus* sp. A and B), the alligator lizard (cf. *Gerrhonotus* sp.), Downs' extinct night lizard (*Xantusia downsi),* skinks (*Eumeces* sp.), and ground lizards and whiptails (subfamily Teiinae) (Norell, 1989).

The closest living relatives of *Pumilia,* animals like the green iguana *(Iguana iguana),* presently inhabit the tropical riparian forests of western Mexico. Since the later types of lizards are indicative of more mesic environments, their presence in the fossil record suggests the spread of grasslands and savanna habitats in the region by 2.5 million years ago (see Sussman et al., this volume, *Paleoclimates and Environmental Change in the Anza-Borrego Desert Region*).

The Helodermatidae (Gila monsters and beaded lizards) are represented by a single specimen identified as *Heloderma* sp. The specimen was recovered through screen washing of sediments from the mid-Pleistocene aged Bautista beds of the Coyote Badlands. Members of the Helodermatidae are the only venomous lizards that exist in the world today and are comprised of only two species; *H. suspectum* (Gila monster) and *H. horridum* (Mexican beaded lizard). This genus is currently evidenced in the Anza-Borrego region by a single boney plate called an osteoderm (Figure 8.9). Individual osteoderms of *Heloderma* are relatively small, less than 2 mm (1/10 in) in diameter, conical, and interlock together under the skin of the lizard to form a type of armor. They are easily noticeable on living examples by their "bumpy" skin texture.

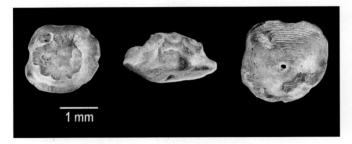

Figure 8.9
Heloderma sp.
Scanning electron microscope photograph of a fossil osteoderm (ABDSP2959/V6260) from a Gila monster or beaded lizard recovered from the Coyote Badlands. (Photograph by Phil Gensler)

The Helodermatidae are very rare throughout the fossil record and were previously unknown in California as fossils. It is unclear which type of *Heloderma* existed in the Anza-Borrego region. Either a skull or jaws will need to be discovered for species level identification.

The Gila monster's range today is from northern Mexico though the Sonoran Desert of Arizona, and into parts of New Mexico, Nevada, southern Utah, and possibly southeastern California. The Mexican beaded lizard's range is throughout the tropical regions of Mexico and perhaps into Guatemala. Neither lizard is found in the Anza-Borrego region today. This may be in part due to the lower amount of precipitation Anza-Borrego receives compared to the tropical regions of Mexico and the summer monsoons that impact the Sonoran Desert. By the end of the Pleistocene, annual precipitation in inland southern California may have decreased making the region unfavorable to the Helodermatidae.

In some instances the absence of a species may have significant paleo-ecological implications. For example, the legless lizard (*Anniella*) is found in

Pliocene and Pleistocene deposits throughout western southern California where it lives today (Bell and Mead, 1993). However, its fossil remains have not been found in the Anza-Borrego region. Modern *Anniella* prefer ground temperatures between 15 to 20°C (59 to 68°F), and specific soil moisture conditions. Apparently, by Pliocene time, the inland regions of southern California were already too hot and/or dry for this small, worm-like burrowing animal.

Five types of fossil snakes have been found in Anza-Borrego. They include a night snake (*Hypsiglena* sp.), the common king snake (*Lampropeltis getulus*), the coachwhip snake *(Masticophis flagellum)*, a garter snake (*Thamnophis* sp.), and a rattlesnake (*Crotalus* sp.). All but *Thamnophis,* which prefers more moist conditions, can be found on the desert floor today.

Conclusions

The fossil freshwater fish from Anza-Borrego are important for several reasons. They represent the first well-documented fossil freshwater fish assemblage from the Colorado River drainage system. Although additional collecting may recover specimens which might represent new records of freshwater fish and allow identification of the *Gila* remains to species, the present lack of fish species common to the Mojave Desert and/or coastal southern California in the assemblage suggests that river drainages into the Salton Trough have been isolated from these regions since the latest Miocene time. Unlike some of the reptile species that are now extinct (Norell, 1989), it appears that all freshwater fossil fish species identified from Anza-Borrego may represent extant or living forms.

Climatic inferences, based solely on the fossil tortoises and turtles, suggest that during the Pliocene and early through middle Pleistocene, Anza-Borrego Desert experienced slightly warmer winters, cooler summers, and an increase in precipitation sufficient to support permanent bodies of water.

The remains of aquatic turtles are found throughout the deposits, whereas the giant tortoises are more abundant in the earlier deposits, and the desert tortoise in the later, suggesting a shift to less mesic conditions. A similar shift from tropical and moist subtropical to more arid conditions is also indicated by changes in the fossil lizard record. Such changes in faunal composition have been interpreted to indicate the spread of grassland savanna habitats about 2.5 million years ago (Norell, 1989).

Unlike the fossil mammals from Anza-Borrego, most of the fossil lizards and snakes belong to the same genera as those present in the region. This suggests that the composition of the desert reptile fauna developed in place rather than through immigration from outside of the region (Norell, 1989). An exception to this is the presence of *Heloderma* sp. which is not found in the region today.

Overall, the fish and reptilian fossil record in Anza-Borrego, while very diverse, is not overly abundant. Tortoises and turtles are an exception because of their larger size. The smaller reptiles typically require extensive screen washing for the recovery of fossils specimens. Further screen washing of sediment, particularly of latter Pleistocene deposits which have not been well sampled, will undoubtedly lead to the recovery of additional reptilian and other microfaunal specimens from Anza-Borrego's distant past.

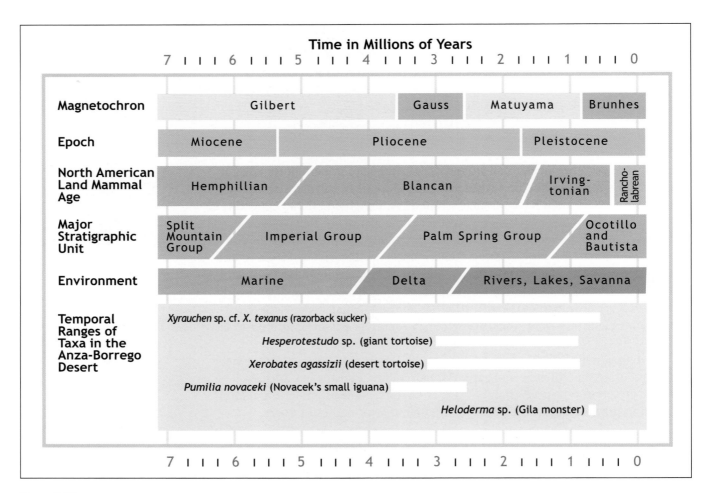

Figure 8.10
The Fossil Lower Vertebrates of Anza-Borrego.
The temperal ranges of these five lower vertebrates, a fish and four reptiles, generally span the latter part of the terrestrial Anza-Borrego record. The iguana, *Pumilia novaceki,* with tropical affinities, is an exception.

Paleolandscape 3: The Shore of Lake Borrego 1 Million Years Ago

INDEX

1. *Aquila chrysaëtos* (golden eagle)
2. *Brantadorna downsi* (Down's gadwall duck, extinct)
3. *Phoenicopterus* sp. (flamingo, extinct)
4. *Rallus limicola* (Virginia rail)
5. *Grus canadensis* (sandhill crane)
6. *Tapirus merriami* (Merriam's tapir, extinct)
7. *Trachemys scripta* (common slider turtle)
8. *Rana* sp. (frogs)
9. *Thamnophis* sp. (garter snake)
10. *Felis rufus* (bobcat)
11. *Sigmodon lindsayi* (Lindsay's cotton rat, extinct)
12. *Fontelicella* sp. (snails)
13. *Typha* sp. (reeds and cat tails)

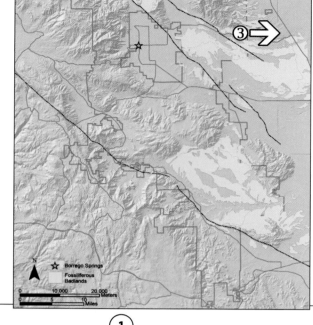

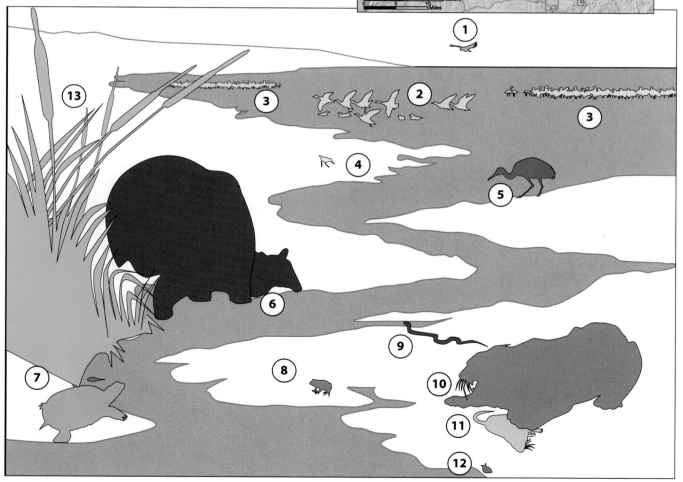

147

This marshy lake margin and open water support a wide variety of waterfowl. A mother *Tapirus merriami* (Merriam's tapir) with young startles a feeding *Felis rufus* (bobcat). The view today is to the southeast from the Truckhaven Rocks turnout. (Picture by John Francis)

The Fossil Birds of Anza-Borrego

Bird bones are extremely fragile, and therefore frequently have not survived the upheavals and metamorphic disturbances to which the older strata have been subjected.

Hildegarde Howard (1947)

Dr. Howard was Chief Curator Emeritus of Sciences at the Natural History Museum of Los Angeles County, first to describe the fossil birds of Anza-Borrego, and a world-recognized specialist in avian paleontology.

George T. Jefferson

151

The Fossil Birds of Anza-Borrego

San Ysidro Mountains as Seen from San Felipe Wash. (Photograph by Paul Remeika)

Figure 9.1
Dr. Hildegard Howard.
In 1938, Dr. Howard measured fossil bird specimens from the Rancho La Brea Tar Pits. (Archival photograph courtesy of the Natural History Museum of Los Angeles County)

Fossil bird remains from Anza-Borrego range from about 3.5 to 0.5 million years in age, and reflect a climatic and biogeographic setting of lakes, ponds, and streams that is strikingly different from the present arid conditions of the Colorado Desert. Over 300 bird fossils have been recovered from the Anza-Borrego Desert State Park. These remains represent at least 40 different kinds of birds (Appendix, Table 3), including such exotic forms as an extinct flamingo, *Phoenicopterus,* and the largest North American raptor, *Aiolornis incredibilis.* Much of our understanding of the fossil birds from Anza-Borrego stems from the pioneering studies of Hildegarde Howard (1963, 1972a, 1972b) (Figure 9.1) (see Marrs, this volume, *History of Fossil Collecting in the Anza-Borrego Desert*), who described seven new species from the Anza-Borrego Desert deposits.

Bird fossils are very rare. Bird bones are fragile, and are hollow with thin walls. This reduces the weight of the bones, and is an adaptation for flight. However, they do not withstand the relatively high energy depositional conditions found in rivers or streams, and usually disintegrate with minimal fluvial (water-borne) transport. As a result, there are over 17 times as many fossil large mammal bones as there are bird bones from Anza-Borrego.

Most of the bird fossils collected from the Hueso Formation and Ocotillo Conglomerate were recovered from finer-grained sediments like silts and clays that are typical of slow-moving or ponded water. As might be expected, these remains are dominated by water birds. This is because the bones of animals that live closest to places where sediment deposition and burial occur are usually better preserved and more numerous. Songbirds (or perching birds, Passeriformes) are the second most abundant group from Anza-Borrego. Although six or more types may be represented in the fossil assemblage, the lack of distinctive skeletal characters makes it very difficult to distinguish individual songbird taxa. Only the crow *(Corvus)* is positively identified.

Water birds comprise 57% of the identified fossils (Paleolandscape 3, Figures 9.2, 9.3). Of these, the ducks, geese, and swans are most common and include a wide variety of forms: the pintail duck (*Anas acuta*), northern shoveller duck (*Anas clypeata*), extinct Downs' gadwall duck (*Brantadorna downsi*), fossil goldeneye (*Bucephala fossilis*) (Figure 9.3), extinct Bessom's stiff-tail duck (*Oxyura bessomi*), Canada goose (*Branta canadensis*), Ross' goose (*Chen rossii*), an extinct large goose (*Anser* sp.), and the extinct Oregon swan (*Cygnus paloregonus*). Other water birds in the fossil avifauna are the western grebe (*Aechmophorus occidentalis*), an eared grebe (*Podiceps nigricollis*) and an extinct *Podiceps* sp., a loon (*Gavia* sp.), coots and rails (*Fulica americana*, extinct *Fulica hesterna, Rallus* sp., and *Rallus limicola*), the sandhill crane (*Grus canadensis*), the surf scoter (*Melanitta persipicillata*), killdeer (*Charadrius vociferus*), a pelican (*Pelecanus* sp.) (see Paleolandscape 1), and an extinct flamingo (*Phoenicopterus* sp.).

Figure 9.2
Ancient Anza-Borrego Lake Shore.
About a million years ago, birds like the extinct flamingo (*Phenicopterus* sp., in distance on water), extinct Down's gadwall duck (*Brantadorna downsi*, landing on water), Virginia rail (*Rallus limicola*, to left), and sandhill crane (*Grus canadensis*, on right) were common on and around the large lakes of the Salton Trough. (Picture by John Francis)

Modern relatives of most of these aquatic birds are migratory and are winter visitors to the lower Colorado River region and the Salton Sea today. They migrate northward in the spring following inland portions of the north-south Pacific coast flyway (Figure 9.4). During Pleistocene time, this route included the now-extinct lakes of the Mojave Desert, including Death Valley, those east of the Sierra Nevada, the western Great Basin of Nevada, and the now-dry lakes of southeastern Oregon. The extinct Oregon swan found in Anza-Borrego was first described from the deposits of Fossil Lake, Oregon. Water birds, such as the surf scoter, killdeer, and pelican are also found along the Pacific coast or on inshore waters and bays.

Extinct species of *Phoenicopterus,* the flamingo, must have been common inhabitants of ancient inland lakes (Figure 9.2). Although we do not know if they were pink-colored or not, flamingo remains have been recovered from Pleistocene deposits throughout the Great Basin, Mojave Desert (Jefferson, 1989), and Pacific coast region. Their presence suggests shallow lake conditions with a rich invertebrate food source.

One water bird is known only from its footprints. The four-toed tracks were found in 3.4 million year old lakeshore mudflat deposits exposed in Arroyo

Figure 9.3
Bucephala fossilis.
This fossil goldeneye and family glide across a Borrego Valley stream about one million years ago. (Picture by John Francis)

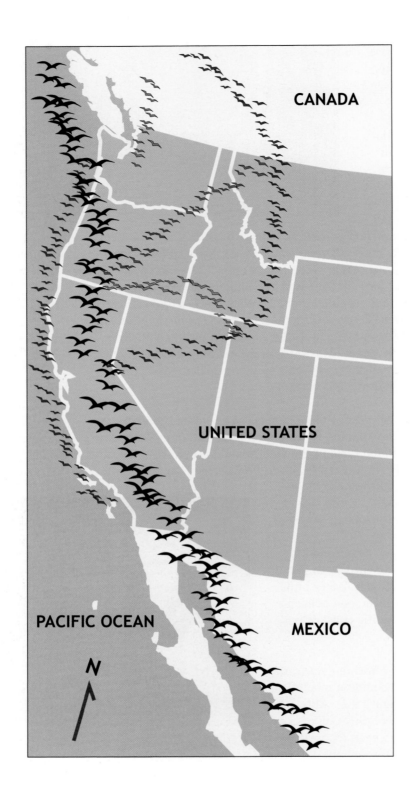

Figure 9.4
Pacific Flyway.
Today many species of birds annually migrate along the Pacific coast of North America. The fossil record from now dry lake deposits indicates that this pattern was followed during the Pleistocene. (Map based on Pacific Flyway Map, U.S. Fish and Wildlife Service.)

Seco del Diablo (Figure 9.5). Accordingly, this extinct shore bird footprint was named *Gruipeda diabloensis,* literally "the Diablo least sandpiper" (Remeika,1999, also this volume, *Fossil Footprints of Anza-Borrego*).

As the types of waterfowl are indicative of both open waters and a variety of nearshore lacustrine habitats, the fossil woodpecker remains from Anza-Borrego (identified only to the family Picidae) imply the presence of local riparian forests (Figure 9.6).

There are 11 extinct species of birds and seven of these are described from remains found only in the Anza-Borrego Desert. Several of the extinct forms are ancestral to living forms or to other extinct species that appear in later Pleistocene (Rancholabrean) age sites in the western U.S. The extinct coot, *Fulica hesterna,* is the probable ancestor of the modern species, *Fulica americana.* The Vallecito neophron, *Neophrontops vallecitoensis,* which is related to the Old World vultures, is the direct ancestor of the extinct *Neophrontops americanus* from the Rancho La Brea "tar pits" in Los Angeles (Howard, 1963). However, the Vallecito neophron is about 15% larger than its later Pleistocene descendant.

Figure 9.5
Fossil Footprints of the Diablo Least Sandpiper.
(Photograph by Paul Remeika)

Unlike *Neophrontops,* the extinct Anza turkey (*Meleagris anza*) (Figure 9.7), is more closely related to the living ocellated turkey of Central America than to the extinct California turkey (*Meleagris californicus*) which is also well known from Rancho La Brea. The remains of quail and the Anza turkey are relatively common, and comprise about 10% of the Anza-Borrego fossil avifauna. Both the California quail and Gambel's quail (*Callipepla californica* and *Callipepla gambelii*) have been reported. The abundance and nesting habitats of such ground dwelling birds suggest the presence of local brushland habitats.

Among the raptors, which include eagles (*Aquila chrysaëtos*) and hawks (*Buteo lineatus*), a falcon (*Falco* sp.), a condor (*Gymnogyps* sp.), and an extinct eared owl (*Asio* sp.), is *Aiolornis incredibilis* (Howard, 1963; Jefferson, 1995; Campbell et al., 1999). The name for this amazing bird is derived from the Greek *Aiolos,* for god of the winds, and translates as the "incredible wind god bird." *Aiolornis incredibilis* is represented by only six specimens, three of which were found in the Anza-Borrego

Figure 9.7
Meleagris anza.
This Anza turkey family forages along a Borrego Valley stream about one million years ago.
(Picture by John Francis)

Figure 9.6
Woodpecker.
The single fossil specimen of this bird suggests the presence of woodlands.
(Picture by John Francis)

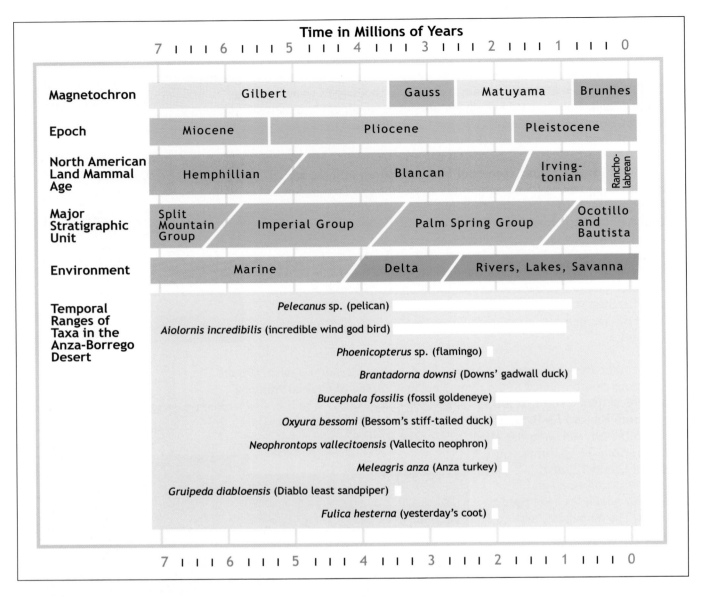

Figure 9.8

The Fossil Birds of Anza-Borrego. Nine of these 10 birds are water fowl. Most are represented by single occurrences. *Pelecanus* and *Aiolornis*, the two largest birds, are exceptions to this preservational pattern.

Desert (Figures 9.9, 9.10). Its bones also have been recovered from Blancan and Irvingtonian age sites in central Mexico and southern California, and from a single Rancholabrean age site in Nevada, a record spanning over 3 million years.

Scaled from measurements of fragments of the major wing bones (Figure 9.9), *Aiolornis incredibilis* probably had a wingspan of between 4 and 5 m (16 to 17 ft). *It was the largest bird ever to soar the skies of the northern hemisphere.* Originally, it was thought that these giant birds were primarily scavengers, like today's condors and vultures. However, based on skeletal features, they are now known to be more closely related to the Old World storks, and are considered active predators (Campbell et al., 1999). Standing about 1.2 m (4 ft) tall, *Aiolornis* must have been an impressive sight as it stalked its prey on the ground.

Figures 9.9A, 9.9B, 9.9C
Aiolornis.
From the upper part of the Huseo Formation.

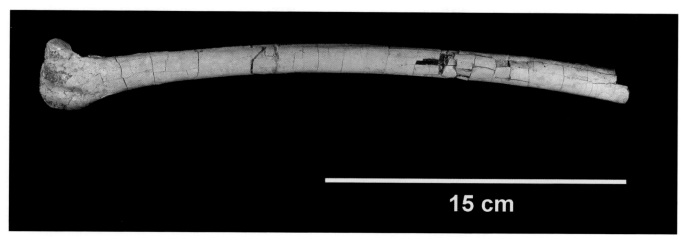

Figure 9.9A
Nearly Complete Radius Bone from the Wing of the Incredible Wind God Bird.
(Photograph by Barbara Marrs)

Figure 9.9B
A Portion of the Shaft of the Ulna Bone from the Wing of the Incredible Wind God Bird.
The two bumps (see arrows) on this crushed specimen provided for the attachment of the primary feathers. By measuring the distance between these bumps or papillae, and comparing them with those in modern raptors, paleontologists can estimate the size of the wing in life.
(Photograph by Barbara Marrs)

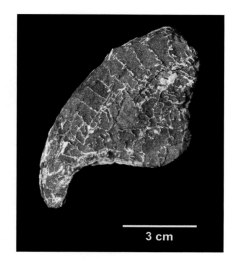

Figure 9.9C
Front Tip from the Beak of the Incredible Wind God Bird.
(Photograph by Barbara Marrs)

Figure 9.10
Aiolornis incredibilis.
About a million years ago, this
incredible wind god bird is about
to land and capture a rabbit.
(Drawing by Pat Ortega)

The Ground Sloths: Invaders from South America

In broader paleontological studies involving faunas of the two continents of the Western Hemisphere, the order Edentata has long been recognized as one which occupies an important position and comprises a number of unusual types.

Chester Stock, 1925

Chester Stock, a colleague of John C. Merriam, worked for much of the first half of the 20th century on the famed Rancho La Brea collection, specializing in fossil mammals. At the end of his career, he was both geology department chairman at California Institute of Technology and Chief Curator of Science at the Natural History Museum of Los Angeles County.

H. Gregory McDonald

The Ground Sloths:
Invaders from South America

Mudhills Along Arroyo Tapiado. (Photograph by Lowell E. Lindsay)

Introduction

Discoveries from the fossil beds of the Anza-Borrego Desert have contributed to our understanding of the evolutionary history of many vertebrate lineages in North America. Among these is one of the more interesting groups of mammals, the xenarthrans (also known as the edentates), the Order that includes living tree sloths, anteaters, and armadillos, as well as extinct ground sloths, glyptodonts, and pampatheres. The traditional name Edentata, meaning "without teeth," is not entirely accurate, since the only members of the group that lack teeth are the anteaters. It has subsequently been replaced by Xenarthra, meaning "strange joint," referring to the presence of extra articulations between the vertebrae of the lower back, a feature shared by all members of the Order.

While a variety of xenarthrans entered North America from South America, the only xenarthran fossils currently known in Anza-Borrego are the extinct ground sloths (Figure 10.1). Studying them has allowed us to incorporate the Anza-Borrego record into our broad understanding of the history of each group of sloths in North America. Their quality of preservation has not only permitted identification to both genus and species, but in some cases, individual fossils provide unique information. An example is the spectacular and rare find of associated dermal ossicles (Figure 10.7) – small bones embedded as armor in the skin of a mylodont sloth, which are usually scattered following death of the animal.

The Story of a Southern Invasion

The Order Xenarthra was the most successful group of South American mammals to enter North America, and ground sloths were its most successful members, in terms of range of distribution, taxonomic diversity, and relative abundance of individuals in faunas. The xenarthrans originated in and most of their evolution took place in South America while it was an island continent, but their dispersal into North America is an equally important part of their history. They entered the North American fauna relatively late in the Cenozoic. Although ultimately a diverse array of xenarthrans was present in North America, they did not all arrive at once, but rather in three stages (Webb, 1985). Fossil finds in Anza Borrego include xenarthrans from all three invasions.

The earliest xenarthrans entered North America in the late Miocene, approximately 9 million years ago, while the last ones arrived during the Pleistocene. Their journey was made possible by the slow creation of the Isthmus of Panama as it connected North with South America, ending the latter continent's millions of years of isolation. In its early stages, the Isthmus was a

Figure 10.1
Paramylodon harlani,
Largest of the Anza-Borrego Ground Sloths.
A pair of Harlan's ground sloths grazes at the savanna edge (see Paleolandscape 5).
Vallecito neophrons
(Neophrontops vallecitoensis)
are seen above.
(Picture by John Francis)

series of islands that served as stepping-stones for only a few forms, including the ground sloths. Later as the dry land connection formed, the armored xenarthrans – glyptodonts, pampatheres, and armadillos, along with other ground sloths and anteaters – expanded their ranges and dispersed northward, to enter North America.

However, the road between North and South America was not a freeway allowing passage between the continents to all animals, but rather a toll road. The toll was paid in evolutionary coin: the ability to survive not only in the tropical habitats of the Panamanian Isthmus, but also the ability to adapt and survive in the more temperate environments to the north. Thus, the Isthmus of Panama acted as a filter that only allowed particular members of each group of animals to pass.

The successful integration of xenarthrans into the North America mammalian fauna suggests that their ecological needs and food requirements were sufficiently different from those of endemic animals that there was little if any competition and they were able to fill "empty" niches. Their success continued until the end of the Pleistocene, when they became extinct, along with other North American herbivores such as mammoths, mastodons, horses, and camels. The exact cause of their eventual extinction is unknown, primarily because we still have much to learn about their paleoecology. One aspect of xenarthran physiology may be relevant – xenarthrans have a low metabolism, with a resultant low body temperature. Environmental temperature, both in terms of seasonal extremes and yearly averages, may have been critical in controlling their distribution and hence their survival.

Anatomical Differences Distinguish Xenarthrans

Xenarthrans are unusual mammals and their unconventionality begins with their teeth. The dentition of xenarthrans is distinctive, unlike that of most mammals in that it is composed of an osteodentine, or bony dentine. Most mammals have teeth covered by enamel, and enamel is still present as a small cap on the teeth of the earliest xenarthrans, but it is absent from the teeth of most members of the Order. Generally, they have no milk teeth; the single set of permanent teeth is open-rooted and grows throughout the life of the animal. While most xenarthrans were herbivores and fed on a variety of vegetation, some modern armadillos are insectivorous or omnivorous and will even eat carrion. In some sloths, the first tooth in the skull and jaw is enlarged. While generally resembling the canine tooth of other mammals, this enlarged tusk-like tooth is not a true canine and is referred to as a "caniniform." The shape of the caniniform varies in different genera of sloths. The cheek teeth, or "molariforms," vary also, and in some sloths may be triangular in cross section, while in others they are square or lobate. The teeth of some sloths have crests on the upper and lower molariforms that would have helped to slice vegetation during chewing. Sloths with this type of dentition are thought to have been browsers. In other sloths, like mylodonts, the top of the teeth tend to be flat, indicating that vegetation was crushed during chewing rather than sliced. Their diet is interpreted to have been that of a grazer.

The xenarthran skeleton also has many distinctive features that readily distinguish members of the Order from other mammals. The preferred name, Order Xenarthra, is derived from the presence of additional bony connections (zygapophyses) between the vertebrae of the lower back (lumbar vertebrae). This "xenarthrous" condition is characteristic of the Order (Gaudin 1999).

The sacrum and pelvis are modified into a very strong and massive structure called a synsacrum, formed by the fusion of the proximal tail vertebrae (caudals) to the posterior pelvic (sacral) vertebrae and additional fusion of these to the bones of the pelvis. Xenarthrans are also distinguished by the presence of a second set of ribs that connect the normal (costal) ribs to the sternum (breastbone). These sternal ribs take the place of the costal cartilages present in other mammals. Each sternal rib articulates with the sternum by a double-headed joint, and is connected to the costal rib by a small amount of cartilage.

Some ground sloths are distinguished by a bizarre modification of the foot. It has become rotated so the animal is actually walking on its outside edge, with the sole facing inward and the claws prevented from touching the ground. This rotation has resulted in a suite of major modifications of numerous bones of the foot. When the foot is thus rotated, the weight of the animal is supported primarily by the fifth metatarsal and the heel bone (calcaneum), which may develop enlarged areas to provide more contact with the ground.

In those sloths with a rotated hind foot, such as *Nothrotheriops* and *Paramylodon* at Anza-Borrego, the anklebone (astragalus) has an enlarged process that fits into a socket on the lower leg bone (tibia) in a peg and groove fashion. As a result of this articulation, the foot could not rock up and down at the ankle as in other animals, but rather pivoted from side to side as the animal walked. Combined with the way the upper leg bone (femur) fit into the hip socket (acetabulum), these sloths must have had a distinctive waddle as they walked.

The only xenarthrans currently known as fossils from Anza-Borrego are ground sloths; each of the three genera represents a distinct stage in the land invasions of mammals from South America (see McDonald, this volume, *Anza-Borrego and the Great American Biotic Interchange*).

Early Invasion – Megalonychid Sloths

A member of the family Megalonychidae was among the earliest representatives of the Order Xenarthra to appear in North America. It is represented by the genus *Pliometanastes*. A mylodont sloth, *Thinobadistes*, also reached North America during this stage, in the Hemphillian (Webb, 1989).

Thinobadistes did not survive into the Blancan, however, and appears to have gone extinct in North America with no descendants. It is known only from a few sites in Florida and Texas. Some ground sloth bones have been found in the Hemphillian age sediments of Anza-Borrego. While it is possible to tell that they are from a megalonychid sloth, they are not diagnostic as to genus; at this time we cannot tell if they are from *Pliometanastes,* or an early species of *Megalonyx.* The recovery of more material may eventually resolve this mystery. Perhaps remains of the other sloth, *Thinobadistes,* may also be unearthed in Anza-Borrego.

Pliometanastes is believed to have evolved into the endemic North American genus, *Megalonyx,* represented in California by *Megalonyx mathisi* in the late Hemphillian. The Blancan to early Irvingtonian species of *Megalonyx* is *M. leptostomus,* first described from Mount Blanco, Texas by E. D. Cope in 1893. The species is widely distributed and is known from Florida, Idaho, Washington, Arizona, Kansas, and Nevada, as well as California.

The Irvingtonian species, *Megalonyx wheatleyi*, was first identified from Port Kennedy Cave, Pennsylvania, and, like *M. leptostomus,* was described by E.D. Cope. It was also, like *M. leptostomus,* widespread, and is present in

Figure 10.2
Megalonyx jeffersonii,
Jefferson's Ground Sloth.
(Drawing by Pat Ortega)

faunas from Florida to California. The terminal species of the lineage is *M. jeffersonii,* first described from a cave in West Virginia. *Megalonyx jeffersonii* is named for Thomas Jefferson, who first described the remains of the animal. Jefferson's ground sloth has the distinction of being the second fossil sloth to be described, and the first discovered in North America.

In Anza-Borrego, *Megalonyx leptostomus* has been identified in the Arroyo Seco local fauna. *Megalonyx wheatleyi* is known from the Vallecito Creek local fauna, and *M. jeffersonii* from the Vallecito Creek and Borrego local faunas (Figures 10.2, 10.3) (see Cassiliano, this volume, *Mammalian Biostratigraphy in the Vallecito Creek-Fish Creek Basin*; Remeika et al. 1995). While the last two species are recorded as being present in the Vallecito Creek local fauna, this does not mean they were contemporaries; *M. wheatleyi* is the direct ancestor of *M. jeffersonii*. Their joint record in the Vallecito Creek local fauna reflects the long span of time, 2.9 to 0.9 Ma, of the sediments from which the fossils have been recovered (Remeika et al. 1995). *Megalonyx* is interpreted as a browser. While widely distributed in North America, it is relatively more common in the eastern U.S. than in the western. Its distribution in the central and western U.S. seems to be closely associated with watercourses and their associated riparian vegetation (McDonald and Anderson, 1983; McDonald, Miller and Morris, 2001).

Compared with the other sloths found at Anza Borrego – *Paramylodon,* which shows up in North America at the next stage of the interchange, and Nothrotheriops, which appears during the final stage – *Megalonyx* probably had the greatest dependency on forest as its preferred habitat. Its enlarged front tooth, or caniniform (canine-like), appears to have served for grabbing leaves and twigs from trees, and the crests of its cheek teeth could have sliced them into small pieces (Figure 10.3). Because the distribution of *Megalonyx* was the greatest of all of the ground sloths in North America, its geographic range overlapped with all of the other species. As a result, *Megalonyx* is sometimes found together with other ground sloths in a fossil assemblage. However, when *Megalonyx* is found along with other sloths, there are usually differences in their relative abundance. If *Megalonyx* is common, for example, then the other small browser *Nothrotheriops* is rare.

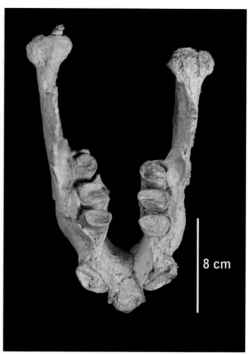

Figure 10.3
Lower Jaw of
Megalonyx jeffersonii.
Occlusal or top view of the lower jaw (mandible ABDSP[IVCM]1027/V3406) of Jefferson's ground sloth. Note the two large oval teeth on either side of the front of the jaw (bottom). (Photograph by Barbara Marrs)

Second Stage – Mylodont Sloths

During the second phase of the Great American Biotic Interchange, the mylodonts, a second group of sloths found at Anza-Borrego, dispersed into North America. This was the second appearance of the family Mylodontidae in North America, following the short-term success of *Thinobadistes*.

These mylodonts are represented by *"Glossotherium" chapadmalense,* a species first described from Argentina. Its assignment to the genus *Glossotherium* is questionable, as the taxonomy of the genus itself is in doubt (McDonald, 1995); hence the use of quotation marks around the name. The earliest record of the species, dated at 2.5 million years, is from Mt. Blanco. The descendant of *"G." chapadmalense* is *Paramylodon harlani*, known from the Irvingtonian and Rancholabrean. Among the evolutionary changes that occur in

Figure 10.4
Paramylodon harlani,
Harlan's Ground Sloth.
(Drawing by Pat Ortega)

this lineage are a pronounced increase in size and a tendency to reduce or lose the caniniform teeth.

In Anza-Borrego, only *Paramylodon harlani* is known from the Vallecito Creek and Borrego local faunas (Figure 10.4) (Cassiliano, this volume; Remeika et al. 1995). Given the richness of Anza-Borrego, it may be just a matter of time before remains of *"G." chapadmalense* are found. Since *"G." chapadmalense* first appears in North America about 2.5 million years ago it would not be surprising for it to be found in Anza-Borrego assemblages above the Gauss Normal-Polarity Magnetochron, which is dated at 2.3 ± 0.4 million years (see Remeika, this volume, *Dating, Ashes and Magnetics*).

The skull of a mylodont sloth is relatively wide for its length, with the rows of cheek teeth diverging anteriorly (forward). The muzzles of the skulls of both *Glossotherium* and *Paramylodon* are much wider than those of *Megalonyx* or *Nothrotheriops.* The molariforms of mylodont sloths are lobate with a flat occlusal surface for crushing vegetation. They are thought to have been grazers that lived in country more open than the forest habitats preferred by *Megalonyx.* This interpretation of their diet is supported by the discovery of preserved dung of a related genus, *Mylodon,* which contains high quantities of grasses and sedges (Moore, 1978). Naples (1989), in an analysis of the skull and inferred locations of the jaw musculature, suggested that *Paramylodon harlani* might not have been a strict grazer but rather a mixed feeder. This assertion was recently confirmed by stable isotope analyses of *Paramylodon* bone from Rancho La Brea by Coltrain et al. (2004).

Features of the ulna and the expanded distal end of the humerus imply that the forearm of *Paramylodon harlani* could generate powerful force. The unguals (claws) are semi-circular in cross section, which further suggests that the animal was capable of digging. The skin contained dermal ossicles (small pieces of embedded bone), which are often recovered. Dermal ossicles are absent in the other sloths found at Anza Borrego. One of the more spectacular Anza-Borrego specimens is a large slab of ossicles still articulated and associated with part of the pelvis of a *P. harlani* sloth (see discussion below, Dermal Ossicles). The discovery of this preserved piece of "skin" shows that the back of the animal was protected from predators by this mosaic of small bony nodules in the skin.

Paramylodon harlani is second only to *Megalonyx* in the extent of its distribution in North America. In California, it is common in numerous paleofaunas and is currently known from 67 localities (McDonald, 1996). It is more common in the western U.S. than *Megalonyx,* but less so in the more heavily forested east where *Megalonyx* is common. Based on its anatomical features, and the pattern of its distribution, its habitat preference was probably more open country than forest.

Final Invasion — Nothrothere Sloths

During the third phase of the Great American Biotic Interchange in the early Irvingtonian, the third type of sloth appeared in North America, and also is found at Anza-Borrego. The earliest species, *Nothrotheriops texanus*, is known only from the Irvingtonian, and the lineage is represented by the descendant species *N. shastensis* in the Rancholabrean. The earliest paleofauna with *N. texanus* is the Leisey Shell Pit in Florida, dated at about 1.7 million years (McDonald, 1995). *Nothrotheriops texanus* was more widespread than *N. shastensis* and is known from California, Florida, Oklahoma, Texas, and northern Mexico. The distribution of *Nothrotheriops* decreased in area during the Pleistocene, with *N. shastensis* being restricted to the western U.S. and Mexico. At Anza-Borrego, this Shasta ground sloth is known from the Vallecito Creek and Borrego local faunas (Figure 10.5) (see Cassiliano, this volume; Remeika et al. 1995).

Nothrotheriops is the smallest of the North American ground sloths, comparable to a small calf in stature, while both *Paramylodon* and *Megalonyx* were similar in mass to an adult cow. *Nothrotheriops texanus* is smaller than *N. shastensis* and there is only a slight increase in body size from the former to the latter. The skull of this genus is longer and narrower than in the other sloths. *Nothrotheriops* also is distinguished from the other sloths at Anza-Borrego by the loss of the first tooth in both the upper and lower dentition. The molariform teeth are rectangular with a prominent transverse crest on the front and back edges of the occlusal surface. These crests on the upper and lower dentition slide past each other during mastication to effectively cut vegetation. The front of the lower jaw has a distinctive spout-like elongation. Lacking the caniniform teeth of *Megalonyx* or *Paramylodon,* it is possible that *Nothrotheriops*, like a giraffe, had a prehensile tongue that was used for gathering vegetation. The tongue would have rested in this prominent spout.

Nothrotheriops has been interpreted as a browser. While no direct evidence for the diet of *N. texanus* is available, dung is known for *N. shastensis.* Numerous dry caves in the southwestern U.S. preserve sloth dung. Analyses of the dung have yielded detailed information on the diet and ecology of the species (Laudermilk and Munz, 1934, 1938; Martin et al. 1961; Hansen, 1978; Thompson et al. 1980). The animal's diet was eclectic and included desert vegetations as well as plants associated with riparian habitats. It appears to have been better adapted than the other sloths for living in arid desert conditions, and may not have been as dependent on water as either *Megalonyx* or *Paramylodon.*

While any of these three species may be found together with one of the other sloths in an assemblage, they are just as likely to be the only sloth in a paleofauna. The joint occurrence of two or more sloth species probably happened only along areas of contact (ecotones) between their preferred habitats.

Figure 10.5
Nothrotheriops shastensis,
Shasta Ground Sloth.
(Drawing by Pat Ortega)

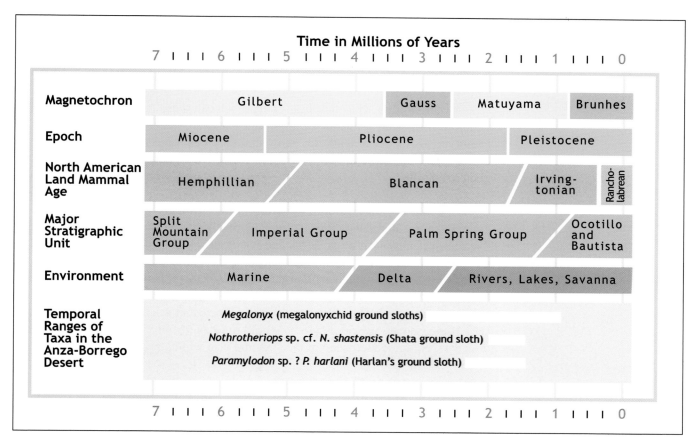

Time in Millions of Years								
7	6	5	4	3	2	1	0	

Magnetochron	Gilbert / Gauss / Matuyama / Brunhes
Epoch	Miocene / Pliocene / Pleistocene
North American Land Mammal Age	Hemphillian / Blancan / Irvingtonian / Rancholabrean
Major Stratigraphic Unit	Split Mountain Group / Imperial Group / Palm Spring Group / Ocotillo and Bautista
Environment	Marine / Delta / Rivers, Lakes, Savanna
Temporal Ranges of Taxa in the Anza-Borrego Desert	*Megalonyx* (megalonyxchid ground sloths) / *Nothrotheriops* sp. cf. *N. shastensis* (Shata ground sloth) / *Paramylodon* sp. ? *P. harlani* (Harlan's ground sloth)

7	6	5	4	3	2	1	0

Figure 10.6
The Ground Sloths of Anza-Borrego.
Temporal ranges for the 3 genera of Anza-Borrego ground sloths apparently are restricted to the terrestrial part of the record.

Dermal Ossicles – "Skin Armor" for Mylodont Sloths

Dermal ossicles, also known as osteoderms, are small nodules of bone embedded in an animal's skin (dermis). They most commonly are found in reptiles including some lizards like the Gila monster (see Gensler, this volume, *The Fossil Lower Vertebrates*), and in the skin of the legs of tortoises, or the larger bony scutes on the backs of alligators. They act as armor and protection against attacks from predators. The only mammals that have developed such armor are members of the Order Xenarthra. Armor is best seen in the living armadillos that are distinguished by having a bony covering or shell. In armadillos, the armor is composed of a series of individual bones that fit together to form a solid carapace. The carapace not only covers the back of the animal but may also be present on top of the head and on the tail as well. Since this bone develops within the skin, it is referred to as dermal armor. It is not equivalent to the shell of turtles or tortoises. The top part of their shells is formed by an expansion and fusion of primarily rib bones.

Armadillos have two extinct armored relatives, the glyptodonts and the pampatheres of South America. The shell of an armadillo is composed of solid front and back parts. These are known as bucklers, and are connected by overlapping and movable bands that allow some but not all armadillos to curl into a ball. The number of bands in the shell varies from three to nine in different types of armadillos. The shell in glyptodonts lacks these bands and is a solid structure formed by a mosaic of large, thick polygonal dermal ossicles. Preserved shells of pampatheres show that they had three movable bands. So their shells were more armadillo-like, and they are sometimes referred to as giant armadillos.

It has been argued that the presence of dermal ossicles in the skin of extinct ground sloths indicates a close relationship with armadillos. However this armor is absent in most sloths and is common only in a number of related genera in the Family Mylodontidae. More importantly, it appears late in the group's history. This suggests that the dermal armor was evolved independently, rather than being inherited from the sloth's common ancestor with armadillos. While the bony armor of the armadillo and its extinct relatives forms an obvious shell, the armor in mylodont sloths would not have been visible. The dermal ossicles were imbedded deep in the skin and also would have been hidden by hair.

Pieces of dried *Mylodon darwinii* skin were found in a southern South American cave. The exact placement of the bony armor within the thick, hair-covered skin can be seen in X-ray images of the specimens. These show that the irregular-shaped dermal ossicles formed a mosaic but that they remained separate and did not fuse together as in armadillos or glyptodonts. Most likely the entire mosaic of small bones covered the entire back of the animal. Any predator, such as a sabertooth cat wishing to take a bite out of the sloth, not only had to get through coarse hair over 15 cm (6 in) long and skin up to 13 mm (1/2 in) thick, but also had to get its teeth past the numerous ossicles embedded in the sloth's skin.

One individual mylodont sloth, like *Paramylodon harlani,* would have been covered by literally thousands of dermal ossicles. Paleontologists often encounter these nodules of bone while screening the sediments that surround a fossil sloth skeleton. Unfortunately, unlike the mummified skin, the original pattern of dermal ossicles and how they fit over the body was lost when the skin holding them together completely decayed. However, on rare occasions an entire piece of skin may become buried and preserved. With careful excavation of such specimens, the original positions of the ossicles may be conserved. Such a specimen was recovered at Anza-Borrego. The entire 1 by 1.5 m (3 x 5 ft) piece (Figure 10.7A.) and the surrounding matrix were carefully removed in a plaster jacket so as not to disturb the mosaic of individual ossicles. Careful preparation in the laboratory provided paleontologists with an exact idea of the density and pattern of how the individual ossicles fit together (Figure 10.7B). An added bonus is that the preserved mosaic of ossicles was associated with a piece of the pelvis, so we know exactly where the layer fit on the animal's back.

2 cm

40 cm

Figure 10.7A
Fossilized skin.
Dermal ossicle layer from
Paramylodon harlani, Harlan's
ground sloth (ABDSP [LACM]
1568/V77700). (Photograph by
Barbara Marrs)
Figure 10.7B (Inset)
Detail of Fossilized
Dermal Ossicles from
Paramylodon harlani.
(ABDSP [LACM] 1568/V77700).
(Photograph by Barbara Marrs)

Why Were These Sloths So Successful?

While all of the North American ground sloths were herbivores, they were quite different in their anatomy, distribution, and habitat preference. A partial explanation of their success in North America is that, while they did not seem to compete with the native herbivores, they also did not compete with each other. Each utilized a different type of habitat: *Megalonyx,* a browser, preferred woody forests and riparian gallery forests; *Nothrotheriops,* another browser, lived in drier and more arid habitats with low scrub; *Paramylodon,* a grazer, fed on grasses in more open savanna. Although each of these sloths was present at Anza-Borrego, each had its own preferred habitat. The presence of all three indicates a diversity of past habitats.

Despite the wealth of knowledge we have concerning these interesting animals, we still have much to learn about their paleoecology and history, both prior to and following the appearance of each of them in North America. More research is needed to understand not only their success but, equally important, what caused their extinction. As this research continues, the rich fossil record at Anza-Borrego will continue to make significant contributions.

Figure 11.0 *Smilodon gracilis*, the Gracile Sabertooth Cat Hunts a Platygonus Family Along the Edge of a Thicket in Pleistocene Anza-Borrego. (Picture by John Francis)

11

The Large Carnivorans: Wolves, Bears, and Big Cats

This tendency for two rather distantly related families of carnivores, Felidae and Nimravidae, to evolve similar ecomorphs in parallel is one of the most remarkable stories in evolution.

Larry D. Martin (1998)

Christopher A. Shaw
Shelley M. Cox

The Large Carnivorans: Wolves, Bears, and Big Cats

Arroyo Tapiado Badlands. (Photograph by Paul Remeika)

Introduction

Plio-Pleistocene meat-eaters, like wolves and bears, whose average weight exceeds about 40 to 45 kg (88-100 lb), are considered large carnivorans. These animals occupy the top of the food chain, exhibit great trophic diversity (wide variety of food preferences), and together with small carnivorans form an ecological guild. Root (1967) defined the ecological concept of a guild as "a group of species that exploit the same class of environmental resources in a similar way" (Van Valkenburgh and Molnar, 2002). During the Pliocene and Pleistocene, carnivorans did partition available food resources which minimized intraspecific competition (Van Valkenburgh, 1988). Nevertheless, species within a guild compete more intensely with each other than with species outside of the guild, and large predatory mammals form a part of the guild in which competition is expected to be relatively intense (Van Valkenburgh and Molnar, 2002). Since the practice of predation is a dangerous business, carcass theft ("kleptoparasitism" of Van Valkenburgh and Molnar, 2002) is a worthwhile endeavor between guild members, so it is to be expected that adaptations to minimize dangerous encounters and to escape predation will occur among less dominant members of the guild (Van Valkenburgh and Molnar, 2002).

Compared with fossil skeletal elements of large herbivorous mammals, those of large carnivorans tend to be rarely preserved in the fossil record and remains of smaller carnivorans tend to be relatively more commonly preserved (see Murray, this volume, *The Small Carnivorans*). Larger members of Pliocene and Pleistocene carnivoran guilds typically include species within the Families Canidae, Ursidae, Hyaenidae, and Felidae.

Dogs: Family Canidae

Two large canid species have been identified in the Anza-Borrego fossil record, one from the Subfamily Borophaginae and the other from the Subfamily Caninae. The bone-eating dog, *Borophagus diversidens*, is represented by several dentaries (one side of the lower jaw) (Figure 11.1), some isolated teeth, and a partially preserved paw. This species is not uncommon in Blancan age sites in North America but it is only rarely identified in Irvingtonian age sites (Kurtén and Anderson, 1980). It is widely distributed in North America, from Washington to central Mexico and California to Florida (Wang et al., 1999). In Anza-Borrego Desert State Park, it has been recovered from 2.1 million-year-old Blancan age deposits of the Hueso Formation.

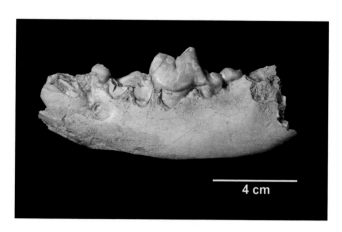

4 cm

Figure 11.1
Borophagus diversidens.
Left dentary (lower jaw) of the bone-eating dog (ABDSP[LACM] 4924/V123906). (Photograph by Barbara Marrs)

Borophagus diversidens (Figure 11.2) weighed 40 kg (84 lb) (Van Valkenburgh, 1991) and was the last species of a diverse borophagine group that dominated North American faunas from Miocene through Pliocene time (Munthe, 1989, 1998; Wang et al., 1999), which became extinct in the early Pleistocene.

It has been suggested that *Borophagus* resided in open-country environments (Matthew, 1930), but their short, robust limbs indicate it was not highly cursorial and that they may have lived in the ecotone boundary zone between grasslands and more heavily vegetated scrub-woodlands (Munthe, 1989; Wang et al., 1999). There are remarkable similarities in the development of the high-vaulted skull and enlarged cheek teeth between *Borophagus* and living hyenas (especially the spotted

Figure 11.2
The Bone-eating Dog,
Borophagus diversidens.
(Drawing by Pat Ortega)

hyena, *Crocuta crocuta*), which Werdelin (1989) found to reflect similar habits in feeding if not mode of life. These adaptations were ideal for "bone-cracking" (termed durophagy by Wang et al., 1999, and ossiphagy by Stefen, 1999) to access nutritious bone marrow and were more highly developed in *Borophagus* than in living hyenas, indicating that they were closer to obligate scavengers than recent hyenas (Munthe, 1989; Werdelin, 1989). Munthe (1989) states that the dietary habits of *Borophagus* may have been comparable to brown and striped hyenas (*Hyaena brunnea* and *H. hyaena*), consisting mostly of carrion, but also including fruits, insects, reptiles, and live-caught mammals.

The demise of *Borophagus* may have been due to a combination of increased competition for food resources by other predators (bears, large cats, and other canids) along with a change in the availability of carrion on which to feed (Munthe, 1989). An extensive area is required for a large scavenger population to support each individual; the *Borophagus* population level was indeed low, as indicated by few and scattered known fossil localities (Munthe, 1998). Ewer (1973) proposed that hyenas evolved when sabertoothed cats were the dominant predator, correctly observing that sabertoothed cats were ill-equipped to process skeletal parts, and felt that this could have been an ideal resource to make bone-cracking adaptations worthwhile. Munthe (1989) entertained the possibility that such a relationship may have existed between *Borophagus* and large felids, and that the *Borophagus* population decreased as the diversity of large felids declined.

Another large canid species has been identified as Armbruster's wolf, *Canis armbrusteri*, based on the large size of three Anza-Borrego Desert State Park specimens. These include a fragmentary scapula and ilium, and a complete left magnum (wrist bone). Until more complete diagnostic material is recovered, however, recognition of the occurrence of this species from Anza-Borrego Desert State Park must remain tentative.

Armbruster's wolf is only known from Irvingtonian sites (Bell et al., 2004). Relatively large samples were recovered from Maryland (Gidley and Gazin, 1938) and Florida (Martin, 1974; Berta, 1995), with tentative identifications made from sites in Arizona, California, Nebraska, Pennsylvania, South Carolina, and Texas (Nowak, 1979; Kurtén and Anderson, 1980). It has been recovered from the Irvingtonian age part of the Hueso Formation in the Vallecito Creek badlands and the Bautista beds of Coyote Creek. Armbruster's wolf weighed 48 kg (106 lb) (Van Valkenburgh, 1991). Nowak (1979) described this species generally as larger than a red wolf (*Canis rufus*) but smaller than a dire wolf (*C. dirus*) and certain similarities of the cheek teeth suggest that *C. armbrusteri* is the direct ancestor of the dire wolf (Nowak, 1979; Tedford et al., 2001). It is thought that Armbruster's wolf is an early Pleistocene immigrant to North America, but from where it came is something of a mystery. Kurtén and Anderson (1980) suggest that dire wolves may have been derived from the South American canid, *Canis nehringi,* before entering North America.

Modern wolves are social and gregarious, habituate in groups, and cooperatively hunt for prey. Merriam (1912) believed that the preponderance of the extinct dire wolves found at the late Pleistocene site of Rancho La Brea suggests that *Canis dirus* "associated themselves in packs, and that groups of considerable size may have assembled to kill isolated ungulates and edentates." Whether such social behaviors can be ascribed to Armbruster's wolf is a matter of conjecture, since the existing fossil record is too poor to predict sociality in this species. If solitary hunters, *C. armbrusteri* was well adapted for predation on medium-sized prey animals commonly found at Anza-Borrego Desert State Park (e.g. tapir, deer, pronghorns); if this species had developed sociality and hunted cooperatively, larger prey items could have been targeted (e.g. llamas and camels or horses).

Bears: Family Ursidae

The bear fossils recovered from Anza-Borrego Desert State Park represent at least two and perhaps three species. Most of the bear remains clearly belong to either of two species in the Tribe Tremarctini, but two specimens are tentatively placed in the Tribe Ursini within the Subfamily Ursinae (see Hunt, 1998, for a taxonomic review). The two species of tremarctine bears include the Florida cave bear, *Tremarctos floridanus*, and the giant short-faced bear, *Arctodus simus*. The ursine bear is tentatively assigned to the genus *Ursus*.

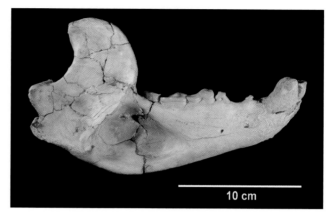

Fossils identified as *Tremarctos* include two dentaries with near complete dentitions (Figure 11.3) and an additional 11 postcranial elements. *Tremarctos floridanus* is a long-lived species, recovered from early Blancan age through early Holocene sites in North America. Blancan records of this species are rare, the oldest of which is found in early Blancan sediments of south-central Idaho; the only other Blancan record is from Anza-Borrego Desert State Park (Kurtén and Anderson, 1980; Cassiliano, 1999). These specimens were recovered from the Hueso Formation and date to about 2.8 million years. Irvingtonian records are equally rare, with remains recovered from sites in Florida, northwestern Sonora, México, and Anza-Borrego Desert State Park, which are from the upper Hueso Formation and are no younger than about 1.1 million years. Rancholabrean age records however are relatively common, and by this time the species had a distinctively southern distribution in North America, from New Mexico to Florida and from Kentucky to Nuevo Léon, Mexico (Kurtén and Anderson, 1980). Unfortunately, there are no Rancholabrean-aged fossiliferous deposits in Anza-Borrego Desert State Park.

Figure 11.3
Tremarctos floridanus.
Right dentary (lower jaw) of the Florida cave bear (ABDSP[LACM] 1454/V4118). (Photograph by Barbara Marrs)

Figure 11.4
The Florida Cave Bear, *Tremarctos floridanus.*
(Drawing by Pat Ortega)

Sexual dimorphism in *Tremarctos floridanus* (Figure 11.4) is pronounced (Kurtén and Anderson, 1980). Van Valkenburgh (1991) lists a body mass estimate at 148 kg (326 lb) for this species, which is an average weight between males and females (Van Valkenburgh, personal communication), and

Guilday and Irving (1967) describe the Florida cave bear as "almost twice the size of the Andean spectacled bear." Kurtén (1967) lists mass estimates based upon total body length (males, 230 to 250 kg [506-550 lb]; females, 115 to 125 kg [253-275 lb]) and femoral cross-section (males, 265 to 285 kg [583-627 lb]; females, 170 to 180 kg [374-396 lb]).

Based upon shared characteristics of the shortened muzzle and the tuberculation ("Z" pattern formed by small ridges on the crown) of the lower second molar, *Tremarctos floridanus* is descended from the middle Hemphillian-early Blancan age *Plionarctos* (Hunt, 1998; Tedford and Martin, 2001). It differs from the living South American species, *T. ornatus*, in its larger size and heavier proportions, its domed forehead and elongated hind molars, longer neck, and its relatively long humerus and femur, features which parallel the condition of European cave bears (Kurtén, 1966; Kurtén and Anderson, 1980). Kurtén (1966) considered this remarkable convergence of characters between the European cave bear and *T. floridanus* indicative of both inhabiting the same niche, but in radically different environments, and felt they were both large, almost entirely vegetarian species. He also believed that tremarctines never evolved the turning-in of the forepaws seen in the gait of living ursine bears (Kurtén, 1966; Hunt, 1998). *Tremarctos floridanus* was a short-footed, plantigrade (walking on the sole of foot) species that probably inhabited more mesic forested coastal environments, as this species is not well-represented in more open savanna or savanna-parkland habitats of the continental interior (Hunt, 1998).

Figure 11.5
The Short-faced Bear,
Arctodus simus.
(Drawing by Pat Ortega)

The giant short-faced bear, *Arctodus simus* (Figure 11.5), is represented in Anza-Borrego Desert State Park by nine isolated skeletal elements (seven foot bones and two vertebrae) and an associated partial postcranial skeleton. The associated skeleton consists of over 48 identifiable bones, including most limb elements and a nearly complete left hind foot (Figures 11.6, 11.7). This species has been recovered from early to late Pleistocene sites in North America; Irvingtonian age records are from California, Kansas, Nebraska, Texas, and Pennsylvania, whereas the Rancholabrean age distribution is much more extensive, from coast to coast and from Alaska and Saskatchewan, Canada, to south-central Mexico

(Kurtén and Anderson, 1980; Agenbroad and Mead, 1986; Richards et al., 1996). The Anza-Borrego Desert State Park partial skeleton was recovered from the Ocotillo Conglomerate of the Borrego Badlands and is about 0.8 million years old.

The genus *Arctodus* is known only from Pleistocene deposits (Bell et al., 2004) and is represented by two species, *A. pristinus* and *A. simus*. *Arctodus pristinus* is generally smaller in size, has a longer muzzle with less crowded anterior lower premolars, and has relatively narrower cheek teeth (Kurtén, 1967; Emslie, 1995). In addition, this species seems to be restricted in distribution to the Irvingtonian of the south-eastern U.S. and may have been ecologically replaced by *Tremarctos floridanus* in Rancholabrean times (Emslie, 1995).

Like *Tremarctos*, sexual dimorphism in *Arctodus* is pronounced, perhaps even more so, as presumed male individuals may be as much as 25% to 47% larger than females (Kurtén, 1967; Nelson and Madsen, 1983). Van Valkenburgh (1991) lists an average mass (between males and females) of 251 kg (552 lb) for *A. simus*. Kurtén's (1967) total body length estimates put females at between

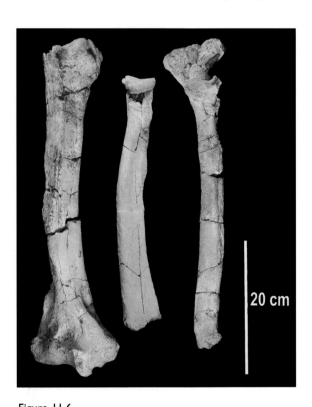

Figure 11.6
Front Limb of *Arctodus simus*.
The right humerus (upper) and radius and ulna (lower) long bones of the short-faced bear are from a single partial skeleton (ABDSP[LACM]68103/V132162). (Photograph by Barbara Marrs)

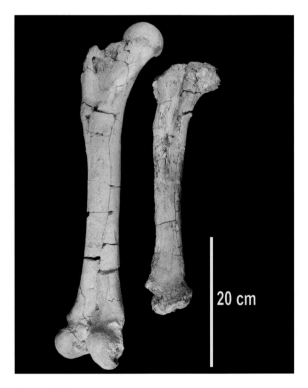

Figure 11.7
Hind Limb of *Arctodus simus*.
The left femur (upper) and tibia (lower) long bones of the short-faced bear are from a single partial skeleton (ABDSP[LACM]68103/V132162). (Photograph by Barbara Marrs)

250 kg and 279 kg (550-614 lb) and males at between 350 kg and 375 kg (770-825 lb), while his femoral cross-section calculations placed females at between 270 kg and 290 kg (594-638 lb). A very large male recovered in Utah was estimated to be between 620 kg and 660 kg (1364-1452 lb) (Nelson and Madsen, 1983). Kurtén (1967) considered that the observed size variation was due to a combination of local, sexual, and individual variation within a single species, and stated that the degree of sexual dimorphism tends to increase with absolute size, both in *Ursus* and *Tremarctos*, so that the largest individuals of the species show excessive size difference between males and females. Even so, he thought that some populations differed enough to recognize subspecies, and assigned the gigantic forms to *A. s. yukonensis* and all remaining, smaller forms to *A. s. simus*. Voorhies and Corner (1982) elevated these subspecies to full specific rank, while Richards et al. (1996) retained the subspecific ranking proposed by Kurtén (1967). Nelson and Madsen (1983) and Gillette and Madsen (1992) remained conservative, relegating their gigantic records to *A. simus* (without designating a subspecies), and attributing size variation to sexual dimorphism. The size range of *A. simus* specimens from Rancho La Brea (Merriam and Stock, 1925; Richards et al., 1996) overlaps both named subspecies and therefore renders suspect the validity of either subspecific designation. This observed variation in size more likely reflects the extremes of sexual dimorphism than specific or subspecific differences. Based on the size of the femur and calcaneum (heel bone) from the partial skeleton, this individual from Anza-Borrego Desert State Park (Figures 11.6, 11.7) was a female.

Like *Tremarctos, Arctodus* was a descendant of *Plionarctos* (Hunt, 1998; Tedford and Martin, 2001). *Arctodus simus* is generally much larger than either *A. pristinus* or *T. floridanus*, and has a shorter and wider muzzle, a high-vaulted calvarium (upper, domed part of the skull or skull cap), short neck, and long limbs. Kurtén (1967) believed these adaptations were remarkably convergent with large felids and concluded that the giant short-faced bear was predominantly carnivorous and may have been capable of bursts of speed exceeding those of the brown bear, without being truly cursorial. However, Emslie and Czaplewski (1985) found this analogy to felids inappropriate. Based upon skull and dental characters and the immense size of this species, they believe that *Arctodus* was mainly herbivorous, feeding on fresh meat or carrion opportunistically like other bears. Gillette and Madsen (1992) found grooves on mammoth bones (see McDaniel, this volume, *Mammoths and Their Relatives*) that were consistent with the size and shape of the canine teeth of the *Arctodus* found associated with them, which is provocative evidence of scavenging by this species. The limb proportions of *A. simus* are distinctly different from cursorial animals, and Esmlie and Czaplewski (1985) suggest this long-limbed adaptation may have improved visibility over tall ground cover or may have had some advantage in tearing and pulling down tall vegetation. In addition, its immense size (the largest of the carnivoran guild) would have facilitated carcass theft to supplement its diet. *Arctodus simus* appears to have been a long-footed, omnivorous carnivoran that inhabited savanna.

If a species of *Ursus* is present in Anza-Borrego Desert State Park, it is indeed a very rare component of the fauna. A first phalanx (toe bone), identified as *Ursus* sp., is typically shaped like a bear, but the size overlaps with that of *Tremarctos*. Another specimen, a pisiform bone from the wrist, also identified as *Ursus* sp., is large enough but lacks typical ursid morphology and could represent a borophagine canid or a hyaenid.

Remains of black bear, *Ursus americanus* (Figure 11.8), are often found associated with those of Florida cave bear (Kurtén and Anderson, 1980), so co-occurrence of these species might be expected in Anza-Borrego Desert State Park even if the two bones tentatively identified as *Ursus* sp. prove to belong to a species in a different Family of Carnivora. Until more complete, diagnostic skeletal material is recovered, the identification of any species of *Ursus* from Anza-Borrego Desert State Park must remain tentative.

Figure 11.8
The American Black Bear,
Ursus americanus.
(Drawing by Pat Ortega)

Hyenas: Family Hyaenidae

No fossils have been recovered yet from Anza-Borrego Desert State Park that can be confidently assigned to the Family Hyaenidae. As mentioned above, a pisiform identified as an ursid may in fact be hyaenid; unfortunately, the only North American hyaenid associated skeletal material does not include this wrist bone. So, a direct comparison is not possible.

The American hunting hyena (Figure 11.9), *Chasmaporthetes ossifragus*, is the only known hyaenid species from North America. This rare species is only documented at about a dozen Blancan and Irvingtonian age localities in the southern U.S. and Mexico (Kurtén and Anderson, 1980; Berta, 1981, 1998). The closest occurrence is in northwestern Sonora, Mexico, approximately 250 km (155 mile) southeast of Borrego Springs, and future recovery of bones representing this species is likely at Anza-Borrego Desert State Park.

Figure 11.9
The American Hunting Hyena,
Chasmaporthetes ossifragus.
(Drawing courtesy of A. Berta)

Chasmaporthetes has been described as a hyaenid of moderate size (Berta, 1998) that weighed 62 kg (124 lb) (Van Valkenburgh, 1991) (Figure 11.9). It is characterized as having a broad muzzle with narrow, sharp teeth, a stepped forehead, and long, slender limbs (Galiano and Frailey, 1977; Kurtén and Werdelin, 1988; Werdelin and Solounias, 1991; Berta, 1998). Kurtén and Anderson (1980) felt that these adaptations indicated a cheetah-like predator. Galiano and Frailey (1977) suggested an ecologic role now filled by wolves and hunting dogs. It is evident that *Chasmaporthetes* occupied a niche that is very different from its living, bone-cracking relatives. The slender limbs suggest a more cursorial habit consistent with active predatory behavior (Berta, 1981). Tooth enamel structure of a European species (*C. lunensis*) indicates that it was capable of bone-cracking, but less efficient in breaking large bones, and from this Ferretti (1999) concluded that this species was primarily an active predator; nevertheless, he felt that scavenging was an important but minor part of its diet. The immigration of *Chasmaporthetes* in the Blancan probably represents the first occurrence of a true pursuit predator on the North American continent (Janis and Wilhelm, 1993; Janis et al., 1998).

Cats: Family Felidae

By far, cats are the most abundantly represented group of large carnivorans in the Anza-Borrego Desert State Park fossil record. Over 45 fossils have been identified that belong to either of two Subfamilies, the Machairodontinae (sabertoothed cats) and the Felinae (true cats). Remains of three felid species have been positively identified from the collections at Anza-Borrego Desert State Park.

The most commonly recovered large carnivoran from the Anza-Borrego Desert State Park is the gracile sabertoothed cat, *Smilodon gracilis* (Figure 11.10). Remains of this species include an associated partial skull and dentary with dentition, an additional skull fragment, four other isolated dentaries, and 15 isolated postcranial skeletal elements. *Smilodon gracilis* has been recorded from late Blancan age sites in Florida and is well represented at early Irvingtonian age localities in Florida and Pennsylvania (Kurtén and Anderson, 1980; Berta, 1987; Bell et al., 2004). The only western North American record of this species occurs at the Anza-Borrego Desert State Park and comes from the late Blancan to middle Irvingtonian Hueso Formation and middle Irvingtonian Ocotillo Conglomerate, dating from 2.1 to less than 1 million years old.

Figure 11.10
The Gracile Sabertooth Cat, *Smilodon gracilis*.
(Drawing by Pat Ortega)

Van Valkenburgh (1991) lists an estimated weight of 80 kg (176 lb) for the gracile sabertoothed cat. Sexual dimorphism is significantly less in *Smilodon* than in living, social felines (Van Valkenburgh and Sacco, 2002). *Smilodon* evolved in North America; *S. gracilis* is the direct descendant of the late Hemphillian age species, *Megantereon hesperus*, based upon shared characters of the skull, lower jaw, and teeth, short and stocky limbs and feet, and shortened tail (Berta and Galiano, 1983; Martin, 1998). *Smilodon gracilis* is the direct ancestor to the more derived Rancholabrean age species, *S. fatalis* (Berta, 1987), the California state fossil. The gracile sabertoothed cat is intermediate in size and morphology between the other two species, sharing characteristics of both (Kurtén and Anderson, 1980; Berta and Galiano, 1983; Berta, 1987). For instance, whereas the long, curved upper canines of *Megantereon* are not serrated on the anterior and posterior edges, and those of *S. fatalis* are strongly serrated on both edges, *S. gracilis* canines have developed very fine serrations that are only visible with the aid of a microscope (Berta, 1987).

Limb proportions of *Smilodon* suggest that they were forest-adapted (Gonyea, 1978), yet most samples come from plains and woodland habitats (Berta, 1987). There has been much controversy over whether these saber-toothed cats exhibited some kind of social structure and behavior that may have included cooperative hunting. Evidence from Rancho La Brea suggests that *S. fatalis* may have been social and gregarious (Gonyea, 1976; Akersten, 1985; Heald and Shaw, 1991), though Van Valkenburgh and Sacco (2002) among others find this unlikely. Whether *S. gracilis* exhibited social behaviors is uncertain due to the relatively meager fossil record. This species was built to prey upon animals that were larger than themselves, and the dental and post-cranial morphology suggests that they developed hunting strategies that relied on stalking and ambush to acquire their preferred prey (Berta, 1987). Martin (1998, 2004) believes that, with their short legs and semiplantigrade feet, they must have been the ultimate ambush predators.

A large feline species from the Anza-Borrego Desert State Park has been identified as jaguar, *Panthera onca*, based on the large size of four fossil bones that include a partial femur and three forepaw elements. Jaguar is the rarest of the large cats found at the Anza-Borrego Desert State Park. Prior to its identification in Anza-Borrego Desert State Park collections, this species was only known from the Pleistocene of North America. The specimens from the Anza-Borrego Desert State Park were recovered from the late Blancan, Pliocene portion of the Hueso Formation and date from 2.5 to 2.2 million years ago. *Panthera onca* had its widest distribution during middle Irvingtonian time, from coast to coast and from Washington, Nebraska, and Maryland into South America (Kurtén, 1973; Seymour, 1989, 1993; Bell et al. 2004). From Irvingtonian to Rancholabrean and into modern times, its distribution steadily retreated southward, and today the historic range of this species is in the southern parts of California, Arizona, New Mexico, and Texas to Argentina (Seymour, 1993).

Panthera onca is also the largest of the felids found at Anza-Borrego Desert State Park. The Irvingtonian forms were estimated by Seymour (1993) to average 106 kg (233 lb) and the largest was 129 kg (284 lb). The size of the Anza-Borrego Desert State Park fossils are even larger than the largest Irvingtonian age samples listed by Kurtén (1965, 1973) or Seymour (1993). Along with the reduction in distribution, there was also a reduction in size of this species through time. Kurtén and Anderson (1980) estimated the size of Wisconsinan (late Pleistocene) jaguars to be 15-20% larger than those in modern populations, and stated that earlier individuals were even larger. Seymour (1993) calculated that the Wisconsinan animals' size may have been up to 25% larger than those in living populations.

Kurtén (1973) felt that the jaguar lineage arose in Eurasia and that they immigrated to the New World in the late Blancan-early Irvingtonian. Although

Kurtén (1973; Kurtén and Anderson, 1980) believed there was late Blancan evidence of jaguar in North America, Seymour (1993) has discounted the evidence for such an early arrival. If the identifications of the Anza-Borrego Desert State Park specimens are correct, then these are the earliest records of this species in North America. The large Irvingtonian age population of *Panthera onca* has most often been given subspecific distinction and called *P. o. augusta* (Simpson, 1941; Kurten, 1973) but others have considered it a separate species, *P. augusta* (McCrady et al., 1951).

Along with the reduction in size through time, the Pleistocene jaguars also exhibited a gradual reduction in limb and paw length, so that the earliest forms were larger and longer-limbed and lacking some of the specializations of the living population (Kurtén, 1973; Seymour, 1993). Modern jaguars are adapted to living in forests (Gonyea, 1978), but the earlier Pleistocene populations may have been adapted to living in more open habitats. Living populations of *Panthera onca* are nocturnal hunters which catch and kill their prey by stalking or ambush, most often by biting through the nape of the neck like most true cats (Leyhausen, 1979; Rabinowitz, 1986). The Pleistocene jaguar, with its longer limbs, could have utilized more of a pursuit phase in its prey capture and could have inhabited less forested, more open habitat.

Bones of a third large felid have been recovered from the Anza-Borrego Desert State Park which are assigned to the American cheetah-like cat (Figure 11.11), *Miracinonyx inexpectatus* (Jefferson and Tejada-Flores, 1995). There are eight bones that definitely represent this species and include a partial humerus, radius, two ulnae, and six foot elements. An additional nine feline

Figure 11.11
The American Cheetah-like Cat,
Miracinonyx inexpectatus.
(Drawing by Pat Ortega)

bones, which are smaller than what might be expected for the size range of this species, may nevertheless also be from this animal. The earliest record of *M. inexpectatus* in North America is about 2.5 million years old, found in the Blancan of Cita Canyon, Texas (Lindsay et al., 1975). Its known distribution includes Blancan and Irvingtonian sites in California, Sonora, Mexico, Arizona, Texas, Nebraska, Arkansas, Pennsylvania, Maryland, Virginia, and Florida, and a Rancholabrean species, *M. trumani*, is well known from Wyoming and Nevada (Kurtén and Anderson, 1980; Van Valkenburgh et al., 1990; Croxen et al., 2003). The eight positively identified skeletal elements from the

Anza-Borrego Desert State Park were recovered from the late Blancan portion of the Hueso Formation, dating at 2.0 million years; those tentatively included with this species are from the same Formation and span 2.2 to 1.3 million years.

Miracinonyx inexpectatus is estimated to weigh 70 to 71 kg (156 lb) (Van Valkenburgh, 1990, 1991). It had extremely elongated and slim limb bones, a lightly built body, and a small head with cranial and dental characters that paralleled similar features found in the Old World cheetah (*Acinonyx*). However, the American cheetah-like cat was not closely related to its Old World cousin, but apparently evolved from a presently poorly understood Eurasian feline that immigrated to North America in the early Blancan (Van Valkenburgh, 1990). The sudden appearance of the Eurasian derived *Chasmaporthetes* at the same time as *Miracinonyx* in Blancan age sites in North America lends support to this suggested immigration event.

The remarkable similarity in body size, limb, and cranial adaptations between *Miracinonyx* and *Acinonyx* indicates convergent modes of predation. *Miracinonyx* undoubtedly employed an extended rapid pursuit to chase down swift prey in the open habitats of a glade or savanna. The preferred prey that these predators exploited in the Anza-Borrego Desert State Park area may well have been the variety of pronghorns present (*Tetrameryx* and/or *Capromeryx*, see Murray this volume, *The Smaller Artiodactyls*).

Paleoecology

The late Pliocene to late-middle Pleistocene large carnivoran ecological guild represented by the Anza-Borrego Desert State Park fossil record includes two canids, two or three ursids, possibly one hyenid, and three felids. From the species represented, one would expect the habitat to include stretches of savanna grading into open woodlands upslope (see Paleolandscape 5). Heavily wooded areas might have been ideal for *Tremarctos*, along with *Ursus* sp. which are often found associated in the same fossil sites in North America (see Paleolandscape 4). Savanna would likely be home to *Miracinonyx, Chasmaporthetes, Arctodus,* and *Canis armbrusteri,* while the rest of the guild (*Smilodon, Panthera onca, Borophagus*) might be found inhabiting the open woodland ecotone between the savanna and heavily wooded areas. The fact is that all large carnivorans have large ranges and home territories, so it is likely that all of them could and would have traversed or inhabited any part of the available area during a 24-hour period.

It is interesting that certain species are often found together in the same fossil localities, like the example of *Tremarctos* and *Ursus* sp. mentioned above. Van Valkenburgh (1990) pointed out that *Miracinonyx inexpectatus, Smilodon gracilis,* and *Panthera onca,* which are often found associated in the same site, are morphologically and ecologically separated from one another. This would aid in avoiding competition and potential injury by partitioning their shared environment and food resources. The only large carnivoran species that is missing from the Anza-Borrego Desert State Park fossil record is the scimitar-toothed cat, *Homotherium serum*. This sabertoothed cat differs morphologically from *S. gracilis* by having longer limbs and shorter upper canine teeth. *Homotherium* is widely distributed in North America, but rarer than *Smilodon*. Nevertheless, they are often found associated in the same sites (Kurtén and Anderson, 1980), and the closest Irvingtonian age occurrence is not far away in Sonora, Mexico. As survey work continues, it seems highly likely that, as with the hyena *Chasmaporthetes*, *Homotherium* also will be recovered from the expansive badlands of Anza-Borrego Desert State Park.

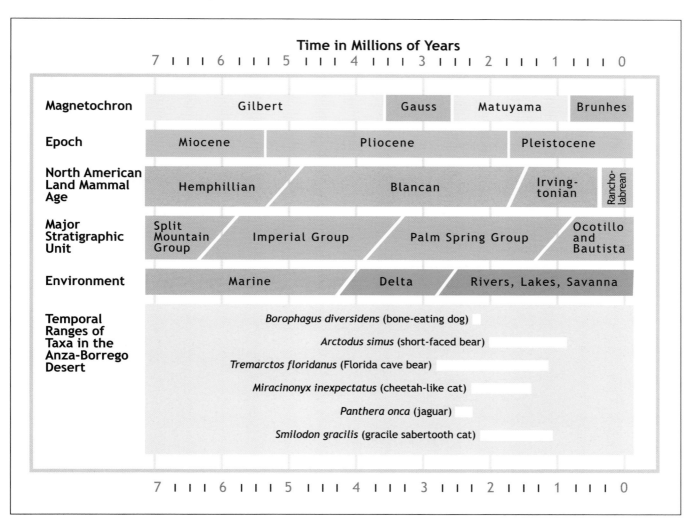

Figure 11.12
The Large Carnivorans of Anza-Borrego.
Remains of *Borophagus* and *Panthera* are restricted to the Blancan part of the Anza-Borrego record. The other four large carnivorans are primarily Irvingtonian in age.

Paleolandscape 4: Borrego Valley Riparian Forest and Stream 1 Million Years Ago

INDEX

1. *Pumilia novaceki* (Novacek's small iguana, extinct)
2. *Washingtonia* sp. (palm)
3. *Tremarctos floridanus* (Florida cave bear, extinct)
4. *Salix gooddingii* (Gooding's willow)
5. *Coendou stirtoni* (Stirton's coendu porcupine, extinct)
6. *Aesculus* sp. (buckeye)
7. *Picidae* (woodpeckers)
8. *Meleagris anza* (Anza turkey, extinct)
9. *Castor canadensis* (beaver)
10. *Capromeryx* sp. (diminutive pronghorn, extinct)
11. *Bucephala fosillis* (fossil golden-eye, extinct)
12. *Kinosternon* sp. (mud turtle)
13. *Lontra canadensis* (river otter)
14. *Salmo* sp. (trout)
15. *Lampropeltis getulus* (common king snake)
16. *Pewelagus dawsonae* (Dawson's small rabbit)
17. *Bufo* sp. (toad)

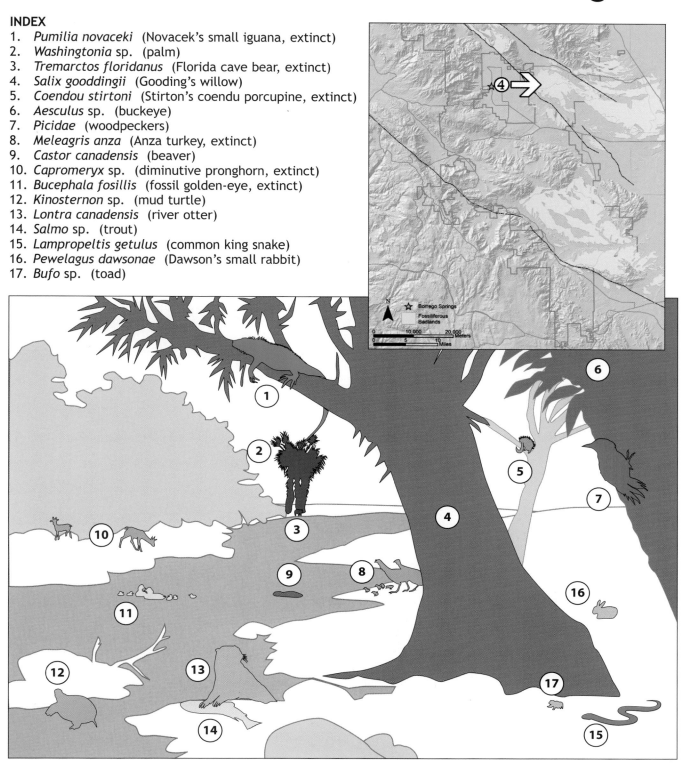

A braided stream and stream bank provide a home for a variety of aquatic and forest adapted animals. The view today is to the northeast, from about 3 km (1.8 miles) east of Borrego Springs (see map, and Landscape 5). The uplifted Borrego Badlands and Fonts Point did not exist at this time. (Picture by John Francis)

195

12

The Small Carnivorans: Canids, Felids, Procyonids, Mustelids

Biologically the species is the accumulation of the experiments of all its successful individuals since the beginning.

H.G. Wells, 1905

Lyndon K. Murray

The Small Carnivorans:
Canids, Felids, Procyonids, Mustelids

Borrego Formation Below Fonts Point in Rainbow Wash. (Photograph by Paul Remeika)

Introduction

The adjective "carnivorous" means flesh eating, or predatory. The related noun "carnivore" is thus applied generally to animals (also insectivorous plants) that naturally prey on animals (e.g., flesh-eating dinosaurs). It applies specifically to the Linnaean Order Carnivora, which includes bears, dogs, cats, raccoons, skunks, weasels, and sometimes walrus and seals. To avoid confusion, and for clarity and precision of meaning, biologists and paleontologists who wish to refer only to flesh-eating mammals of the Linnaean Order Carnivora use the term "carnivoran."

There are about 300 carnivoran specimens from over 230 localities in the Anza-Borrego Desert State Park (ABDSP) collection of over 14,000 numbered vertebrate fossils. Approximately half of these specimens have been identified to the smaller (less than 40 kg or 88 lb) carnivoran taxa such as coyotes and bobcats. Less than a third represent the larger carnivorans like wolves, bears, or sabertoothed cats (see Shaw and Cox, this volume, *The Large Carnivorans*). The remaining specimens are indeterminate fossils of the order Carnivora and the families Canidae (dogs) and Felidae (cats). Due to the fragmentary nature of much of the fossil material from the Anza-Borrego collection, many specimens cannot be identified to genus or species and identifications are often limited to the more inclusive but less precise order or family ranks. Coyote-like dogs and small wolves of the genus *Canis* are the most abundant small carnivorans represented, with specimens recovered from about 60 sites. This is slightly more than the next most abundant group, bobcat-size cats of the genus *Felis* (around 50 sites). Many of the indeterminate Carnivora, Canidae, and Felidae specimens undoubtedly represent additional fossils of the smaller species of *Canis* and *Felis*. All other small carnivoran taxa total between 1 and 5 specimens each. The latter groups include gray fox, spotted skunk, weasel, otter, grison, badger, ringtail, coati, and raccoon. With the exception of otter, grison, and coati modern representatives of all the small carnivorans have been recorded within Anza-Borrego Desert during historic times.

Disproportion of Carnivores to Herbivores

Environments of sediment deposition in the terrestrial portion of the Anza-Borrego stratigraphic sequence are primarily streams and lakes (see Dorsey, this volume, *Stratigraphy, Tectonics, and Basin Evolution in the Anza-Borrego Desert Region*). These tend to accumulate vertebrate remains at random, preserving a fossil cross-section of all the animals that died within the stream drainage boundaries. That is, any animal that dies within the local drainage system has the potential of being washed down slope into a site where deposits are being laid down. Ultimately, its remains become entombed in the sediments that are eroded off the local highlands and deposited into the streams and lakes. This slope-wash action may displace animals from their local micro-habitats, such that a creature known only from hillside environments may become a fossil in a lakeshore deposit. Ancient local drainage areas were reasonably small, approximately several hundred square kilometers (often much less). This means that if a desert dweller is found in the fossil deposits, there was very likely a desert environment within the ancient drainage. Likewise, if a tree dweller, a burrower, and a rock dweller are all found in the same deposit, there were probably trees, diggable soil, and rocky habitats all within the same ancient drainage area.

Under such conditions, the sediments tend to preserve the remains of various animals in the same relative numbers as when they were alive. This is a broad generalization, since different bone sizes and shapes are affected differently by sediment grain size, stream flow, and abrasion. Nevertheless, bones of similar size and shape will react in a similar way during water transport, regardless of taxon. After death, not all vertebrate bones end up on the surface of the ground to be washed down slope. Some animals are eaten by others and many of their bones destroyed, some die in underground burrows or in rock crevices, become encased in sediment, and are discovered as fossils exactly where they died. In general, though, remains of fossil animals of similar size will be found in quantities relative to their living numbers. In a living environment, herbivorous animals far outnumber predatory animals. So in stream and lake deposits, the fossils of terrestrial herbivores should be much more numerous than the fossils of their predators (see Jefferson, this volume, *The Fossil Birds of Anza-Borrego*).

In contrast, trapping environments of deposition, such as the tar pits of Rancho La Brea in Los Angeles, California, tend to preserve a relatively high number of predators and scavengers, which are attracted by the other trapped and dying animals (Stock and Harris, 1992). Animals that live in environments differing from that in which the tar seeps exist will not normally be caught in the trap. Therefore, the relative numbers of different animals trapped is not the same as those living in the area at the time the remains were deposited.

In the Anza-Borrego fossil beds, herbivorous taxa are about a hundred or more times as abundant as carnivorous taxa. Fossil carnivoran remains are thus comparatively rare, and identifiable specimens are very rare.

Dating the Specimens

The great majority of terrestrial vertebrate fossils recovered from the Anza-Borrego Desert come from sediments deposited roughly between 4.0 Ma (million years ago) and 0.5 Ma. Sediments of that age fall into two globally recognized divisions of time or "epochs": the earlier Pliocene (5.3 to 1.8 Ma), followed by the Pleistocene (1.8 to 0.01 Ma). When the exact age of a specimen cannot be determined it may be described, for example, as being of Pliocene age, indicating it was incorporated into sediments deposited during that epoch. The Pliocene-Pleistocene epochal boundary is linked globally to sediments deposited during the topmost portion of the geomagnetically polarized chron C2n, also called the Olduvai subchron, and determined to be 1.77 Ma. This date is often rounded to 1.8 Ma (Berggren et al., 1995; Pasini and Colalongo, 1997; see Remeika, *Dating, Ashes, and Magnetics*; also Cassiliano, *Mammalian Biostratigraphy of the Vallecito Creek and Fish Creek Basins*, this volume). A more detailed division of Cenozoic time was developed specifically for North America and is comprised of a series of 19 ages (North American Land Mammal Ages or NALMAs) (Woodburne, 1987), extending from the end of the Cretaceous (65 Ma) to the end of the Pleistocene (10,000 years ago) (see Cassiliano, this volume). Chronologically, the last four NALMAs (Hemphillian, Blancan, Irvingtonian, and Rancholabrean) include the entire Pliocene and Pleistocene time span. The Anza-Borrego Desert terrestrial vertebrates fall mostly within the Blancan and Irvingtonian NALMAs.

Theodore Downs, of the Natural History Museum of Los Angeles County, established a transect overlay system for placing fossil sites in stratigraphic levels of the Deguynos and Hueso Formations (see Dorsey, this volume) of the Vallecito Creek-Fish Creek Badlands in the southern Anza-Borrego Desert (see Cassiliano, this volume). The transect allows sites to be located relative to one another stratigraphically and to be dated by correlation to the Geomagnetic Polarity Time Scale (Cassiliano, 1999; Remeika, this volume). This system divides the strata into horizontal Collecting Units (CU), each representing approximately 61 vertical meters (200 ft) of sediments. For example, CU29 is located at the transitional boundary between the Gilbert and Gauss Chrons at 3.6 Ma, while CU56 is near the top of the Olduvai (C2n) subchron at 1.8 Ma (Opdyke et al., 1977; Cassiliano, 1999; Remeika, this volume).

Not all Anza-Borrego localities can be dated with Downs' CU transect system. Fault blocks and geoclines (folded strata) have not yet been fully correlated into the main stratigraphic transect, so the relative position and age of specimens found in those areas can only be estimated (Cassiliano, 1999). Also, the collection areas in the north part of the Park, such as Clark Lake or the Truckhaven Rocks, are not presently correlated with the Geomagnetic Polarity

Time Scale, with the exception of the Borrego Badlands (see Remeika, this volume).

The Collecting Units provide only a stratigraphic organization of localities and the specimens they produce. That is, they show the sequence in which the specimens were buried, but not actual dates of burial. Dates associated with the chronologic ages of the specimens are provided by correlating the CUs with the Geomagnetic Polarity Time Scale. The most accurate date that can be assigned to a specimen is actually a range of dates based on the known timing of the geomagnetic reversals. For example, the bottom of the Olduvai (C2n chron) was measured at 1.95 Ma and the top was at 1.77 Ma (Berggren et al., 1995). An Anza-Borrego Desert specimen, discovered in sediments known to have been deposited during the Olduvai, is no older than 1.95 Ma and no younger than 1.77 Ma. This is a chron date range and is applied throughout this chapter to describe the range of time during which a specimen may have lived (see Murray, *The Smaller Artiodactyls,* this volume). The oldest and youngest dates of the range are merely boundaries for the known chron and should not be confused with the still unknown absolute dates of the specimens.

This distinction is important when discussing oldest and youngest known appearances of species in Anza-Borrego Desert, North America, or the world. Refinement of the Anza-Borrego Desert paleomagnetic sequence, sedimentation rate, and digital mapping of localities is ongoing, and age resolution for specific localities and specimens will undergo continuous improvement.

The Canidae: Dogs and Their Relatives

Three distinct sizes of *Canis* are present in the fossil collection. While a multivariate biometric analysis will no doubt provide more accurate taxonomic distinctions, most of the specimens can be grouped easily by visual comparison with modern osteological material. Skeletal elements of the small *Canis* species are very close in size to the living coyote (*C. latrans*). The large Canis specimens vary from slightly smaller to larger than modern gray wolf (*C. lupus*). The medium *Canis* is midway in size between the two living species, *C. latrans*, and *C. lupus*.

Figure 12.1
Canis priscolatrans.
This hunting pair of wolf coyotes overlooks a herd of *Camelops* one million years ago.
(Picture by John Francis)

The small canids probably belong to the taxa *C. latrans* and *C. lepophagus* (Johnston's wolf); whereas the medium size canids are likely *C. priscolatrans* (wolf coyote) (Figures 12.1, 12.2); and the large Anza-Borrego Desert

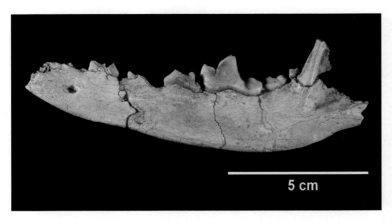

Figure 12.2
Canis priscolatrans.
Partial left dentary (lower jaw)
of the wolf coyote
(ABDSP[IVCM]1595/V6250).
(Photograph by Barbara Marrs)

Canis may belong to *C. armbrusteri* (Armbruster's wolf – see Shaw and Cox, this volume). The oldest *C. latrans* are known from Irvingtonian sites in North America. Multiple records at Porcupine Cave, Colorado are at least 600,000 years old and probably exceed 780,000 years old (Anderson, 1996; Barnosky and Bell, 2004). *Canis priscolatrans* is reported from Blancan through Irvingtonian age sites in North America (Kurtén and Anderson, 1980). *Canis lepophagus* is known primarily from Blancan localities (Munthe, 1998; Bell et al., 2004). Previous published faunal lists for Anza-Borrego Desert include *C. latrans* and *C. edwardii* (Edward's wolf) or *C. priscolatrans,* with no mention of *C. lepophagus* (Remeika et al., 1995; Cassiliano, 1999), although Kurtén and Anderson (1980:167) suggested that *C. lepophagus* "probably also occurs at" Anza-Borrego Desert State Park. Kurtén (1974) stated that *Canis edwardii* is a junior synonym of *C. priscolatrans,* however, both names are still used regularly (Tedford et al., 2001; Nowak, 2002; Bell et al., 2004). *Canis edwardii* has been compared with the modern red wolf (*Canis rufus*). While recent phylogenetic studies show no direct genetic relationship between the latter two species, there may be an ecomorphic relationship between their relative sizes and their favored habitat (Nowak, 2002). Specimens of the small *Canis* that date to the middle Irvingtonian part of the Anza-Borrego Desert stratigraphic sections or higher (see below) likely represent *C. latrans* rather than *C. lepophagus.*

One lineal progression proposed by Kurtén (1974) extends from *Canis lepophagus* through *C. priscolatrans* to *C. latrans.* Nowak (2002) suggested an evolutionary sequence from *C. lepophagus* or some related small species through *C. priscolatrans, C. armbrusteri,* and *C. dirus* (dire wolf) (see Shaw and Cox, this volume). Nowak noted that while *C. lepophagus* may be an ancestor common to both *C. latrans* and *C. priscolatrans,* the latter two species probably arose through separate lines of descent. A preliminary report by Tedford et al. (2001) from a larger work in progress identified *C. lepophagus* and *C. edwardii* (*C. priscolatrans* of Kurtén [1974] and Nowak [1979, 2002]) as sister groups, with *C. latrans* possibly arising out of *C. edwardii,* and *C. dirus* derived from the immigrant *C. armbrusteri.*

Based on 20 *Canis* specimens from the Vallecito Creek-Fish Creek Basin that can be dated within the Downs CU system, the small *Canis* (probably *C. lepophagus*) appears to have no time-stratigraphic overlap in the southern part of Anza-Borrego Desert with the medium *Canis* (probably *C. priscolatrans*). The occurrence of small *Canis* begins during middle Blancan, between 3.58 and 3.33 Ma and extends to late Blancan, disappearing before about 1.95 Ma. Both the medium and large *Canis* (probably *C. armbrusteri*) (see Shaw and Cox, this volume) first appear stratigraphically above the last dated small *Canis* and extend well into the Irvingtonian, possibly later than 1.07 Ma.

A small *Canis* partial skull from Clark Lake, in the northern part of the Park may represent *C. latrans*. The age of the Clark Lake sediments is about middle Irvingtonian or younger. Several canid specimens from the nearby Coyote Badlands and Borrego Badlands are medium size canids (probably *C. priscolatrans*). The stratigraphic and age relationships of these fossiliferous areas are under study.

The gray fox, *Urocyon* has been recovered from three localities in the Anza-Borrego fossil beds. Two sites are dated well into the Blancan, between 3.58 and 3.33 Ma. The third site, at 1.95 to 1.77 Ma is closer to the Blancan-Irvingtonian boundary. *Urocyon* is a small fox, with skull and mandible readily distinguished from other canids. The oldest of the three specimens is distinctly larger than modern comparative specimens of *Urocyon* obtained from local populations. This indicates the older fossils probably belong to the Blancan progressive gray fox (*U. progressus*), which was larger than modern gray fox (*U. cinereoargenteus*) and may have been a lineal ancestor to it (Kurtén and Anderson, 1980; Anderson, 1984). *Urocyon progressus* is not known from the Irvingtonian age. *Urocyon cinereoargenteus* was reported elsewhere from early to middle Irvingtonian localities, with a possible record from Kansas between 2.01 and 1.2 to 1.5 Ma, based on volcanic ash dates (Kurtén and Anderson, 1980; Bell et al., 2004). The youngest of the three Anza-Borrego Desert specimens probably represents the Recent species or a transitional form of *Urocyon*.

The Felidae: Cats

Various authorities present differing views regarding the taxa that make up the genus *Felis*: whereas some placed most living cats in *Felis*, including the African lion (*Panthera*) and lynx (*Lynx*); others limit members of *Felis* to mountain lion (subgenus *Puma*) and smaller taxa; yet other workers split out multiple genera from *Felis* (Kurtén and Anderson, 1980). A detailed study of extant felid phylogeny (history of ancestor-descendant relationships) revealed *Lynx* to be a distinct group closely related to *Felis*. However, none of the other North American cats is closely related to *Felis* or *Lynx* (Salles, 1992). To reduce confusion in the following section, the taxonomic terminology represents all small cats as *Felis* with a parenthetic 'subgenus' classification where various authors disagreed on taxonomic assignment, e.g. *Felis (Lynx) rufus*.

Figure 12.3
Felis rexroadensis.
This Rexroad cat has caught its dinner. (Drawing by Pat Ortega)

The small Anza-Borrego Desert felid fossils appear to represent at least two different species based on size differences. The smaller specimens are near the size of modern bobcat *Felis* (*Lynx*) *rufus*. The noticeably larger skeletal elements are comparable to *F.* (*Lynx*) *rexroadensis* (Rexroad cat) or *F.* (*Puma*) *lacustris* (lake cat) (Kurtén and Anderson, 1980; Werdelin, 1985; Martin, 1998) (Figure 12.3). At least one Anza-Borrego Desert felid lower jaw was compared to a large sample of measured jaws and referred to *F. rexroadensis* (Werdelin, 1985) (Figure 12.4). The Anza-Borrego Desert specimens also should be compared with a multitude of small native cat species recorded from southern U.S. and northern Mexico, including jaguarundi (*F.* [*Herpailurus*] *yagouaroundi*), margay (*F.* [*Leopardus*] *wiedii*), ocelot (*F.* [*Leopardus*] *pardalis*), and *river cat* (*F. amnicola*). Detailed measurement and study of the fossils may resolve some specimen identifications.

The ages of small *Felis* specimens, as determined by CU numbers, are not very informative. The oldest dated Anza-Borrego specimen appeared between 4.18 and 3.58 Ma, and the youngest was found in sediments slightly younger than 0.76 Ma (Gensler, 2002). These are both larger size specimens. The ages of the smaller specimens lie between the earliest and latest dates for the larger specimens.

Felis (*Lynx*) *rexroadensis* is known from North American sites of Hemphillian through late Blancan age; *F.* (*Puma*) *lacustris* is known from early to late Blancan; *F.* (*Herpailurus*) *yagouaroundi* is recorded from Irvingtonian to present; *F.* (*L.*) *rufus* is recorded from Blancan to the present; other small felid species are known only from Rancholabrean age or younger localities (Kurtén and Anderson, 1980; Werdelin, 1985; Martin, 1998). It is very possible several species of small cats are represented in the Anza-Borrego Desert collection.

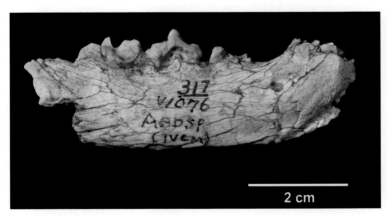

Figure 12.4
Felis rexroadensis.
Partial right dentary (lower jaw)
of the Rexroad cat
(ABDSP[IVCM]317/V1067).
(Photograph by Barbara Marrs)

The Mustelidae: Badgers, Grisons, Otters, Skunks, and Weasels

The third most abundant Anza-Borrego fossil small carnivoran is the badger, *Taxidea* (Figure 12.5). *Taxidea* is known from both the Vallecito Creek-Fish Creek and Borrego Badlands. Of five identified specimens, only one is currently datable (1.95 to 1.77 Ma). The single species *Taxidea taxus* has been recorded in North America from early Blancan age to the present (Kurtén and Anderson, 1980). Fossil badger localities have been reported from the western and southern U.S., while modern badgers range through northern Mexico, western and Midwestern U.S., and western Canada (Hall, 1981). Extant badgers have a total length 520 to 875 mm (20.5 to 34.4 in) and weight 4 to 12 kg (8.8 to 26.5 lb). They have a distinctive flattened, stocky shape with short tail, powerful forelimbs, and large digging claws (Nowak, 1991).

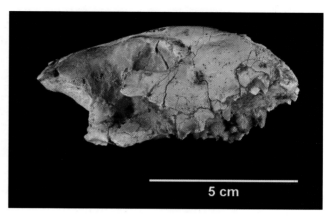

Figure 12.5
Taxidea taxus.
Partial skull (in profile, tipped) of the badger (ABDSP[IVCM]1666/V4597). (Photograph by Barbara Marrs)

At least two specimens (lower jaws and a partial pelvis) of the grison (*Trigonictis*) were recovered from Anza-Borrego in 2.58 to 1.77 Ma aged sediments. *Trigonictis macrodon* and a smaller form, *T. cookii* comprise the two known grison species recorded from North American Blancan sites. Specimens from Anza-Borrego Desert were referred to the former species (Anderson, 1984). Trigonictis may be ancestral to the extant grison (*Galictis*) and tayra (*Eira*) both found from central Mexico to South America. Modern grisons and tayras have long bodies with short legs, and a body shape similar to weasels, otters, and martens. The larger of two *Galictis* species, *G. vittata* has a total length of 600 to 760 mm (23.6 to 29.9 in) and weight 1.4 to 3.3 kg (3.1 to 7.3 lb) (Presely, 2000; Yensen and Tarifa, 2003a, b). *Galictis vittata* is about 2/3 the size of *T. macrodon* (Kurtén and Anderson, 1980).

The Anza-Borrego Desert collection contains at least one specimen (lower jaws) of the spotted skunk (*Spilogale*), and one specimen each of otter (*Lontra or Satherium*) and weasel (*Mustela*). The spotted skunk is from late Blancan sediments, about 1.95 to 1.77 Ma. The species represented by these fossils may be *Spilogale rexroadi* (Rexroad skunk), which is known elsewhere almost exclusively from Blancan age localities, or *S. putorius*, known from late Blancan through present day. Spotted skunks are the smallest of the modern North American skunks, total length 185 to 565 mm (7.3 to 22.2 in), weight 0.2 to 1.0 kg (0.44 to 0.97 lb) (Nowak, 1991). Both of the Blancan and Irvingtonian species were smaller than modern *Spilogale* (Kurtén and Anderson, 1980; Anderson, 1984). Fossil *S. rexroadi* were recorded from Texas and Kansas while Blancan *S. putorius* was found in Kansas and Arizona (Baskin, 1998). Modern distributions of *S. putorius* include most of continental U.S. and northern Mexico (Hall, 1981).

Figure 12.6
Lontra canadensis.
This fish-eating river otter catches a large trout (*Salmo*) from a Borrego Valley stream. (Picture by John Francis)

The otter specimen, a partial skull, was recovered from late Blancan sediments, age 2.58 to 1.95 Ma and probably represents the Blancan otter (*Satherium piscinarium*) (Anderson, 1984). The other genus of otter, *Lontra* (some authors still use *Lutra*) (Figure 12.6) is known primarily from Irvingtonian to recent North American localities (Larivière and Walton, 1998). River otters have long cylindrical bodies with short legs, webbed feet, a powerful tail, and a flattened skull, all adaptations to an aquatic lifestyle. *Satherium* was larger than extant *Lontra*, which varies between 900 to 1,150 mm (35.4 to 45.3 in) in length and weighs 5 to 14 kg (11.0 to 30.9 lb) (Larivière and Walton, 1998). *Lontra* may be derived from *Lutra licenti,* a Chinese otter (Kurtén and Anderson, 1980). The modern distribution of river otters in North America covers most of Alaska, Canada, and the Great Lakes, mountain, and coastal regions of the U.S. Although they are absent from the southern half of California (Hall, 1981), they are found on the lower Colorado River (Hoffmeister, 1986).

The Anza-Borrego Desert weasel specimen, an isolated canine tooth, is similar in size and morphology to modern specimens of the long-tailed weasel (*Mustela frenata*). It was recovered from sediments aged 1.95 to 1.77 Ma. M. frenata is often the most abundant small carnivoran in paleofaunas, with the longest chronostratigraphic range and widest distribution of North American weasels (Anderson, 1984). However, there are several extant *Mustela* species very close in size to *M. frenata,* and more Anza-Borrego Desert fossils must be found and studied to determine which species is present. The Blancan age Rexroad weasel (*M. rexroadensis*) is a possible ancestor of *M. frenata* and is also an excellent candidate for the fossil.

Weasels and ferrets are the smallest of the mustelids, with an elongate body, enabling them to pursue burrowing rodents and other ground-dwelling animals into their tunnels and nests. *Mustela frenata* ranges in total length from 280 to 420 mm (11.0 to 16.5 in) and weighs 0.08 to 0.45 kg (0.17 to 0.99 lb). Long-tailed weasels are found today from southern Canada to northern South America. However, they are absent from most of the Mojave and Sonoran Deserts (Sheffield and Thomas, 1997).

The Procyonidae: Coati, Raccoon, and Ringtail

Two records each of raccoon (*Procyon*), ringtail (*Bassariscus*), and coati (*Nasua*) are identified in the Anza-Borrego fossil collection, all from different localities. One of the raccoon specimens is a robust lower jaw, considerably larger than modern local raccoon. It compares favorably with published descriptions of the Blancan age Rexroad raccoon and was referred to *Procyon sp.* cf. *P. rexroadensis* by Murray and Jefferson (1996). The ages of the Anza-Borrego raccoons lie between 2.58 and 1.77 Ma. Extant raccoons show a regional size variation: height 228 to 304 mm (about 9 to 12 in) and weight 2 to 12 kg (about 4.4 to 26.5 lb). They are found from central Canada to Panama, but are absent from many high mountain ranges of the western U.S. (Hall, 1981).

The Anza-Borrego Desert ringtail specimens (lower jaws) were identified as *Bassariscus casei* (Case's ringtail) (Anderson, 1984), known in North America only from Blancan age localities. The fossils were recovered from sediments between 3.04 and 1.95 Ma. *Bassariscus casei* is probably ancestral to extant ringtail (*B. astutus*) (Kurtén and Anderson, 1980). The modern ringtail, also known as "ringtailed cat" and "cacomistle," is about the size of a small domestic cat with a head and elongate body shape similar to *Martes americana* (American marten), with a long tail. Today its range extends from southwestern U.S. to central Mexico (Poglayen-Neuwall and Toweill, 1988).

The fossil specimens previously identified as *Nasua* (Murray and Jefferson, 1996) consist of three dentaries (lower jaws) with incomplete dentition and a partial humerus. The two localities that produced them fall outside the measurable portion of the Downs' CU transect. However, Murray and Jefferson (1996) estimated a time range of 2.5 to 1.6 Ma based on the general position within the upper Hueso Formation. "Coati" is the proper common name for all members of the genus *Nasua*. Another popular term for these animals, "coatimundi" derived from a Brazilian vernacular name 'coati monde,' and properly refers only to solitary males (Gompper and Decker, 1998). The Anza-Borrego Desert coati specimens compare well with modern *Nasua narica* jaw size and morphology. However, the measurements overlap with fossil *Procyon* specimens, plus most of the diagnostic features that readily separate the two taxa are missing from the Anza-Borrego Desert fossils. Further study of these specimens may resolve the taxonomic identities. The fossil record for *Nasua* is very poor – a single tooth from a Blancan (4.5 to 2 Ma) locality in Texas (Dalquest, 1978) and report of a possible *Nasua* sp. from a late Miocene locality in Florida (Baskin, 1982; Gompper and Decker, 1998). *Nasua* and *Procyon* may be descended from the common ancestor *Paranasua* in the late Miocene (Baskin, 1982). Extant coatis are the size of a small dog, with tail about the same length as the body. *Nasua narica* modern range extends from southern Arizona, New Mexico, and Texas to Panama (Gompper, 1995).

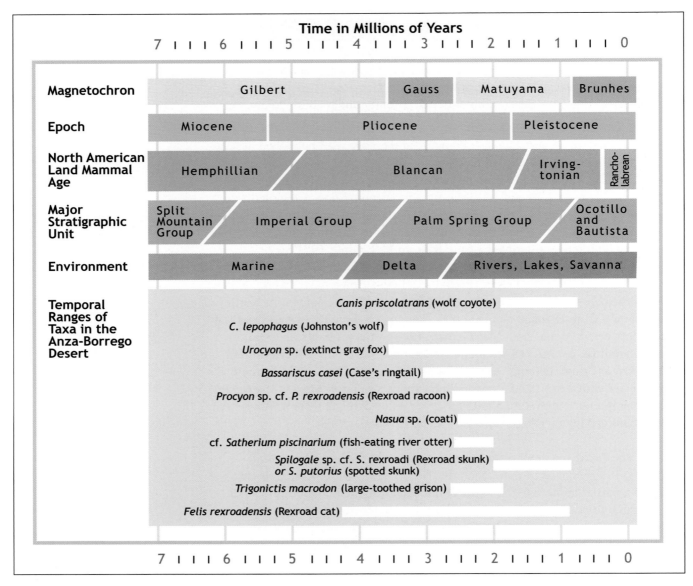

Figure 12.7
The Small Carnivorans
of Anza-Borrego.
The temporal ranges of ten
smaller Anza-Borrego carnivorans
are plotted here. They span from
the first Colorado River delta
deposits through the largely
floodplain and riverine Hueso
and Ocotillo formations.

Small Carnivorans from Older Deposits

A felid specimen and a possible canid specimen were recovered from one of the oldest terrestrial vertebrate localities in the Anza-Borrego Desert, Hawk Canyon. Ages for these sediments are currently under study but appear to be between 9 and 7 Ma. The felid specimen is a partial lower jaw, slightly larger than "bobcat" size range. Although the specimen was previously referred to the Clarendonian/Hemphillian age cat *Pseudaelaurus,* this identification can not be confirmed. The possible canid specimen is a partial shoulder blade and is "coyote" size.

A small procyonid dentary was recovered from the 4.5 Ma marine sediments near the top of the Deguynos Formation. The jaw has no preserved teeth but otherwise is comparable to specimens of *Nasua* or *Procyon.*

Questionable Identifications

A fourth fox-like specimen, an isolated canine tooth from strata that date to between 2.58 and 1.95 Ma, appears to be more like the canine of a small red fox (*Vulpes vulpes*) than a large *Urocyon*. However, it is incomplete and may never be fully identified. The oldest record of *Vulpes* in North America is from Porcupine Cave, Colorado, at a maximum age in excess of 780,000 years old (Anderson, 1996; Barnosky and Bell, 2004).

Similarly, a badly weathered and fragmentary distal femur (the part of the leg just above the knee) compares better with modern wolverine (*Gulo*) than any other known Anza-Borrego Desert mammal in the same size range. Wolverines are known from eastern North America in the early Irvingtonian, but do not appear in the west until the latest Pleistocene (Kurtén and Anderson, 1980).

While these Anza-Borrego Desert specimens present intriguing possibilities for reporting chronological and geographic range extensions of important carnivoran taxa, neither specimen has sufficient diagnostic features to make a definitive claim. This is an unfortunate aspect of many of the fragmentary Anza-Borrego Desert fossils. Happily, in some cases, minimally diagnostic bones can be assigned to a genus or species with a reasonable amount of confidence, when several positively identified representatives of that genus or species are already recorded from the same stratigraphic horizon. Until more complete specimens of red fox or wolverine are found in the Anza-Borrego Desert fossil beds, preferably a complete skull or skeleton, these single, isolated, incomplete, and weathered specimens can be identified only to "Canidae, genus and species indeterminate" and "Carnivora, family indeterminate" respectively. Comments in the database records for the specimens provide reminders to researchers to keep on the lookout for red fox or wolverine, and to keep checking for other taxa that may have skeletal elements that resemble these specimens.

Carnivoran Diets

Paleontologists make the common assumption that fossil taxa interacted with their local environment in the same way their modern relatives do. The accuracy of this assumption may never be known with scientific certainty for every fossil species. Yet, it is a useful working tool with which we address the questions about the lives of organisms that left the fossilized remains we study. With this tool we reconstruct past environmental conditions based on the assumed requirements of the taxa found in the sedimentary record. This method works well for general behavior. All carnivoran species from Anza-Borrego Desert were carnivorous, the flesh of other animals being one of their main sources of food. This is readily seen by examining their teeth. Carnivoran teeth are designed for capturing and consuming other animals. It is more difficult to match a given predator with its primary prey, e.g. *Canis lepophagus* (the name *lepophagus* means "leporid-eater" or "rabbit-eater") probably preyed on the local rabbit and hare populations in much the same way as *C. latrans* does today. However, that is not necessarily the case, and may be impossible to prove.

Large numbers of fossil rodents and rabbits are represented in the Anza-Borrego fossil collection (see White et al., this volume, *The Small Fossil Mammals*). Many of these specimens were recovered from coprolites (fossilized feces) of carnivorans. As with their modern relatives, the Pliocene and Pleistocene small carnivorans were preying on the local rodents and rabbits. Several bird specimens have been found in coprolites. Bird fossils are not as abundant as rodents, partly because their bones are more delicate and do not preserve as well (see Jefferson, this volume). Some invertebrates and digested plant material are also identifiable. Determination of the taxon that deposited a particular coprolite is usually very difficult. So, although these fossil packages are direct evidence of predation on the organisms found in the coprolite, they still do not provide definite proof of which carnivoran was feeding on which food item. With no further information, the best we can do at present is to make the assumption that the ancient Anza-Borrego taxa had dietary habits very similar to their living relatives.

Reported observations of the dietary ranges of living representatives of the carnivoran genera show general trends that prove useful in understanding the environmental conditions of the Anza-Borrego during the Blancan and Irvingtonian ages. Rodents and birds are the only prey common to all carnivorans, while reptiles, insects, carrion, and fruit comprise at least part of the diet of most carnivorans. Canids and felids appear to be the only carnivorans that hunt and kill ungulates or hoofed mammals such as deer and horses as well as camels. Only otters and raccoons exploit aquatic food resources regularly. Various forms of vegetation provide a large percentage of the diet of procyonids, canids, and spotted skunks. The life histories cited here, of bobcat, badger, and weasel, list no utilization of vegetation. Carnivorans also prey on each other. Larger animals tend to be regular predators on carnivorans smaller than themselves, especially those slightly smaller. Bobcats and coyotes are prey for mountain lions and wolves, and in turn eat raccoons, foxes, and otters. Large and small cats eat coatis. Coyotes eat badgers and foxes (although coyotes and badgers sometimes hunt cooperatively – the coyote sniffing out the prey and the badger digging it up). Raccoons, bobcats, and coyotes eat ringtails. Foxes and bobcats eat weasels, and big weasels eat smaller weasels. Small carnivorans and their young also fall prey to raptorial birds (hawks, eagles, owls, and vultures) and snakes (Mech, 1974; Hoffmeister, 1986; Poglayen-Neuwall and Toweill, 1988; Nowak, 1991; Gompper, 1995; Sheffield and Thomas, 1997; Gompper and Decker, 1998; Larivière and Walton, 1998; Presley, 2000; Verts et al., 2001; Yensen and Tarifa, 2003b).

Small Carnivoran Habitats

The preferred habitats of different taxa are often described based on the types of the largest or most abundant plants covering the ground and the density and extent of vegetation (pine forest, oak woodland, grassland, desert), the presence or absence of water sources (lakes, streams, and seas), and elevation and description of the ground (flat, slope, cliff, sandy, rocky, broken). Some major habitat types include: open areas covered by grasses or other low, scattered vegetation, or devoid of vegetation; partially open areas with scattered trees and sparse ground vegetation; brushy areas with various plants about 1 m (39 in) in height or more; open woodlands with widely scattered trees interspersed with grasses or low vegetation; forest areas with a more or less continuous canopy of overhanging branches and leaves, with sparse to dense understory plants; riparian areas where vegetation is lush, with trees, bushes, and low plants relatively dense on both sides of a stream; rocky slopes and hillsides, usually with stunted, sparsely scattered trees, bushes, and low vegetation. The ways animals move through their environment are often used to describe the taxa: burrowing or digging (fossorial); running (cursorial); climbing (scansorial); tree-living (arboreal); leaping (saltatorial).

Some carnivorans are generalists, such as wolves and coyotes, and may be found in almost any habitat, while others are very restricted to specific conditions, such as badgers and otters. The presence in the fossil record of a species with restricted habitat requirements provides definitive evidence for those conditions existing locally during the time the taxon lived in the area. Likewise, if a taxon avoids only one or two habitat types the presence of the fossil animal may indicate a local absence or reduction of those unfavorable environments.

Coyotes avoid dense forests and territory hunted by wolves, but range most other places, especially open woodlands, grasslands, and brushlands, including arid regions and deserts. Coyotes prefer terrain broken up by rocks, bushes, trees, and other visual blinds. Other arid-loving taxa include gray foxes, badgers, and spotted skunks. The badger is restricted to ground that can be excavated easily and inhabits areas with loose sandy soil such as that found on alluvial fans at the base of mountains. Otters require perennial open water, while raccoons, ringtails, and long-tailed weasels are usually found near water, preferring riparian areas. Coatis are also found in riparian areas but are equally at home in desert areas with abundant cactus and other fruit-bearing plants. Bobcats frequent river bottoms. Open brushy and grassy areas are favored by weasels and badgers, while bobcats and gray foxes, prefer more wooded brushy areas. Gray foxes also prefer brushy areas on rocky and broken terrain of hillsides, along with spotted skunks and ringtails. Gray foxes avoid coyotes that share the same habitat. Grison and tayra favor tropical rainforests, dry forest, shrub woodlands, and open fields. Otters are swimming specialists, while raccoons, weasels, grisons, and tayras are also good swimmers. Canids are good at digging prey out of burrows, however badgers are fossorial specialists able to completely bury themselves in seconds. Ringtails are truly scansorial, able to scale vertical cliffs at ease. Raccoons, cats, weasels, and grisons climb trees, while tayras are so agile they are almost arboreal (tree living). Foxes and coyotes are cursorial. The latter can run long distances, often covering several kilometers in a night, and can reach speeds up to 64 kilometers per hour (about 40 miles per hour), nearly as fast as the the fleetest extant North American carnivoran, the gray wolf (Jameson and Peters, 1988; Hoffmeister, 1986; Nowak, 1991; Gompper, 1995; Larivière and Walton, 1998; Verts et al, 2001; Murray, Artiodactyls, this volume).

Environmental Inferences

The sparse numbers of most taxa of small carnivorans from the Anza-Borrego are not statistically informative enough to make detailed environmental syntheses of the natural history of Anza-Borrego Desert based on carnivorans alone. However, a few observations may be made. The sharp increase in size of *Canis* between 2.58 and 1.95 Ma may reflect a locally manifested environmental change. Such a change allowed a larger canid either to evolve out of the smaller form, or physically to displace the smaller form from the region. This happened during the time it took for approximately 43 vertical meters (140 ft) of sediment to be deposited. While speciation is not suggested by any recent canid phylogenetic studies (Nowak, 2002; Tedford et al., 2001), either scenario implies an environmental change favoring the ecological requirements of the medium size *Canis* over the small *Canis*.

During the tenure of *Canis lepophagus*, beginning between 3.58 and 3.33 Ma and ending after 2.58 Ma, the majority of the landscape was probably arid open woodland, grassland, or brushland. *Urocyon,* another inhabitant of dry, brushland was present during the beginning of that time period and well after the disappearance of *C. lepophagus. Bassariscus* was the only water-loving taxon around throughout the transition from small to medium *Canis*. The presence of ringtail indicates perennial springs were probably active in the rocky cliffs and steep drainages being formed by the rising Peninsular Ranges. By the time *Canis priscolatrans* arrived the landscape may have been more forested and wetter. The one otter specimen is also found at about the time of the introduction of *C. priscolatrans*, as well as grison, raccoon, and (somewhat later) weasel, indicating an increase in permanently available water within the ancient drainage basin. However, the appearance of badger during the time of *C. priscolatrans* indicates that part of the drainage was still open and covered by grassy to brushy vegetation, most likely on sandy slopes or washes forming at the bases of the emerging mountains.

The relative numbers of fossils of each of the small carnivorans recovered (more canids and felids than procyonids and mustelids) points to at least two conclusions: canids and felids were moving through a wider range of habitats within the areas of sedimentary deposition than the other groups; and/or canids and felids were consistently more abundant. This can be seen particularly when comparing modern taxa that are restricted to a unique habitat, such as the otter, dependent on perennially flowing water, and raccoons and ringtails, also relying on a nearby water source. Also, the range of diet allows canids and felids to travel virtually anywhere rabbits, rodents, and small prey may be found, while taxa that rely on digging into soft soils for arthropods and burrowing rodents are necessarily barred from living on rocky ground or substrates that may be devoid of food resources. The physical size and morphology of the various carnivoran taxa may also be a hint as to their relative abundance and physical distribution in the fossil record. Canids and felids are generally larger than the other taxa, with the exception of raccoons, coatis, otters, and badgers, which may be close to or exceed some of them in body mass. Canids and felids, however have long legs (relative to body length) for greater range and speed of travel, lacking in raccoons, otters, and badgers. While coatis have longer legs, they travel in large herds of 10 to 30 individuals, foraging as they move, and spending some time in trees. This mode of travel does not lend itself to rapid coverage of large distances. Small felids are mostly solitary, while canids may be solitary or travel in small cooperative hunting groups.

Several of the small fossil carnivorans – *Canis* between 2.58 and 1.07 Ma, *Urocyon* between 3.58 and 1.77 Ma; and *Procyon* between 2.58 and 1.77 Ma – were considerably larger than modern forms. Since smaller modern representatives of all of these genera are found in the Park today, the size change is an indicator that some aspect of the regional environment changed subsequent to their last appearance in the Anza-Borrego Desert fossil record. Phenotypic (physical appearance) responses of individual taxa to environmental change may have multiple causes, including adaptation to changes in topography (rise of mountain ranges), climate (reduced precipitation due to mountains blocking storm systems), or vegetation (increased desert conditions and reduction of trees and grasses) as well as adaptation to a general turnover in faunal content. A reduction in carnivoran body size may be related to an overall reduction in the prey size. The local large *Odocoileus* (deer) and *Tetrameryx* (extinct pronghorn) were replaced by smaller modern forms, while the large peccary, *Platygonus* and the diminutive pronghorn, *Capromeryx* disappeared without replacement, along with the loss of large gophers (*Geomys*). Generally, larger body size costs more energy to maintain than the same body shape in a smaller form. And, since reduction in prey size provides fewer calories per kill, the parallel reduction in a predator's body size may improve its overall ability to survive in the changed environment.

The general types of small carnivorans present in the Anza-Borrego fossil beds have changed very little in the last three to four million years. Felids and canids were present at least seven million years ago, and probably considerably longer. With the exception of taxa requiring perennial flowing streams, such as otter and the locally extinct grison, most of the small carnivorans may still be seen within the Park boundaries. They include the ubiquitously present coyotes, as well as bobcats and gray foxes using dense vegetation for cover, badgers digging in deep loose sand, raccoons, ringtails, spotted skunks, and weasels restricted to habitats near canyons and the bases of mountain ranges.

Mammoths and Their Relatives

13

Elephants, as the largest living terrestrial animals, have always fascinated people. The two extant proboscidean species, Loxodonta africana *from Africa, and* Elephas maximus *from Asia, are endangered species. As such, they need a determined policy for conservation. Once a formidable group, found on all continents except Australia and Antarctica, the proboscideans now are restricted to these two species.*

E. Shoshani and P. Tassy, 1993

Shoshani and Tassy are co-editors of the latest comprehensive reference *"The Proboscidea: Evolution and Paleoecology of Elephants and Their Relatives,"* 1996.

George E. McDaniel, Jr.

Mammoths and Their Relatives

Borrego Formation Vista del Malpais. (Photograph by Paul Remeika)

Introduction

It is difficult to imagine that, in the not-too-distant past, herds of elephants roamed an area that today is a desert. What now is Anza-Borrego Desert State Park (ABDSP) once was the home of several species of elephant-like proboscideans. The skeletal remains of these lumbering beasts are among the most important of the fossils recovered from the Park, due in part to the scientific impact they have had on the field of vertebrate paleontology.

Over eighty localities yielding proboscidean remains have been recorded in ABDSP. This makes it one of the most productive proboscidean fossil localities in North America. Although most fossils are very fragmentary, some sites have yielded nearly complete skeletons and tell us much about the living conditions and the environment that supported life and brought death in ancient Anza-Borrego.

The proboscideans from ABDSP are represented by two separate groups, the gomphotheres and the mammoths. The earliest gomphothere remains from the Park (*Gomphotherium*) may be as old as 9 million years (Jefferson, 1999), the latest (*Stegomastodon*) 1.4 million (McDaniel and Jefferson, in press). Thus, the gomphotheres persisted in this area for about 7.5 million years. Mammoths (*Mammuthus*) crossed the land bridge from Siberia and arrived in North America about 1.8 million years ago, but their earliest recorded appearance in the Park is 1.4 million years ago (McDaniel and Jefferson, in press).

Proboscidean Family Tree

The proboscideans originated in Africa with the appearance of the primitive genus *Moeritherium* forty to thirty million years ago (Shoshani et al., 1996). This was a small semi-aquatic animal the size of a large pig, weighing up to 225 kg (500 lb). It had a large nose, which was destined to grow even longer. The Order name Proboscidea refers to this elongated nose or proboscis (trunk). All of the front teeth except the upper second incisor eventually would be lost as evolution progressed, and this remaining tooth would develop into the elephant's tusk. The earliest proboscideans had four tusks, while later species, including some of the gomphotheres, mammoths, and the elephants, would lose the

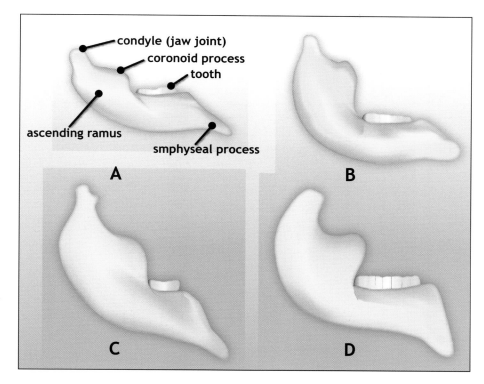

lower tusks and retain only the uppers. The skin of the proboscideans would become quite thick, hence the term pachyderm (Greek for thick skin). Some rather strange forms would arise and subsequently disappear, such as *Amebelodon* with short spoon-like lower tusks and short upper tusks, *Platybelodon* with an extended lower jaw with wide shovel-shaped lower tusks, and the spoon-billed *Gnathobelodon* from the Pliocene of Nebraska. None of these odd beasts occur in the Park.

The early proboscideans had a long, low skull, a long mandible (lower jaw), with a low ascending ramus (rear portion of jaw bone) that projected posteriorly (back) at an obtuse angle (Figure 13.1). The cheek teeth (premolars and molars) had a few thick plates (see Proboscidean Dentition, Figure 13.3). When the elephants (Family Elephantidae) split from the gomphotheres (Family Gomphotheriidae), elephant cheek teeth changed from having thick plates to being composed of many thin lamellae. Mammoths became even more progressive, with more numerous and thinner lamellae.

The proboscideans produced families such as mastodons (Mammutidae), gomphotheres (Gomphotheriidae), and elephants (Elephantidae) which include mammoths, *Loxodonta,* the African elephant, and *Elephas,* the Asian elephant. The gomphotheres are the main stem of the proboscideans from which the mastodons and elephants branch. The mastodons split off early in the evolution of the Proboscidea, while the elephants split off much later (Figure 13.2). The

Figure 13.1
Evolutionary Stages of Elephantid Mandible.
Images of the lower jaw, A through D, progress from primitive through the most evolved. Note that from the oldest to youngest (A-D), the condyle and ascending ramus move forward and become more vertically oriented.
Explanation:
A = *Gomphothere;*
B = *Mammuthus subplanifrons;*
C = *M. meridionalis;*
D = *M. columbi*
(not to scale).

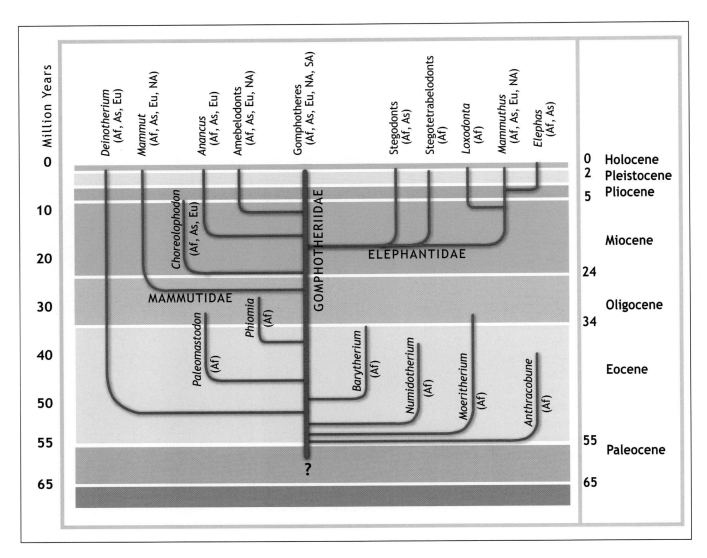

Figure 13.2
Simplified Family Tree of the Order Proboscidea.
Geologic Epochs are on the right and age in millions of years marks the major branches.
Geographic distribution of proboscidean taxa:
Af = Africa;
As = Asia;
Eu = Europe;
NA = North America;
SA = South America.

gomphotheres are a very diverse group, as every proboscidean that is not classified as a mastodon or elephant (including mammoths) is classified as a gomphothere. The gomphotheres are further split into three groups.

Mastodons have been reported from coastal southern California and the Mojave Desert (Tedford et al., 1987) to the north of ABDSP, but have not been identified from Plio-Pleistocene deposits of the Salton Trough. Since mastodons generally have been associated with conifer forests, we do not expect to find their fossils here. However, at one time, mammoths and gomphotheres were common to Anza-Borrego.

Proboscidean Dentition

Dentitions are one of the most important features that paleontologists use to distinguish one taxon from another, and mammoths are no exception. Like most mammals, a typical proboscidean tooth has a core of dentine covered by a layer of harder enamel. The root of the tooth is held in the alveolus (bony socket) by cement. However, elephantid cheek teeth (premolars and molars) differ greatly from those of other mammals. In these teeth, the enamel has developed numerous transverse folds with cementum (not to be confused with the cement of the roots) deposited on the outer surfaces of the crown between the folds. This folding increases the length of the cheek teeth in elephants and mammoths many times over that found in their ancestors.

Early mastodons and gomphotheres had molars with three thick transverse ridges or lophs (trilophodont) while later species are tetralophodont (with four lophs). With continued folding, the lophs develop into tall plates (Figure 13.3). In the Elephantidae, these plates, referred to as lamellae, are thin and numerous. The number of plates is relatively consistent within a species, but varies greatly between species (Hildebrand, 1974). Different mammoth species have different numbers of lamellae in their last or third molars: *Mammuthus meridionalis*, 11-14; *M. columbi*, 16-24; and *M primigenius*, 20-28 (premolars and first and second molars have fewer lamellae).

As elephantid teeth wear, the enamel wears slower than the softer dentine and cementum. This forms ridges of enamel on the occlusal surface (where upper and lower teeth meet) of the teeth. A worn lamella appears as a long transverse loop across the occlusal surface. This rough surface facilitates the grinding of food. The occlusal surfaces of cheek teeth are flat, and provide a broad area for grinding. To extend the life of the tooth, the roots are deep in the jaw of the young animal, and the crown (that part of the tooth above the roots) is tall, or hypsodont (high crowned).

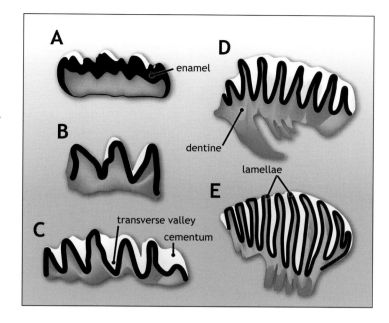

Figure 13.3.
Stages in the Evolution of Elephantid Teeth.
The teeth are depicted in cross section. A typical mammal tooth has a core of dentine (gray) covered by a layer of harder enamel (black). As elephantid teeth evolved from a simple set of cross lophs or cusps into numerous plates or lamellae, the enamel was folded with the dentine remaining in the core of the tooth. On the crown of the tooth, the spaces between the lamellae filled with cementum (white). As the animal eats, the softer dentine and cementum wear more easily than the harder enamel, leaving enamel ridges. Teeth with numerous hard ridges are more suited to grinding grasses, or grazing. Grasses are tough and cause more wear on teeth because they have higher levels of silica than leaves or other browse.
Explanation:
A = *Gomphotherium*;
B = *Stegotetrabelodon*;
C = *Primelephas*;
D = *Mammuthus meridionalis*;
E = *M. columbi*;
black = enamel; gray = dentine; white = cementum
(not to scale; after Maglio, 1973).

Mammoths, like elephants today, have six successive teeth (three premolars and three molars) in each quadrant of the mouth at any one time during their life. The first and second cheek teeth are in place one behind the other in the upper and lower jaws (maxillae and dentaries respectively) (Figure 13.4). They erupt vertically, the first followed later by the second. The other four cheek teeth form sequentially in the posterior region of the upper and lower jaws, subsequently move forward, erupting in the process one at a time (Todd and Roth, 1996). Each successive tooth is larger and has more lamellae than the

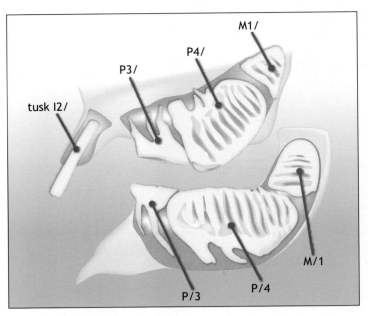

Figure 13.4.
The Process of Tooth Development and Replacement in Elephantids.
A typical elephant tooth (white) is held in an alveolus or socket (gray) in the bone (yellow). Shown here, the third and fourth premolars are in wear, and the first molars are developing in the germination cavities.
Explanation:
I2/ = upper second incisor or tusk;
P3/ = upper third premolar;
P4/ = upper fourth premolar;
M1/ = upper first molar;
P/3 = lower third premolar;
P/4 = lower fourth premolar;
M/1 = lower first molar
(after Cornwall, 1956; from Saunders, 1970).

preceding tooth. The teeth lie in a groove, which is deep posteriorly and slopes towards the occlusal surface. They erupt into the posterior part of upper and lower jaws at an angle to the occlusal surface, and thus the anterior lamellae come into wear first. The teeth are worn obliquely. As each tooth moves forward in the maxilla or dentary, a replacement tooth erupts from behind, and begins to wear. The cheek teeth are very long, and usually only one is fully erupted and in wear in each quadrant at any given time. The tooth continues to wear until only a small shallow-rooted remnant remains. It then is swallowed or falls out of the mouth. This displacement and replacement continues throughout the life of the animal until only the last molars (sixth cheek tooth or third molars) remain.

The proboscidean tusk developed from the upper second incisor (Figure 13.5). All of the other incisors have been lost in both mammoths and elephants. Originally, the outside of the second incisor was covered in enamel. As evolution progressed, this layer was lost in most species. However, some of the later gomphotheres retained an enamel band on the tusks, while others and the Elephantidae did not. The tusks of some proboscideans are straight, while others curve to various degrees. The curvature of the tusk is one criteria used to distinguish species (Figure 13.12).

Using the Craig scale (Haynes, 1991), based on which cheek tooth and the percentage of the tooth that is in wear, we can estimate the age of an elephantid at the time of its death. It is felt that the life span of mammoths is close to that of the African elephant, *Loxodonta africana,* approximately 70 years.

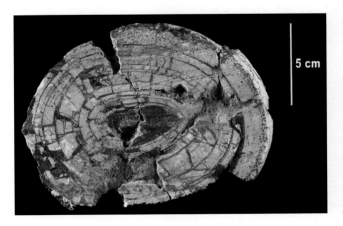

Figure 13.5.
Cross Section of a Mammoth Tusk.
Tusks are modified incisors. As they evolved, becoming longer, the enamel covering the dentine was lost. Some intermediate species retain an enamel band on the tusks, which distinguishes species. As the tusk grows, dentine is laid down inside the base or root of the tooth as long, forward-pointing cones. These nested cone-shaped layers appear as rings in cross section (right I2/ABDSP[IVCM]964/ V4056.28). (Photograph by Barbara Marrs)

Gomphotheres

Two members of the family Gomphotheriidae have been found in the Park, *Gomphotherium* (Figure 13.6) and *Stegomastodon* (Figure 13.7). *Gomphotherium* is a short-legged, hippopotamus-sized, water-loving animal that lived in marshy areas. The skull is low and the mandible low and long. Four tusks, two uppers and two lowers are present, the upper tusks being larger and longer than the lowers. The chin extends far forward so that the tips of the shorter lower tusks extend almost to the tip of the larger upper tusks.

Figure 13.6
Gomphotherium.
(Drawing by Pat Ortega)

The gomphotheres arose in east Africa from primitive proboscideans (*Moeritherium* mentioned above) in the early Miocene (Tassy, 1996). Several species spread through Africa and Eurasia. Four genera are reported from North America. The earliest of these, *Gomphotherium,* arrived via the Bering Land Bridge (see McDonald, this volume, *Anza-Borrego and the Great American Biotic Interchange*) during the Barstovian North American Land Mammal Age (NALMA) a little over 15 million years ago (Saunders, 1996). The upper tusks have well-developed enamel bands (Lambert, 1996). The tusks of later proboscideans, including the gomphothere *Stegomastodon* and the elephants (Elephantinae) are composed only of dentine. The elongated chin is straight, a feature that distinguishes *Gomphotherium* from some other taxa. The other three North American gomphothere genera, including *Stegomastodon,* which is found in the Park, originated from this first New World genus. *Gomphotherium* became widespread throughout North America, while the geographic distribution of the other three genera was more restricted. Why only two of the four genera of the New World gomphotheres occur in the Park and the others are absent is not understood.

The earliest proboscidean specimens recovered from the Park are approximately 9 million-year-old tusk fragments from Hawk Canyon (Jefferson, 1999). Identifiable only to the family Gomphotheriidae, the genus cannot be

Figure 13.7
Stegomastodon.
(Drawing by Pat Ortega)

determined from this meager material. The next oldest is a 3.7 million-year-old *Gomphotherium* specimen recovered from the upper Deguynos Formation of the Vallecito Creek/ Fish Creek basin in the southern part of the Park (Jefferson and McDaniel, 2002). Only a few fragments of the skull and mandible, including parts of the alveoli (tooth sockets) of the upper and lower tusks were found, but this was enough to identify the remains. Thought to have become extinct about 4.5 million years ago, this specimen extends the time range of the genus by nearly one million years.

Stegomastodon lived in North America from the early Blancan to early Irvingtonian NALMA, approximately 4.5 to 1.2 million years (Fisher, 1996), although it continued to exist in South America much longer (Lambert, 1996). The name is somewhat confusing, as *Stegomastodon* is a gomphothere, not a mastodon. It is unknown east of Texas, occurring only in the western part of the U.S. Most gomphothere molars have three or four large tooth plates, occasionally five. *Stegomastodon* has five to seven, or more plates. The tusks of *Stegomastodon* are distinctive among gomphotheres by having no enamel bands. It has a long, low skull and a long mandible (lower jaws) like other gomphotheres.

The geographic distribution of the genus suggests that *Stegomastodon* may have been more specialized for grazing than other gomphotheres, and occupied grasslands habitats. It may have competed with mammoths for some of the same foods. Mammoth teeth, having more numerous lamellae and enamel ridges, are better adapted for chewing grasses than are those of *Stegomastodon*. When mammoths arrived in North America, about 1.8 million years ago, *Stegomastodon* apparently could not compete with them, and shortly thereafter became extinct (Saunders, 1996; Lindsay et al., 1984). To date, a single 1.2 million year old locality, in the southern part of the Park, has produced specimens identified as *Stegomastodon*. One large incomplete molar was recovered along with large fragments of vertebra, ribs, and leg bones.

Figure 13.8
Southern Mammoths.
A condor (*Gymnogyps*) soars above this southern mammoth family at the edge of a Borrego Valley riparian woodland (see Paleolandscape 4). (Picture by John Francis)

Mammoths

Mammoths are medium to large-sized elephants (Maglio, 1973). They originated in southern and eastern Africa about 4 million years ago (Lister and Sher, 2001), with the appearance of *Mammuthus subplanifrons*. Subsequent species spread later through Europe, Asia, and North America, but did not reach South America. The first mammoth to reach the New World from Siberia via the Bering Land Bridge was the southern mammoth, *M. meridionalis*. Four species of mammoths are recognized in North America: *M. meridionalis, M. columbi* (Columbian mammoth) (Agenbroad, 2003), *M. exilis* (pigmy mammoth of the California Channel Islands) and *M. primigenius* (wooly mammoth). The well-known wooly mammoth, a late comer to this continent, was limited primarily to Alaska, Canada, and the northeasternmost U.S.

By far the most common proboscidean recovered from ABDSP is the mammoth. Almost fifty localities have yielded mammoth remains, nearly all in the Borrego Badlands of the northern part of the Park. Two species are identified: *Mammuthus meridionalis* and *M. columbi* (Agenbroad, 2003; McDaniel and Jefferson, 2003). ABDSP is the only place where fossils of both are found together (Jefferson and Remeika, 1994, 1995, McDaniel and Jefferson, 2003).

Mammuthus meridionalis

Mammuthus meridionalis evolved in Africa and first appeared in Europe during the middle Pliocene (Figure 13.8). It spread throughout Eurasia, giving rise to the steppe mammoth, *M. trogontherii,* in Siberia about 1.2 million years ago (Lister and Sher, 2001). In North America, early specimens of *M. meridionalis* are dated at around 1.8 million years, and current evidence indicates that it persisted here for over 1.5 million years and then went extinct. The southern mammoth is a medium-sized elephant, the males measuring 3 to 4 meters (9 to 12 ft) at the shoulder. The tusks are long and curved in a gentle spiral (helix).

Fragmentary evidence of *Mammuthus meridionalis* from the southern part of the Park indicates that this species was present in the region from at least 1.4 million years ago. Several partial skeletons of southern mammoths have been recovered from the Borrego Badlands. These range in age from 1.1 to about 0.8 Ma.

Although not assembled, the most complete skeleton of *M. meridionalis* from North America also was recovered from the Borrego Badlands (Figures 13.9, 13.10) (McDaniel and Jefferson, 2003). Almost 70% of this skeleton (with skull, mandible, one tusk 3.5 meters (10.8 ft) long, a scapula (shoulder blade), most of the vertebrae and ribs, and leg bones) was recovered from near the base of the Ocotillo Conglomerate, a horizon dated to about 1.1 million years (Jefferson and Remeika, 1996; Remeika and Beske-Diehl, 1996). The specimen was discovered by the late George Miller and his college class in December 1986 (Miller et al., 1991) (see Marrs, this volume, *History of Fossil Collecting in the Anza-Borrego Desert Region*). Some of the bones were lying on the surface, and others were protruding from the side of a hill. A little over two years

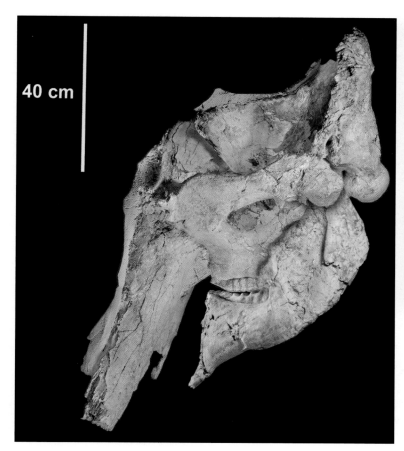

40 cm

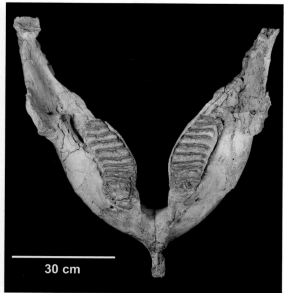

30 cm

Figure 13.10
Jaw of *Mammuthus meridionalis* (southern mammoth).
(ABDSP[IVCM]1277/V5126.02)
(Photograph by Barbara Marrs)

Figure 13.9
Articulated Skull and Jaw of *Mammuthus meridionalis* (southern mammoth).
(ABDSP[IVCM]1277/V5126.01 and V5126.02) (Photograph by Barbara Marrs)

was spent painstakingly excavating the skeleton, and final cleaning and conservation of the fossil took several more years. Three to 4 meters (10 - 12 ft) of sediment overburden was removed in order to expose the bone bed, and only small hand tools such as dental picks and small brushes were used (see Taphonomy below). Because of the completeness of the skeleton, this is one of the most exciting finds from ABDSP.

Mammuthus columbi

The second mammoth to arrive from Siberia, *Mammuthus columbi* (Figure 13.11), is the largest North American elephant, but its origin is not clear at this time. Some paleontologists feel that it was derived from Eurasian *M. meridionalis,* while others, including the author, believe that the morphology of the Columbian mammoth more closely compares to that of *M. trogontherii,* the Steppe mammoth. Males reached a height of over 4.5 meters (14 ft), measured at the shoulder. The curvature of the tusks (Figure 13.12) is much greater than in *M. meridionalis,* but less curved than in the smaller wooly mammoth, *M. primigenius.*

Figure 13.11
Mammuthus columbi
(Columbian mammoth).
(Drawing by Pat Ortega)

In the past, the imperial mammoth (*Mammuthus imperator*) was considered a valid species and regarded as the immediate ancestor of the Columbian mammoth (*M. columbi*). Most paleontologists now feel that the two taxa are the same species (Agenbroad, 2003), and that the imperial mammoth is the early evolutionary stage of the Columbian mammoth lineage. All of the specimens of *M. columbi* found in the Park belong to this early evolutionary stage. A specimen representing the later stage of *M. columbi* was found immediately east of the Park near the Salton Sea, where the sediments are much younger.

Many localities in the Borrego Badlands have yielded significant skeletal remains of this elephant. The oldest *Mammuthus columbi* was recovered from the same horizon as the nearly complete skeleton of *M. meridionalis,* discussed above, and is 1.1 million years old. The youngest (see below) was recovered from sediments about 0.7 million years old.

One of the most interesting fossil localities in the Park, dubbed Elephant Hill, is in the Borrego Badlands where the remains of three Columbian mammoths were removed: a juvenile individual; a subadult 22 to 23 years old at the time of death; and an old individual with bone cancer. Along with these are the bones and teeth of horse, camel, deer, and several small mammals. The skeletal remains of the subadult are the most numerous and include the skull and mandible (Figures 13.12, 13.13) in excellent condition, tusks, vertebrae, ribs, scapulae, pelvis, bones of the front and back legs, and a few foot bones. The bones of the juvenile include many ribs and vertebrae, one nearly complete scapula, and a patella (knee cap). The bone bed lies at the base of greenish-gray clays and silts indicative of a shallow water, marshy nearshore lake environment. An occasional fresh water snail was found during excavation.

Figure 13.12.
Comparison of Tusk Curvature in *Mammuthus columbi* **and** *Mammuthus meridionalis.*
The degree of curvature in Columbian mammoth (*Mammuthus columbi*) tusks (upper) is much greater than that in southern mammoth (*M. meridionalis*) tusks (lower). This is one of the characteristics that differentiates these species. Tusks continue to grow throughout life, but also are worn away at the tip through use. An estimated 1.5 to 2 meters (4-6 ft) is worn from the tip of the *M. meridionalis* tusk. The *M. columbi* right tusk (ABDSP[IVCM]964/V4056.28) is from a 22-23 year old individual, and the *M. meridionalis* right tusk (ABDSP[IVCM]1227/V5126.03) is from a 55-60 year old individual. (Photographs by Barbara Marrs)

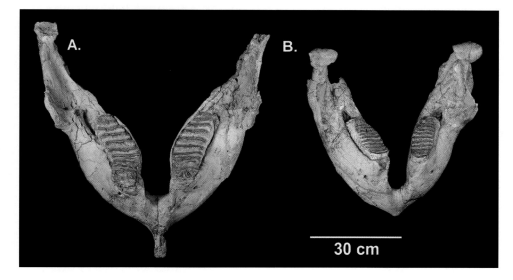

Figure 13.13
Mandibles of *Mammuthus meridionalis* **and** *Mammuthus columbi.*
A. 55-60 year old female southern mammoth (ABDSP[IVCM]1277/V5126).
B. 22-23 year old subadult Columbian mammoth (ABDSP[IVCM]964/V4056).
The ages of these individuals was determined using the Craig scale. The teeth in the older southern mammoth are fully worn last or third molars. Those in the Columbian mammoth are partially worn second molars. Note that, although the female southern mammoth (left) is an adult, it is about the same size as the subadult Columbian mammoth (right). Male elephants are larger than females, but it is impossible to determine the sex of this young Columbian mammoth. However, it is clear that should had it reached adulthood, it would have grown considerably larger than this female southern mammoth. A distinguishing feature between *Mammuthus meridionalis* and *M. columbi* is the thickness and number of lamellae in the cheek teeth. *M. meridionalis* has fewer and thicker lamellae. These differences are clearly visible. (Photographs by Barbara Marrs)

Taphonomy: The Crime Scene Investigation of a Pleistocene Site

One of the most intriguing areas of paleontology is taphonomy, the study of what happens to the remains of an animal after it dies. This also might be described as paleo-forensics. A classic taphonomic study was done on the *Mammuthus meridionalis* skeleton from the Borrego Badlands (McDaniel and Jefferson, 1997, 2003) discussed above. When such a site is discovered, we first thoroughly study the types of sediments and local stratigraphy. From this information, we know that the mammoth died on the lower part of an alluvial fan. Sediment grain size and textures indicate that he small streams on the fan were braided, meaning that they branched then came back together to form a network. By measuring individual bed thicknesses and lateral continuity, we know that individual stream channels were no more than one meter (3 ft) deep. From the position of the bones, it is apparent that the carcass had not been moved far from the site of death. There was sufficient sediment to bury the bones, but the stream currents were not strong enough to transport most of them very far.

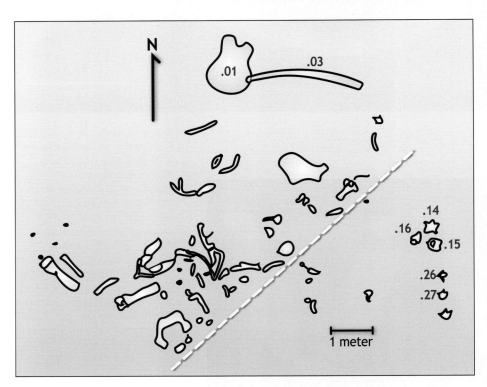

Figure 13.14

Bone Bed Map of the *Mammuthus meridionalis* Excavation Floor.
Numbers refer to catalogued skeletal elements of ABDSP(IVCM)1277/V5126: e.g. .01 = skull, .03 = right tusk; .14, .15, .16 = three consecutive cervical vertebra removed from the carcass by scavengers; and .26, .27, .28 = three consecutive lumbar vertebrae also removed by scavengers Note that the southeastern margin of the bone distribution (lower right corner) is truncated along a line (white dashed) that is parallel to the current direction (see Figure 13.15). This is the edge of the active stream channel. Specimens to the far right, those moved by scavengers, lie outside of this paleochannel. (Map redrawn from Miller et al., 1991; McDaniel and Jefferson, 2003)

As each bone was uncovered, its location and orientation were recorded, and a sketch of the specimen was drawn on a map of the bone bed (Figure 13.14). We know from the distribution of skeletal elements on this map that the mammoth died with its right side down, and its head pointed upstream. Notably, most of the foot bones are missing, having been washed away. In a stream current, rounded bones, such as those in the feet, will scatter, long bones will move perpendicular to the direction of flow, and long bones with one heavy end, will move parallel to the direction of flow with the lighter end pointed downstream. By plotting these on a diagram (Figure 13.15), we can see that the stream flowed from northeast to southwest at the time the bones were buried. Today, water flows from west to east, nearly a complete reversal, and is eroding sediment due to uplift and folding of the badlands.

The mammoth carcass was processed extensively by scavengers. Potential large predators in the Borrego local fauna that may have been responsible for this bone modification include *Borophagus diversidens* (bone-eating dog), *Arctodus simus* (short-faced bear), and *Smilodon gracilis* (graceful saber-tooth cat) (Remeika and Jefferson, 1993, 1995; McDaniel and Jefferson, 2003; Appendix, Table 3; see Shaw and Cox this volume, *The Large Carnivorans*). Many of the bony processes (projections) on the vertebra were chewed off, and many ribs exhibit carnivore "bite and drag" tooth marks. These are the result of gnawing by carnivores in a manner similar to the way you chew corn-on-the-cob. Three consecutive neck vertebrae, as well as three consecutive lumbar vertebrae were removed from the carcass as pieces, drug a few meters to the east, out of the main stream channel, and scavenged.

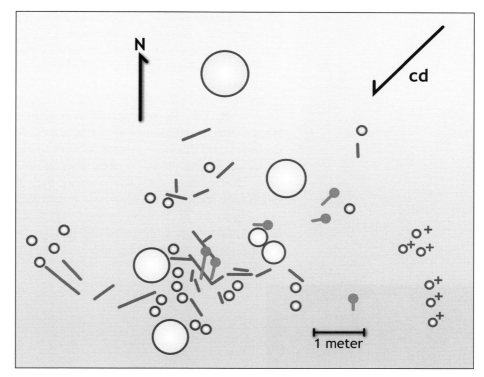

Figure 13.15
Orientation and Hydrodynamics of Skeletal Elements.
The orientation of bones in this stream deposit at locality ABDSP(IVCM)1277 directly reflects the direction water was flowing (cd) at the time the bones were buried. Spherical-shaped (short and round) bones will scatter and quickly roll down stream, limb bones will roll with their long axes perpendicular to the flow of water, and long bones with one heavy end will move parallel to the direction of flow with the lighter end pointed downstream. Explanation: open circle = spherical-shaped bone; open circle with + = moved by carnivores or scavengers; rod = long bone, ends equal in size and weight; rod with knob on end = long bone with one heavy end (knob); cd = inferred paleocurrent direction (from McDaniel and Jefferson, 2003).

The skull is missing the top of the cranium, and the cranial cavity was filled with sediment. We can conclude from this that the skull was partially buried initially, that the top of the cranium weathered away during a period of non-deposition, then the skull subsequently was buried completely. This documents two separate depositional events, possibly months apart, before the mammoth skeleton was completely covered.

Generally, we can tell male elephants from females by characters on the pelvis. Unfortunately, this bone in the Anza-Borrego specimen is incomplete. However, the diameter of the base of the tusk is much larger in male elephants than in females. The small size of the tusk indicates that this mammoth is a female. The amount of wear on the cheek teeth provides an age of about 55 to 60 years at the time of death (Figure 13.16). Arthritis, as evidenced by very irregular surfaces in the jaw joints, probably made it very painful for the mammoth to chew, and may have led to her being malnourished and an easy target for predators.

A thorough taphonomic study requires careful and detailed observations. Ancient sediments can reveal past depositional environments. Their study provides data that, in conjunction with fossil bone distribution and orientation maps, allow interpretation of the conditions of carcass transport and burial. A close examination of skeletal remains may reveal subtle clues as to the cause of death, age and health. The processes of carcass disintegration and scattering by scavengers are discernable in the fossil record. All of these are taphonomic factors, that when understood together, permit the reconstruction of paleo-environments and can bring to life events that took place millions of years ago.

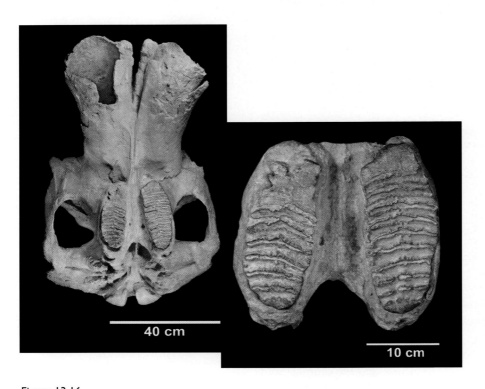

Figure 13.16
Occlusal Surface of the Upper Molars in *Mammuthus meridionalis*.
The upper third molars in the skull of this southern mammoth (ABDSP[IVCM]1277/V5126) are in full wear, and are smooth (the lamellae are completely worn away) in front (top of right image). The estimated age of the animal at the time of death is approximately 55 to 60 years. (Photographs by Barbara Marrs)

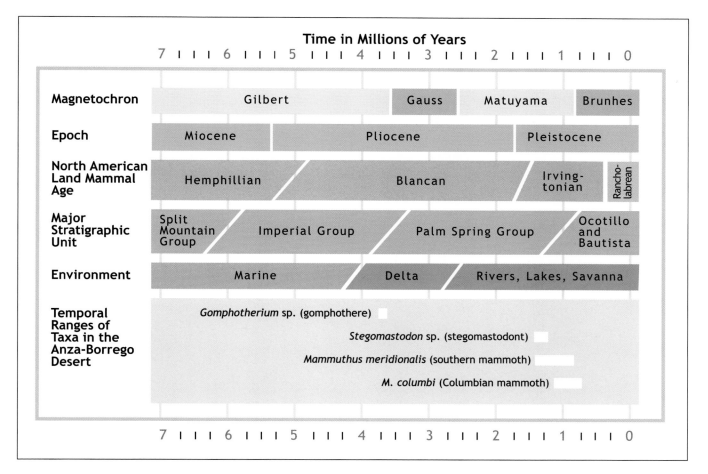

Figure 13.17
Mammoths and Their Relatives in Anza-Borrego.
The time ranges of the four Anza-Borrego Proboscideans are plotted. Note that *Stegomastodon* disappears shortly after the earliest *Mammuthus meridonalis* in the Stratigraphic section.

14
The Small Fossil Mammals: Rodents, Rabbits, and Their Relatives

The completed study of faunal or community associations at each "level" in this entire section together with sedimentary interpretation may reveal environmental changes primarily reflecting the provincial tectonic activity or geologic history of this unique area in Southern California.

Ted Downs and John White (1968)

John A. White
Hugh Wagner
George T. Jefferson

The Small Fossil Mammals:
Rodents, Rabbits, and Their Relatives

Mudhills Along Arroyo Tapiado. (Photograph courtesy Colorado Desert District Stout Research Center Archives)

Is Bigger Better?
What Small Fossils Tell Paleontologists

Bigger is better, so they say, and the best-known fossils usually fall into this category. From the ever-popular dinosaurian titans to the post-Cretaceous mammalian "oddities" such as *Brontotherium* or "thunderbeast" (a 5 meter-tall rhinoceros-like animal with large bony horns on its snout) and *Moropus* (a huge, long-necked herbivore related to horses, only with claws instead of hooves), on to the more "modern" mammoth elephants, giant camels, and ground sloths, the fossils of the gargantuan beasts of the past have long fascinated and captivated the interest and imagination of all – from school children to scientists.

Though these are undoubtedly spectacular discoveries, the large size of these forms has some drawbacks. From a paleontological point of view, large animals usually provide insufficient data for in-depth studies about their past habitats. In addition, small animals, particularly small mammals, are a key, and provide essential data necessary for reconstructing specific paleoenvironments. The small guys help to paint an overall picture of where and how they, and their larger counterparts, lived. Without these little "fossil critters," much would be lost in the fossil record, and therefore, not as much would be known about the larger denizens of the past.

Small mammal fossils, often called microfossils or microvertebrates, are excellent in providing such a wealth of paleoenvironmental data because:

1. Small mammals often produce offspring at an age of less than two or three years and have a very short gestation period, while large

mammals take longer to mature, have a longer gestation period, and often produce fewer offspring at any one birthing; thus small mammals, over time, produce more offspring, and therefore there is the potential to yield more microfossils;

2. Specimens of small fossil mammals are therefore usually more numerous than those of large fossils. It is possible to concentrate large numbers of these specimens by screen washing sediments through various sized meshes, catching the small bones and teeth;

3. The larger number of fossils also makes it possible to determine the degree of morphological variation within each species;

4. Small mammals have a much smaller home range than larger mammals. A mouse, for example, may have a home range of perhaps 1/3 of a hectare, whereas horses have home ranges of hundreds of square kilometers. Therefore the ecological setting of a fossil mouse is better constrained and defined than that of a fossil horse.

Figure 14.1
Microfossil Field Collecting.
Sediments containing the fossils are excavated and bagged. Then the sediment is washed through screens and the residue is laid out to dry. Seen in the photographs are Drs. Ted Downs and John White, and Laboratory Preparator Leonard Bessom of the Natural History Museum of Los Angeles County (see Marrs, this volume). (Archival photographs courtesy of the Natural History Museum of Los Angeles County)

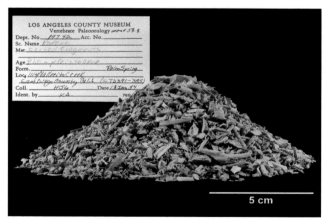

Figure 14.2
Microfossil Sample.
Screen washed microfossils from locality ABDSP(LACM)1114, from the Hueso Formation, await identification.
(Photograph by Barbara Marrs)

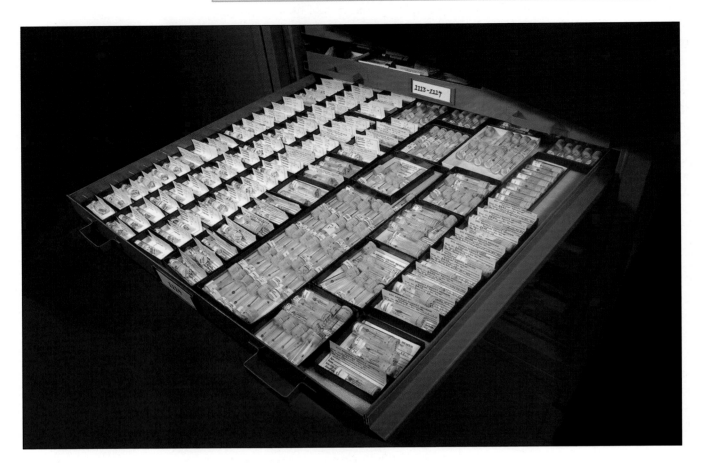

Figure 14.3
Identified and Catalogued Microfossils.
A typical drawer of curated microfossils. Each glass vial contains a single tooth, mounted on a pin for study under the microscope. The fossils are from locality ABDSP(LACM)1114.

The remains of small animals from Anza-Borrego consist of teeth, limb bones, vertebrae, and jaws and parts of skulls. Whole skulls are known from Anza-Borrego but are extremely rare, possibly because these relatively delicate structures have been damaged by stream action. Microvertebrates are commonly found in fluvial stream deposits, sandstones, and siltstones of the Hueso Formation within the Vallecito Creek region of the southern part of the Park. They occur through about 2.7 km (9,000 ft) of sediments in an incredible continuous stratigraphic sequence (see Dorsey, this volume, *Stratigraphy, Tectonics, and Basin Evolution in the Anza-Borrego Desert Region*) representing some 2.5 million years of time. The number of identifiable specimens collected exceeds 3800, and constitutes diverse and highly significant fossil assemblages (Downs and White, 1965a, 1965b).

Many of the small mammals are extinct, but some of the microfossils represent extant species. Some of these living species, that are found as fossils, presently do not inhabit the Colorado Desert region. These are called extralocal (outside the area) taxa, and reflect environmental and habitat changes that through time have forced these animals to shift their geographic range. By looking at the modern conditions in which these animals live, we can infer past habitats for Anza-Borrego. Also, to complete the picture of lifestyle and habitats, paleontologists can infer the ecological requirements of extinct small mammals from the known habitats and food preferences of their closest living relatives.

Fossil remains recovered from Anza-Borrego, designated type specimens, have been used for the definition of nine new species of small mammals. Insectivores, shrews and moles; Chiroptera, bats; and Lagomorpha, cottontails and jackrabbits, are present, and more than 40 different types of Rodentia alone have been identified (Appendix, Table 3). These include porcupines, ground squirrels, chipmunks, pocket gophers, beavers, pocket mice, kangaroo rats, kangaroo mice, pygmy mice, grasshopper mice, deer mice, cotton rats, harvest mice, pack rats, and voles. Ecological and behavioral data presented in the following discussions are largely from Burt and Grossenheider (1976), Hoffmeister (1986), Ingles (1965), Jameson (1988), Nowak (1991), and Walker (1968).

Insectivoria

Shrews: Family Soricidae

Two species of shrews have been recovered from the Hueso Formation strata: *Sorex* sp., the red-toothed shrew, and *Notiosorex jacksoni,* Jackson's extinct desert shrew. The genus *Sorex* includes some of the smallest (as small as 4 g weight, 0.14 oz) mammals in North America and they are found in a variety of habitats. Although *N. jacksoni* is extinct, based on what is known of the living species, *N. crawfordi*, one can postulate that its habitat would have been similar. *Notiosorex crawfordi* lives in the deserts of the southwestern U.S. and northern Mexico, and is found in Anza-Borrego today. They live under dead agave plants, under piles of sheet metal and lumber, among debris in village dumps, in pack rat middens, and in other sheltered places where cover is ample. They need no water other than that obtained from their insectivorous prey (Hoffmeister, 1986).

Moles: Family Talpidae

The moles are represented by a single fossil skull, identified as the extinct *Scapanus malatinus* (Hutchison, 1987). Living moles of the genus *Scapanus* occur in grasslands near forests. These animals are highly modified for life underground; the eyes are greatly reduced in size, the humerus of the forearm is broad and flat with several large places for muscle attachment, an adaptation to digging. The animals literally "swim" through the top level of the soil; they also excavate tunnels and leave the removed dirt in rounded piles on the surface. Based on the habitats of the living species of *Scapanus* in California, *S. latimanus* and *S. townsendi,* more moist and cooler climates in the past can be inferred for Anza-Borrego.

Chiroptera

Bats

Two genera and one identified species of fossil bats are recognized from the Hueso strata: *Myotis,* the mouse-eared bat, and *Anzanycteris anzensis* (White, 1969), the Anza bat. Living species of *Myotis* are known throughout North America, including the Anza-Borrego Desert. They catch insects in flight.

Anzanycteris anzensis is related to the pallid bat, *Antrozous pallidus,* as demonstrated by similar measured skull proportions. Pallid bats have relatively large eyes and fly and hunt close to the ground. It is probable that *A. anzensis,* having a skull structure similar to that in *A. pallidus,* had similar life habits. Like the modern pallid bat, which is known from southern California deserts, the Anza bat probably caught insects on the ground and on the trunks of trees, but seldom in flight.

Lagomorpha

Rabbits and Hares: Family Leporidae

Two groups of rabbits and hares are found in the Hueso strata: the completely extinct but advanced rabbits (Subfamily Archaeolaginae), and the advanced and living rabbits and hares (Subfamily Leporinae).

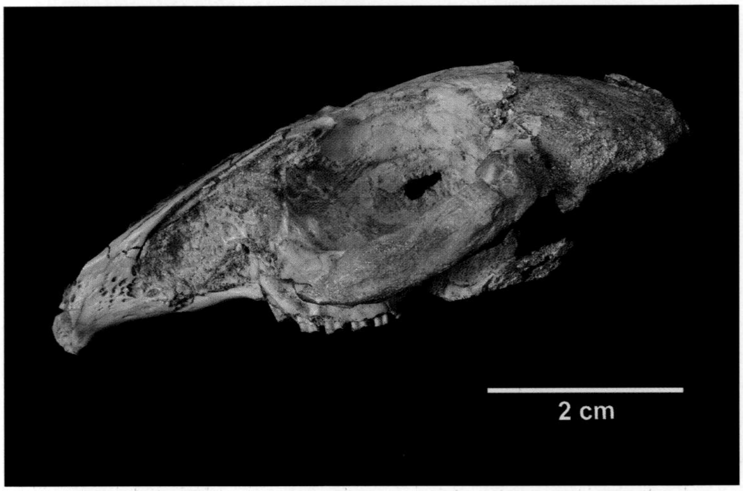

Figure 14.4
Sylvilagus hibbardi.
Hibbard's cottontail rabbit, view left side of skull (ABDSP[LACM]1301/V3479). (Photograph by Barbara Marrs)

Representing the Archaeolaginae are *Hypolagus vetus, H. edensis,* and *Pewelagus dawsonae. Pewelagus dawsonae,* Dawson's dwarf rabbit, was described using Anza-Borrego fossils (White, 1984): a partial skull of an adult specimen and two partial skulls of immature specimens. It was about a third the size of a large black-tailed jackrabbit, *Lepus californicus.* The outstanding feature of *Pewelagus* is the large-sized tympanic bullae, the bubble-like structures on the base of the skull that house the middle ear. The comparable structure in a large jackrabbit for example, is about the same size, suggesting that *Pewelagus* had a hearing capacity similar to larger rabbits. The enamel pattern on the lower third premolar also is distinctive. Because of many morphologic similarities between *Pewelagus* and *Ochotona,* the living pika or rock rabbit, *Pewelagus* probably was ecologically similar to *Ochotona.* These two genera therefore are called ecomorphs (White, 1984).

Hypolagus vetus has almost identical skull proportions as the living Florida cottontail, *Sylvilagus floridanus*. This suggests that *H. vetus* may have been a functional cottontail, or ecomorph, and occupied a similar ecological niche and exhibited similar food preferences. However, because of dental characteristics, it is classified in the Archaeolaginae, a separate group not closely related to the Florida cottontail. *Hypolagus vetus* disappears from the stratigraphic record about 3.2 Ma, shortly before the arrival of *Sylvilagus hibbardi* (Figure 14.4). This is a probable case of ecological replacement (see Cassiliano, this volume, *Mammalian Biostratigraphy in the Vallecito Creek-Fish Creek Basin*).

Hypolagus edensis was a small rabbit about the size of the modern brush rabbit, *Sylvilagus bachmani*, that occurs today in coastal San Diego County. *Hypolagus edensis* is characterized by the structure (folded enamel pattern) of the front-most, third premolar or chewing tooth in the jaw.

Nekrolagus, known only from a few specimens recovered from the Blancan, older part of the Hueso Formation, also occurs in similar aged deposits in southern Arizona. It may have been ancestral to later Leporinae, including both *Lepus* and *Sylvilagus*. The jackrabbit, *Lepus*, was reported previously from the Hueso Formation (Cassiliano, 1999). However, current studies of these fossils (Murray et al. pers. comm., 2005) failed to confirm the presence of *Lepus*. The few remains in question can not be confidently assigned to the genus *Lepus*, and could represent either *Lepus* or *Nekrolagus*.

Rodentia

Porcupines: Family Erethizontidae

Porcupines are relatively large rodents, and *Erethizon dorsatum* is the only living member of the family in the U.S. (Frazier, 1981). They prefer woodland habitats, though they are highly adaptable and have been found in open range land, tundra, and desert environments, but usually stay within a riparian habitat. Their diet ranges from the bark of trees to leaves, fruit, flowers, and shoots. They are good climbers and will nest either in trees or on the ground. Porcupines are immigrants to North America from the South American tropics (see McDonald, this volume, *Anza-Borrego and the Great American Biotic Interchange*).

Coendou stirtoni, Stirton's coendu (White, 1970), the extinct North American porcupine (Figure 14.5), was originally described from Anza-Borrego fossils. Within the Hueso stratigraphic section, it ranges from about 2.5 to 1.0 Ma. It is a close relative of the living *C. mexicanus*. This modern prehensile-tailed porcupine lives in the tropical forests of southern Mexico and Central and South America.

Figure 14.5
Coendu stirtoni.
This extinct porcupine, related to the Mexican form, has found a comfortable place to nap in the riparian forests of Borrego Valley one million years ago. (Picture by John Francis)

The skull of *Coendu* can be distinguished from that of the living porcupine, *Erethizon dorsatum*, because its upper tooth rows are almost parallel to one another, whereas they are strongly convergent towards the front of the skull in *Erethizon.* The Anza-Borrego collection includes a specimen of *C. stirtoni* with the tail vertebrae preserved, which, when compared with that of the living *C. mexicanus,* indicates that it probably had lost its prehensile function.

If *Coendu stirtoni* had ecological requirements similar to its closest living relative, *C. mexicanus,* then it probably was restricted to the wetter habitats of the ancient Anza-Borrego landscape. The presence of *Coendu* supports a tropical Mexican climatic aspect to the Hueso sequence. Its extinction about 1.0 Ma may have been a result of increasing regional aridity due to rain shadow effects from the rising Peninsular Ranges (see Dorsey, this volume).

Squirrels: Family Sciuridae

Squirrel fossils are rare in Hueso strata. Based on the few teeth and some jaws recovered, they can be identified to the genus *Spermophilus,* subgenus (*Otospermophilus*), the common ground squirrel. No tree squirrels are known.

The common ground squirrel is primitive with low-crowned cheek teeth. The upper cheek teeth have a cusp pattern characterized by a pi-shape, the Greek letter π, and the lowers are square with four cusps. They most closely resemble living *Spermophilus* (*Otospermophilus*) *beecheyi,* the rock squirrel. Today, common ground squirrels inhabit prairies, steppes, tundra, rocky country, open woodlands or desert mountain ranges. Their diet consists of fruits, seeds, nuts, grains, roots, bulbs, mushrooms, green vegetation, insects, and small invertebrates. They often are colonial. Banks (1964) reports *Spermophilus* as being rare in the Park, but confined to higher elevation rocky terrain.

They differ from the antelope ground squirrel, *Ammospermophilus harrisi,* that is smaller, has slightly higher-crowned teeth, and prefers open dry sandy desert areas including the desert floor of Anza-Borrego. The difference between these two types of ground squirrels is indicative of different climatic and environmental conditions. The dominance of *Otospermophilus* strongly suggests a region with significantly more vegetation than is present today. The absence of *Ammospermophilus* indicates that the arid conditions present today had not fully developed.

Chipmunks are represented by a single skull, referable to the western North American genus, *Eutamias.* The chipmunk, a small ground squirrel, is a forest animal and an excellent climber that inhabits western North American spruce, fir, redwood, and pine forests, but also sagebrush plains, brush-covered mountains, and dense temperate-zone forests. It prefers rugged or brush-covered land. *Eutamias* excavate a burrow system and live mostly on the ground. Their diet consists mainly of fruits and seeds of various trees with some herbs, mushrooms, insects, and other animal foods.

The presence of *Eutamias* indicates a moister environment more typical of higher elevations. Chipmunks live in varied habitats, but always where there is significant vegetation from shrub to forests, which is not present in the Salton Trough today.

Pocket Gophers: Family Geomyidae

Pocket gophers have fur-lined cheek pouches on each side of the mandible for the storage and transport of seeds and food items. The western pocket gophers, genus *Thomomys*, are rare in Anza-Borrego, represented only by a jaw and a few teeth, found in the younger part of the Hueso Formation. The other pocket gophers in the collection are *Geomys anzensis,* the Anza gopher, and *G. garbanii* (Figure 14.6). Both of these extinct species were named from Anza-Borrego specimens, *G. garbanii* for Harley Garbani, who has found the greatest number of significant gopher specimens in the Park (see Marrs, this volume, *History of Fossil Collecting in the Anza-Borrego Desert Region*). Because pocket gophers burrow, when they die their remains are "pre-buried." Fossil skulls and jaws of gophers are the most common of all rodent fossils from Anza-Borrego.

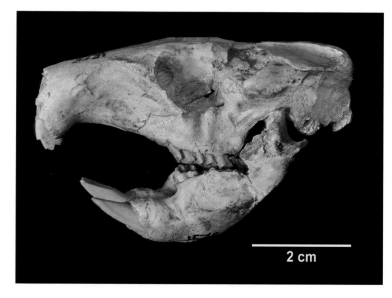

Figure 14.6
Geomys garbini.
Garbani's pocket gopher,
view left side of skull
(ABDSP[LACM]1192/V3231).
(Photograph by Barbara Marrs)

Living species of *Geomys* occur in the Great Plains, but only as far west as White Sands National Monument, New Mexico. There are also extinct *Geomys* in the Curtis Ranch local fauna in Cochise County, Arizona, from sediments about the same age as those in Anza-Borrego. The presence of *Geomys* supports a tropical Mexican type of environment, but in addition it indicates a rather stable tectonic environment, i.e. no large alluvial fan development associated with orogenic activity. Living species of *Geomys* only borrow into fine-grained soils, and are never found in coarse gravely deposits. This suggests an open area not in close proximity to a rapid uplift or factors that would contribute a coarse-soil component.

Thomomys prefer less humid, more diverse climatic conditions, and occur throughout the western U.S., with an eastern limit literally at the western edge of *Geomys* range. *Thomomys* burrow not only in fine-grained soils, but also in coarse gravely deposits. It appears elsewhere in California in Hemphillian and Blancan age assemblages. Although not as abundant as *Geomys*, the appearance of *Thomomys* in the Hueso stratigraphic section at 1.9 Ma suggests a dispersal and colonization of the Anza-Borrego region by *Thomomys* due to environmental changes, either a shift in weather conditions to a more temperate climate, a change to coarser soils due to uplift of local mountain ranges, or both.

Pocket Mice, Kangaroo Rats and Kangaroo Mice: Family Heteromyidae

The pocket mice, genus *Perognathus* (Figure 14.7), are named for their fur-lined cheek-pouches. There are over 160 skulls, jaws, and teeth in the Anza-Borrego collection. Some of these remains have been referred to *P. hispidus,* the hispid pocket mouse, which presently inhabits the central and southern Great Plains and central Mexico, an environment very different from today's Anza-Borrego Desert.

The kangaroo rats, genus *Dipodomys*, also have fur-lined cheek pouches for the transport of seeds. *Dipodomys* is especially suited for hopping and is bipedal with a long tail. They are nocturnal and live in complex burrow systems under bushes and other objects. Many species are adapted to arid conditions and live in deserts to coastal shrublands throughout western North America. Their food consists primarily of dried seeds and vegetation, and many species can derive all their water by metabolic processes from the solid foods they eat. Two species of kangaroo rats presently inhabit the Anza-Borrego Desert region.

Figure 14.7
Perognathus sp.
Pocket mouse, palatal view of skull (ABDSP[IVCM]290/VI135).
(Photograph by Barbara Marrs)

Dipodomys is well represented by over 150 specimens from the Hueso strata, including a nearly complete skull, and several jaws. Five species have been recognized from Anza-Borrego fossil deposits, one living form, *D. compactus*, the gulf coast kangaroo rat. The other species, *D. hibbardi,* Hibbard's kangaroo rat, *D. minor,* the dwarf kangaroo rat, and two yet-to-be-described species (Cunningham, 1984) are extinct. Interestingly, *D. compactus* presently inhabits the gulf coast of Texas, suggesting Anza-Borrego was warmer and more humid.

The kangaroo mouse, *Microdipodops,* also has an enlarged bubble-like structure or bulla that houses the middle ear. Today, this genus with two species occurs primarily in Nevada where they inhabit a wide variety of arid and semi-arid environments. The Anza-Borrego specimens of *Microdipodops* consist of only a few teeth.

Beavers: Family Castoridae

Castor canadensis is the largest extant North American rodent. They are semi-aquatic, either living in constructed structures in ponds or holes in stream banks. They prefer streams and small lakes with growths of deciduous hardwoods such as willows, aspen, poplar (cottonwood), birch, and alder (see Remeika, this volume, *Ancestral Woodlands of the Colorado River Delta Plain*). *Castor* eats bark, twigs, leaves, and roots of deciduous trees and shrubs and various aquatic plants such as water lilies. They build dams and impound running water making small ponds, and are an important factor in development of stream pond ecologies. They are essentially nocturnal, beginning activities at sunset. Beavers do not hibernate but during winter months live off stored food stockpiled on the bottom of ponds when the surface is frozen. *Castor* has been reported from the lower Colorado River area, but this is in its extreme southern geographic distribution (Hoffmeister, 1986).

The genus *Castor* first appears in the fossil record in late Miocene deposits of western North America as a primitive species with lower crowned teeth. By the Pleistocene it is well established on the continent. The Anza-Borrego collection contains three beaver specimens, including a partial lower jaw and teeth. Two of the specimens are from the Hueso Formation and date to about 2 Ma. The third is from the Coyote Badlands and is younger than about 700,000 years. The occurrence of beavers indicates a cooler and wetter climate and the presence of permanent-flowing streams lined with riparian vegetation.

Mice, Rats, and Voles: Family Cricetidae
Packrats or Woodrats, White-footed Mice, Pigmy Mice: Subfamily Sigmodontinae

Woodrats are the most common and the largest cricetid rodents in North America. There are 20 described species that live in a variety of habitats from low deserts to humid jungles, to rocky slopes above timber line. Some species build elaborate nests of vegetation and other objects such as bones, and have been known to take objects and leave a replacement. In the Anza-Borrego Desert, they build large nests, up to 1.8 meters-tall (6 ft), from twigs and branches usually in rocky crevices (Ingles, 1965). Their diet consists almost entirely of plant tissues such as roots, stems, leaves, seeds, and some invertebrates. Their geographic range extends from as far north as northwestern Canada to as far south as Nicaragua and Honduras.

As a group, the woodrats have undergone a major evolutionary radiation in southern North America, primarily in Mexico and the tropics. They are important in interpreting paleoenvironmental conditions as the most common fossil types in the Hueso Formation are found in southern Mexico today.

Nearly 300 specimens of *Neotoma* have been recovered from the Hueso Formation. The Anza-Borrego *Neotoma* include at least three different forms, but have not been studied in detail or assigned to species (Cunningham, 1984). *Neotoma* are of interest because their ancestry can be traced back to a large species of a *Peromyscus*-like animal in the late Miocene. They appear to diverge into several lineages, represented by subgenera such as *N.* (*Paraneotoma*), and *N.* (*Parahodomys*) that both occur in the deposits of Anza Borrego. They are an endemic North American rodent group valuable for biochronologic interpretations.

Repomys is an extinct pygmy woodrat (or packrat). Specimens from Anza-Borrego were tentatively referred to the living genus *Nelsonia,* a pygmy woodrat from coastal tropical western Mexico. However, *Repomys* is now regarded as a very rare Mexican-Arizona-California taxon.

Only three teeth of the pygmy mouse, genus *Baiomys,* have been identified in the Anza-Borrego collection, and there are no living *Baiomys* in California. However, they are found as fossils in the Blancan part of the San Timoteo Formation (Albright, 1999) in Riverside County to the west of Anza-Borrego. The living pygmy mouse is the smallest rodent in North America. They do not occur in western North America today, but are restricted to the central southwest south into Mexico to Nicaragua and are associated with a warm humid environment. These minute *Peromyscus*-like rodents prefer cover in dense grass, shrubs and cultivated fields. *Baiomys* is currently extending its range north across Texas.

Calomys, the vesper mouse, is related to *Peromyscus.* It is extinct in North America, but relatively common in southwestern and central U.S. paleofaunas during latest Hemphillian and Blancan time. It can be distinguished from other *Peromyscus*-like mice by unique features on the first molars and by a well developed anterior masseteric crest, a ridge on the cheek side of the lower jaw. Today *Calomys* is restricted to Central and South America. It lives in a variety of habitats including montaine grasslands, brushy areas, and forest fringes. *Calomys* find shelter in bunch grass, in holes in the ground, and in rotting tree stumps and rock piles (Walker et al., 1999). It climbs well and often hops on its hind legs like a kangaroo rat.

Species of white-footed mice in the genus *Peromyscus* are probably the most numerous in numbers and kinds in today's North American rodent fauna. However, this distribution may have been the result of a relatively recent radiation. If so, *Peromyscus* is a more boreal animal that is expanding its range south. It is not a common genus in Anza-Borrego assemblages, with fewer than 75 identified fossils. *Peromyscus maniculatus,* the deer mouse, is the most abundant and widespread living mammal in California, and presently is found in the upland areas of the Park (Bond, 1977).

Onychomys, the grasshopper mouse (see lead photograph, this chapter), is very rare in the Anza-Borrego fossil record, with only three catalogued specimens. This is not unexpected because it is a carnivore (see Murray, this volume, *The Small Carnivorans: Canids, Felids, Procyonids, Mustelids*). Currently, species of *Onychomys* live throughout the western U.S., including Anza-Borrego. The grasshopper mouse is a small *Peromyscus*-like rodent that prefers dry prairie and desert scrub. It preys primarily on insects, but will consume other small rodents including *Peromyscus, Perognathus,* and *Microtus.* They even bay like a tiny wolf with a whistle! *Onychomys* are very territorial and live in low densities as do other predators. They inhabit practically any shelter they can find, are nocturnal, and active throughout the year.

Reithrodontomys sp., the American harvest mouse, is a small *Peromyscus*-like rodent represented by approximately 20 specimens from Anza-Borrego. It resembles the common house mouse, *Mus,* but has more hair on its tail and has grooved incisors, the front gnawing teeth. Its habitat varies from salt mashes to tropical forests, and it usually lives in stands of short grass. Its food consists primarily of seeds and small green shoots of vegetation. *Reithrodontomys* is one of the few mammals that can drink salt water, so it can survive in areas along coastlines with no fresh water, and around saline lakes in arid environments. Although there is no evidence that it likes desert environments, it can tolerate them. The most common western species is *R. megalotis*, which has been identified in many middle and late Pleistocene paleofaunas of North America.

The cotton rat, *Sigmodon*, is generally considered to be from a group of rodents that is distantly related to the Subfamily Arvicolinae mainly confined to Central and South America. Cotton rats prefer grassland and shrubby areas (Martin, 1979). They are the most abundant rodents in areas where they occur in the southern U.S., south through Mexico and Central America. Their range today in California is confined to the Colorado River area, where they make tunnels in the grass in their quest for seeds (Hoffmeister, 1986). They don't thrive where winter temperatures drop below freezing. They are omnivorous, feeding on vegetation, insects, and other small animals. The genus can be traced back into the late Miocene to its ancestor *Prosigmodon*, recovered from deposits in northern Mexico and Arizona.

The genus is abundant in the fossil record, and several species of *Sigmodon* have been described from the Hueso Formation, including *S. curtisi, S. lindsayi,* and *S. medius/minor.* According to Cassiliano (this volume), *Sigmodon* occurs throughout the Anza-Borrego section, from strata that are about 4.2 Ma to less than 700,000 years old. This is significant because the presence of *Sigmodon* indicates warm climates, and apparently, Anza-Borrego remained relatively warm throughout the past 4 million years. However, lemmings and voles (below), animals that today live in cool climates, also occur in the Anza-Borrego fossil record. The remains of warm and cool adapted animals found together as fossils makes interpreting the climatic and paleoenvironmental conditions at Anza-Borrego difficult (see Sussman et al., this volume, *Paleoclimates and Environmental Change in the Anza-Borrego Desert Region*). It is possible that climatic conditions shifted back and forth on a scale that is presently beyond the resolution of the fossil record.

Muskrats, Lemmings, and Voles: Subfamily Arvicolinae

The living muskrat, *Ondatra zibethicus*, occurs in northern Baja California and throughout most of Canada and the U.S. They are found in a wide range of aquatic habitats, fresh and salt-water marshes, lakes, ponds, rivers, and sloughs, where they build houses using twigs, grass, and other vegetation. Muskrats eat all types of vegetation with some animal material, but prefer cattails and bulrushes. Muskrats are not good climatic indicators as they apparently occur wherever there is water regardless of the climate.

Ondatra is represented in the fossil record by several extinct species including *O. idahoensis*, the extinct Idaho muskrat, that is found in the Hueso Formation of Anza-Borrego. *Ondatra idahoensis,* is distinguished principally by its tooth structure or tooth enamel pattern. Its habits were probably similar to the living species *O. zibethicus*, which suggest the presence of permanent bodies of water in Anza-Borrego.

Pliopotamys minor, the pygmy muskrat, is an extinct form with habits that may have been somewhat like *Ondatra*. *Pliopotamys* appears to be ancestral to *O. idahoensis* and to the extant *O. zibethicus,* and the North American ancestry of the muskrat can be traced back to *P. minor* in the middle Blancan, approximately 3.5 Ma. Evolutionary trends through time are an increase in size and an increase in crown height of cheek teeth best recognized in the lower first molar as added reentrants forming additional triangles on the labial (lip) and lingual (tongue) sides of this tooth. With the increase in crown-height the production of dentine exceeds that of the enamel resulting in the development of dentine tracts. Measurements of the lengths of these tracts on the margins of the triangles have been used to determine relative ages of populations of muskrats such as *Pliopotomys* and *Ondatra* (reviewed by Albright, 2000). The length of the lower first molar and the number of reentrant angles and associated number of anterior triangles has been used to differentiate the species of *Ondatra,* specifically *O. idahoensis, O. annectens*, and *O. zibethicus.*

Lemmings and voles are small to medium-sized mice with short tails, short ears, and short legs. In general these rodents prefer an environment close to water. They are herbivores, feeding principally on grasses, and some like living in burrows in bogs while others are almost entirely arboreal adapted to eating conifer needles. They are generally considered to be northern latitude rodents that prefer to live above the frost line.

Stratigraphic distribution of the fossil Arvicolinae remains from the Hueso Formation indicate that there were a number of late Blancan dispersals of these rodents from east Asia into the southern U.S., including the *Mimomys* group, the lemmings and *Microtus*. A thorough study of the early *Mimomys* group and subsequent genera and species of lemmings and voles in the Anza-Borrego stratigraphic section will greatly enhance our knowledge of when cool, wet spells in the climate permitted them to travel this far south. However, it is evident that Plio-Pleistocene climates were significantly wetter and cooler than in the Salton Trough today.

Synaptomys (*Mictomys*), bog lemmings, live in moist areas throughout the humid Pacific northwest, and are commonly found near rocky cliffs. In the southern parts of their geographic range they are limited mainly to isolated cold bogs and springs, like the Dismal Swamp of Virginia and North Carolina. They feed primarily on plants, mainly the green parts of low vegetation and to a lesser extent on slugs, snails, and small invertebrates.

Synaptomys (*Mictomys*) *anzaensis,* the extinct Anza bog lemming, is described from Anza-Borrego specimens. It is present in other early Irvingtonian paleofaunas of approximately 1 Ma (Repenning, 1987), and in the Hueso stratigraphic section (Cassiliano, 1999) between 2 and 1 Ma. Permanent water and overall cooler, more mesic (moderately wet) climates in Anza-Borrego's past are suggested by the fossil remains of these animals.

Lasiopodomys deceitensis, the Cape Deceit water or bank vole, occurs late in the record, near the top of the Hueso Formation in late Irvingtonian age deposits younger than 700,000 years. *Lasiopodomys* is often treated as a subgenus of *Microtus.* Today this vole inhabits the steppes of eastern Asia, in meadows in mountainous regions around lakes. Its diet consists of grass, other green vegetation, and roots. They are extinct in North America, and their occurrence in Pleistocene assemblages signals a climate change to cooler, wetter conditions. This allowed them to extend their range west and south into North America.

Fossils of *Microtus californicus,* the California meadow vole, occur in the Hueso Formation at several stratigraphic intervals less than 1.7 Ma. Living *M. californicus* are usually found in grassy areas near water. Voles establish runways, connecting their nests with foraging sites. They are not desert animals, are restricted to areas of permanent water and lush grass, or permanent marshy ground (Banks, 1964), and do not live in the Park today. Voles indicate the presence of permanent water in Anza-Borrego's past.

Mimomys is a small extinct vole ancestral to many later Microtinae including lemmings and all other voles. *Mimomys* has been given multiple names since its discovery in fossil localities of Blancan and Irvingtonian age in North America and Eurasia. *Mimomys* (*Cosomys*), first found in Pliocene deposits near the town of Coso, California, and *M.* (*Ophiomys*) are two subgenera that immigrated to North America approximately 4.8 Ma. Both *M.* (*Cosomys*) and *M.* (*Ophiomys*) *parvus,* the Snake River vole known from late Blancan deposits along the Snake River in Idaho, are found in the Hueso Formation of Anza-Borrego. *Mimomys* (*Ophiomys*) *parvus* is restricted to assemblages that occur between 2.5 and 2.0 Ma.

The genus *Pitymys,* the pine mouse, according to some authors is placed as a subgenus under *Microtus* (see Bell et al., 2004). Here, it is considered a valid genus following Walker (1964), and Repenning (1983, 1987). In North America, *Pitymys* ranges from Texas to Wisconsin and eastward to the east coast and south into east-central Mexico. It is also found in Europe and Asia Minor. *Pitymys* live in varied habitats from open field to tamarack swamps, and have adapted to a semi-fossorial, subterranean, existence with reduced eyes, ears, and close pelage. Its front claws are enlarged for digging, and its diet consists of seeds, leaves, nuts, fruits, and succulent roots and tubers.

Pitymys meadensis (*Microtus meadensis* of Bell et al. [2004] and *Terricola meadensis* of Repenning [1992]), the extinct Mead, Kansas, vole, is recovered from the Ocotillo Conglomerate. The species is a typical Irvingtonian

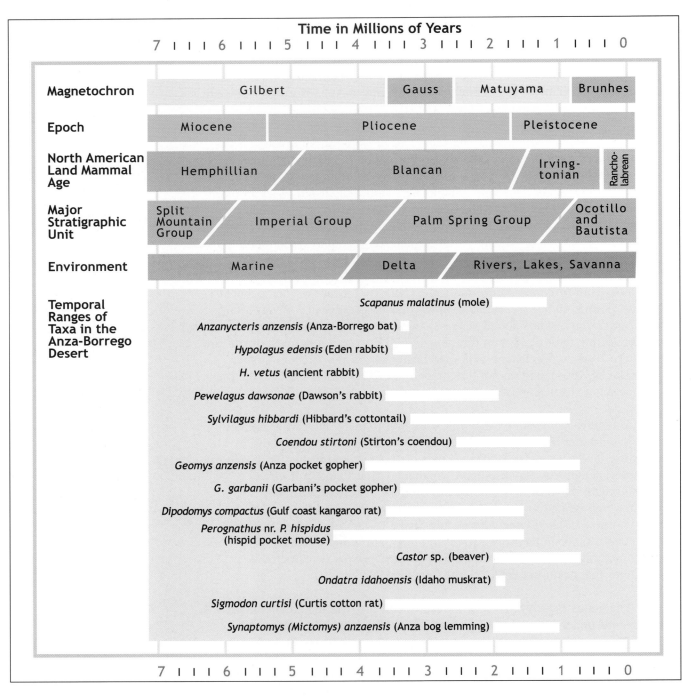

Figure 14.8
The Small Fossil Mammals of Anza-Borrego.
The temporal ranges of fifteen small mammals are plotted, and they span almost 4 million years.

taxon (Repenning, 1987), and occurs in assemblages associated with the Bishop Tuff, less than about 760,000 years old (see Remeika, this volume, *Dating, Ashes, and Magnetics*). Deposits at the type locality in Cudahy, Kansas, are considered to have accumulated during a time of dry hot summers, before and during the Nebraskan glacial period (Repenning, 1987). The genus *Terricola* is presently limited in its geographic distribution to a relict population located in the eastern slopes of the Sierra Madre Orental in central Mexico (Repenning, 1992).

Conclusions

Many of the small mammals found as fossils in Anza-Borrego appear to have southern or tropical affinities, based on the geographic ranges of either extant species or the closest living relatives of the extinct forms. This suggests that, for much of the time represented by the Hueso Formation, the Anza-Borrego region was dominated by a warmer, moister environment with less seasonal changes in temperature, more tropical vegetation, and permanent water sources. It also would imply the absence of cold winters that resulted in frost or snow. This paleoenvironment may have been the result of different global climatic conditions combined with the lack of a mountainous rain shadow to the west of the Salton Trough. However, it is the earlier part of the record that is more tropical, not the later. This pattern is also seen in the stratigraphic distribution of the tropical fossil lizards and the giant tortoises (see Gensler et al., this volume, *The Fossil Lower Vertebrates*).

Fewer, but still a significant number of small mammal taxa (again based on the geographic ranges of either extant species or the closest living relatives of the extinct forms) suggest a cooler environment with temperate plants, and seasonal changes broad enough to result in frosts and snow during the winter. Again as with the more tropical forms, the presence of water in streams or lakes is indicated. Animals adapted to these types of conditions seem to be more abundant and or restricted to the younger part of the Hueso Formation. It is interesting that both *Eutamias* and *Peromyscus*, two cooler climate forms, first appear about 3.3 Ma, suggesting the beginnings of a climate shift at this time. The Arvicolinae in general suggest at least periods of cooler conditions, moving away from the earlier warm moist tropical environment typical of the earliest part of the Anza-Borrego fossil record.

Extinct Horses and Their Relatives

In those days animals thought nothing of growing a new limb or organ to adapt themselves to conditions. We seem to have lost the knack.

Will Cuppy (1941, "*How To Become Extinct*")

Eric Scott

Extinct Horses and Their Relatives

View from Fonts Point to Borrego Mountain. (Photograph by Paul Remeika)

Introduction

The sedimentary sequence exposed in the Anza-Borrego Desert State Park (ABDSP) has yielded abundant fossil remains of extinct animals dating from the late Miocene into the middle Pleistocene Epochs (see Cassiliano, this volume, *Mammalian Biostratigraphy in the Vallecito Creek-Fish Creek Basin*). Common in this assemblage are the remains of perissodactyls (commonly termed "odd-toed ungulates," although this informal designation is somewhat incorrect). Ungulates are large herbivorous mammals, often bearing hooves. Perissodactyls are ungulates lacking the first and fifth digit (toe) on the hind foot and the first digit on the fore foot; the central third digit is larger than the digits flanking it. In horses, this condition is extreme; the classic horse hoof is the massive central digit, while the accessory splint bones on either side are all that remains of the lateral toes.

Present-day perissodactyls include the equids (members of the family Equidae: "true" horses, zebras, asses, and hemionines or "half asses," and their ancestors), tapirs, and rhinos. Both equids and tapirs were present during the time period represented by the sediments in Anza-Borrego; equids are among the most commonly recorded animals from the region, while tapirs are much rarer.

Horses: "Evolution in Action"

Horses are native to North America, originating on this continent approximately 57 million years ago during the Eocene Epoch. Although they went extinct in North America approximately 11,000 years ago, at the end of the Pleistocene Epoch, until then equids were well represented in North American fossil faunas. At various times in their long history, equids have successfully emigrated from their home continent to Europe, Asia, Africa, and South America, becoming relatively prosperous and widespread on each of these continents as well (Simpson, 1951; MacFadden, 1992).

The dramatic sedimentary outcrops within Anza-Borrego capture a relatively recent period in the long evolutionary history of the horse. While the extinct horses from Anza-Borrego are assigned to different species, and in two cases to a separate genus from *Equus*, they all share the following general characteristics:

1. They are generally relatively large, bigger than present-day burros, and similar to extant zebras in body size, although smaller than most breeds of domestic horse.

2. They possess one massive central digit flanked by two vestigial digits (also termed "splints") on each foot. This reduction in lateral digits, also observed in extant equids, is an evolutionary adaptation to running, wherein all of the force generated by the limbs is driven through a single point rather than dispersed through several digits.

3. The elbow, "knee," and "fetlock" of the equine forelimb as well as the "stifle," "hock," and "fetlock" of the hind limb are ginglymus joints. Joints of this type move in only one plane, with no appreciable rotation. In the horses from Anza-Borrego, as well as present-day horses, this lack of rotation is taken to the extreme. Articular surfaces at the elbow, hock, and fetlocks are deeply grooved, permitting very little lateral play, and all the joints are tightly bound by ligaments and tendons, further restricting rotation. In addition to these adaptations, the paired bones of the lower limbs have also undergone substantial modifications reducing or eliminating rotation. In the forelimb, the ulna is reduced in size and solidly fused to the radius, while in the hind limb the fibula is much reduced relative to the tibia. These features are also adaptations for running. (Note: the elbow, knee, and fetlock of the equine forelimb are equivalent to the elbow, wrist, and finger joints of the human forearm, respectively. The "stifle," "hock," and "fetlock" of the equine hind limb are analogous to the knee, ankle, and toe joints of the human leg, respectively.)

4. The vertebral column of these horses, like that of present-day horses, is rigid and heavily reinforced by muscles, tendons, and ligaments. This relatively inflexible spine is necessary to support the massive intestinal tract, wherein the grass, browse, and other abrasive fodder ingested by these grazing animals is slowly digested. (Although only bones and teeth of extinct horses have been recovered, and soft parts such as intestines are not preserved, nevertheless the structure of the vertebral column indicates that the extinct North American horses possessed large, heavy intestines like those of their present-day cousins.)

5.   The relatively tall or "high-crowned" teeth of the horses from Anza-Borrego are adapted to eating coarse, gritty plants such as grasses, rather than leaves. Grass and related plants can be very tough and fibrous, with high concentrations of silica preserved in their cells.

Grasses also grow in regions where wind-blown dust is a factor, and so these plants can often have a considerable coating of residual dust. Animals with short or "low-crowned" teeth would wear them down very quickly eating this tough, gritty plant matter. High-crowned or hypsodont teeth, in contrast, can last much longer, and so extend an animal's life. Additionally, horse teeth are reinforced with cementum, a substance that covers the outer surface of the teeth as well as filling in the interior spaces of each tooth. This cementum both strengthens the teeth and helps them sit more firmly in the jaw.

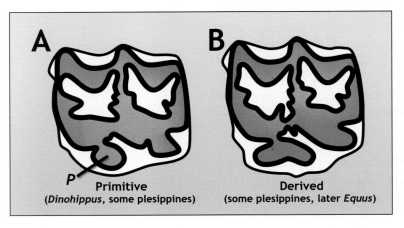

A

B

P

Primitive
(*Dinohippus*, some plesippines)

Derived
(some plesippines, later *Equus*)

Figure 15.1
Horse Upper Cheek Tooth Anatomy.
Wear-surface view of upper left third premolars, black = enamel; grey = dentine; white = cementum. The front of the jaw is to the left of the diagram.
A: primitive condition observed in *Dinohippus* and some *Equus* (*Plesippus*) species. The proto-cone (P) is usually small and rounded, with little or no forward or rearward projection.
B: derived condition observed in later *Equus* species. The protocone is generally more elongate, with a forward "heel" and a long backward projection (towards the right of the diagram). These features are often age-dependent, and can change as a tooth wears.

6.   The enamel of the cheek teeth does not cover the outer surface of the teeth, as in many other mammals (including humans), but rather forms complex columns running from top to bottom of each tooth, surrounding the softer dentine and surrounded by cementum. As a tooth wears down during normal use, the enamel columns are, in effect, cross-sectioned to reveal the inner dentine, much the way slicing a banana reveals the pattern of the inner fruit. The enamel appears on the wear surface of each cheek tooth in an elaborate pattern that comprises a large part of the chewing surface (Figures 15.1, 15.2). Enamel does not wear down as quickly as dentine or cementum, and so a sharp enamel ridge is always exposed for effective grinding of grass and other fodder.

Equid Dental Anatomy

Students of Plio-Pleistocene horse evolution classify extinct equids and argue about them largely in terms of their dental features. This has engendered a complex and potentially intimidating descriptive terminology. Since this terminology is essential to the language of the debate, a brief discussion of some of these terms and their significance is presented in the caption for Figures 15.1 and 15.2.

The enamel patterns observed on the wearing surfaces of horse teeth are of particular interest to paleontologists, as variations in these patterns can help to distinguish one species from another. Such determinations are by no means hard and fast, as the observed enamel patterns can differ not only among species, but also among individuals and even in single animals, depending upon which tooth is being considered and how worn it is. For example, there are subtle differences between the enamel patterns of premolars and molars. In addition, because equid teeth are high-crowned, the enamel patterns seen on the wear surface of the teeth can change noticeably as the tooth wears down through the animal's life.

While the potential variability of equid cheek tooth enamel patterns is widely recognized by paleontologists (and has been since J.W. Gidley in 1901 diagnosed the potential for and degree of such variation), nevertheless scientists have for decades attempted to assign species names to isolated equid teeth – or even to name new equid species based upon teeth alone. This is partly because vertebrate teeth are more common than bones in the fossil record, as teeth are harder than bones and so are generally preserved more frequently and in better condition. Many equid species have been named based upon fossil teeth, without full consideration of the potential for variation within a species or during the development of an individual. There are literally dozens of published species names for North American Pleistocene horses that are invalid, usually because the type fossil material was technically insufficient for correctly diagnosing a given species. North American Pleistocene horses are historically a taxonomic and systematic quagmire, and as of this writing, there is no universally accepted systematic classification for these equids – despite more than a century of fossil collecting and analysis.

Some paleontologists have proposed that other morphological features than teeth should be used to define equid species. Such features include body size (estimated from measurements of the skeleton), the anatomy of the skull and jaws, and the relative sturdiness (or slenderness) of limb elements (Bennett, 1980, 1984; Groves and Willoughby, 1981; Eisenmann and others, 1988; Downs and Miller, 1994; Azzaroli 1995, 1998). In fact, dental characters have been considered by some authors (Groves and Willoughby, 1981; Winans, 1989) to be useful only in corroborating specific identifications using cranial (skull) or postcranial (behind the skull; essentially, the rest of the skeleton) characters. However, because of the nature of the fossil record, only very rarely are all the requisite skeletal and dental elements preserved together, in unequivocal association, to enable such identifications.

The variability of dental characters can be reduced when only like elements are compared (e.g., molars compared with molars, premolars compared with premolars), and when individuals of similar ontogenetic (developmental) age are examined. Further, although dental traits may be poorly diagnostic when employed in isolation, such traits can help discriminate among fossil equids when used in combination with other morphologic features such as the overall size and relative robustness of the metapodials (foot bones).

For example, present-day horses and their relatives can be distinguished to some extent by the shape of a fold of enamel, the ectoflexid, on the wear surface of each of the lower molar teeth (Skinner, 1972; Dalquest, 1978, 1988; Eisenmann and others, 1988; Downs and Miller, 1994; Hulbert, 1995b) (Figure 15.2). Then other characters, when used in conjunction with ectoflexid

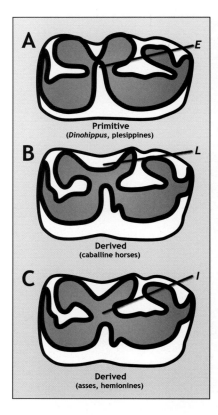

Figure 15.2
Horse Lower Cheek
Tooth Anatomy.
Wear-surface view of lower left first molars, black = enamel; grey = dentine; white = cementum. The front of the jaw is to the left of the diagram.
A: primitive condition observed in *Dinohippus* and *Equus* (*Plesippus*) species. The ectoflexid (E) is elongate and projects completely through the center of the tooth.
B: derived caballine condition, with a reduced ectoflexid and a linguaflexid (L) that is generally (but not always) broad and U-shaped.
C: derived condition observed in asses and hemionines, with a generally V-shaped linguaflexid and a short ectoflexid that usually does not enter the molar isthmus (I). Again, all of these features are often age-dependent, and can change as a tooth wears.

morphology, can help to clarify specific distinctions. These include: the shape of the protocone (a cusp on the upper cheek teeth) (Figure 15.1), and the depth and shape of the molar linguaflexid (a fold of enamel on the lower molars, Figure 15.2) (Forsten, 1986; Eisenmann and others, 1988; Dalquest, 1988; Dalquest and Schultz, 1992; Downs and Miller, 1994; Hulbert, 1995b).

In addition to the cheek teeth discussed above, the incisors – the teeth in the front of the mouth – can also prove useful for classifying fossil horses. The appearance, patterns, and presence or absence of enamel-lined cups, or infundibulae, in the lower incisors varies among different species of horses, both living and extinct (Cope, 1892; Gidley, 1901; Hoffstetter, 1950; Bennett, 1980; Eisenmann and others, 1988; Downs and Miller, 1994; Hulbert, 1995b; Scott, 2004).

The Equids of Anza-Borrego

The family Equidae in Anza-Borrego State Park is currently represented by seven taxa (see Appendix, Table 3). The following section will provide an overview of some of them as they are presently understood, and in general is confined to those fossils that have been assigned to species. The naming and classification of each of them has been accompanied by vigorous scientific debate, not only over the identification of individual specimens, but also about how the various equids are to be organized into a system that includes them all.

In the course of identifying and classifying our seven species, problems with invalid taxa have figured in the history of the taxonomic names. For example, the earliest published record of extinct horses from Anza-Borrego, Bowers' mention of Alverson's 1901 recovery of "remains of the fossil horse (*Equus occidentalis*)" (see Marrs, this volume, *History of Fossil Collecting in the Anza-Borrego Desert Region*), was too sketchy to permit confirmation, with no listing or description of fossils included. Furthermore, the species itself, initially described by the renowned paleontologist Joseph Leidy in 1865 based on three isolated teeth, is now considered invalid. The three teeth, having been duly compared with already-classified specimens, were ultimately not thought to be sufficiently diagnostic to warrant specific distinction (Miller, 1971; Winans, 1985; Scott, 2004).

Another difficulty concerns the classification of equids into subgenera. Some species of *Equus* reported from Anza-Borrego (Downs and Miller, 1994; Cassiliano, 1999) were classified based on their proposed subgeneric affinities. However, the subgenera employed were nearly all defined on extant Old World equids. There is uncertainty about how many equid species were present (and how these are defined) in North America during the Pliocene and Pleistocene Epochs. Furthermore, considering that the phylogenetic relationships of these animals to each other and to equids from the Old World have yet to be satisfactorily resolved, it seems somewhat presumptuous to assign virtually all

Plio-Pleistocene *Equus* from Anza-Borrego to extant Old World subgenera. This review will therefore employ only one subgeneric assignment – the North American subgenus *Plesippus* (American zebra-like horse) as appropriate. Other equids will be treated at the species level with no subgeneric assignments.

Some species reported from Anza-Borrego by Downs and Miller (1994) were assigned to the subgenus *Dolichohippus* Heller, 1912, an extant zebrine subgenus whose type species is the living Grevy's zebra, *Equus grevyi*. Where previous studies employed the name *Dolichohippus* or its derivatives in reference to North American equids, this paper will employ *Plesippus*, the American zebra-like horse. The apparent clinal (gradual but continuous) nature of equid evolution from *Dinohippus* through *Equus* (Hulbert, 1989) renders *Plesippus* difficult to define. For convenience, the present review considers *Plesippus* a valid subgenus of *Equus,* but acknowledges that more work needs to be performed in order to resolve this taxonomic conundrum.

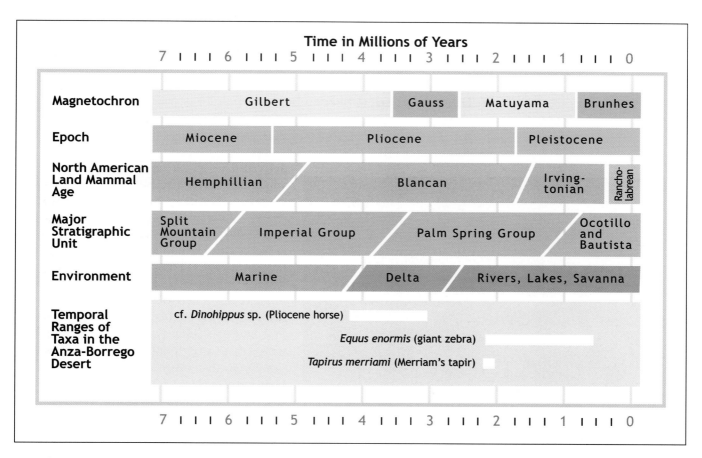

Figure 15.3
Extinct Horses and Their Relatives in Anza-Borrego.
Until more specimens of *Equus simplicidens* and *E. scotti* are identified, plotting Anza-Borrego ranges for these taxa is not possible. Also, even though we know that the only *Hippodean* remains were recovered from the upper Hueso Formation, the site is bounded by faults and can not be precisely correlated with dated horizons.

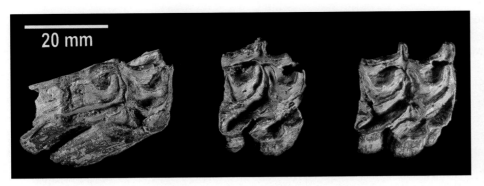

Figure 15.4
Upper Cheek Teeth
Assigned to *Dinohippus*.
The front of the mouth is
to the right, occlusal view.
(ABDSP[IVCM]177/V2257.01,
ABDSP[IVCM]177/V2257.02,
ABDSP[IVCM]177/2257.03.
(Photograph by Barbara Marrs)

It should be noted, also, that the first appearance in North America of the subgenus *Equus* (*Equus*), the subgenus that includes the modern caballine domestic horse, may have been recorded in the rocks of Anza-Borrego. Cassiliano (1999) combined the criteria of Forsten (1986, 1988), who proposed separating equids into two groups based on dental characteristics, with the taxonomy of Downs and Miller (1994), who further subdivided the equids based on other morphological features. Cassiliano reviewed specimens from Anza-Borrego that had not been discussed by Downs and Miller, and determined that upper cheek teeth with long protocones, and lower cheek teeth with U-shaped linguaflexids, were present low in Anza-Borrego strata; these specimens were interpreted to represent early records of *Equus* (*Equus*). This interpretation must be viewed with some caution, because other subgenera than *Equus* (*Equus*) also exhibit similar dental features, but it is intriguing and suggestive.

cf. *Dinohippus* sp. (Pliocene horse)

The equid genus *Dinohippus* is generally considered immediately ancestral to *Equus*. Until the 1950s, most paleontologists believed that *Equus* was derived from an earlier form, *Pliohippus*, which had become something of a wastebasket taxon after its initial description by the renowned vertebrate paleontologist Othniel C. Marsh in 1874. Any specimens (particularly teeth) considered more advanced than earlier forms, but not as advanced as *Equus*, were consigned therein, and the genus came to include so many distinctly different equids that substantial confusion ensued. Paleontologist John Lance, in 1950, determined that some "pliohippines" were more like *Equus* than others, and in 1955, J.H. Quinn erected the genus *Dinohippus* to accommodate them, which resolved at least some of the confusion.

Figure 15.5
Dinohippus.
This herd of Pliocene horses
forages on the ancient Colorado
River delta (see Paleolandscape 2).
(Picture by John Francis)

The geologically oldest equid fossils known from Anza-Borrego are a right upper last molar from the Deguynos Formation (early Pliocene), three isolated upper molars from the "Imperial Group - Palm Spring Group transitional formations" (early Pliocene), and a partial mandible (jawbone) (Figures 15.4, 15.6) from the Hueso Formation (middle to later Pliocene) (Downs and Miller, 1994). They have been assigned to cf. *Dinohippus* sp. (see Paleolandscape 2,

Figure 15.5), based on primitive features of the specimens, compared with the younger genus *Equus*. For the cheek teeth, these include small tooth size, low height of the tooth crowns, simple enamel patterns, and small, rounded protocones (Downs and Miller, 1994; Lundelius et al., 1987).

Primitive features of the partial mandible from the Hueso (Figure 15.6) are more equivocal. Some aspects of the incisor infundibular "cups" and the molar ectoflex- ids are also seen in later equids, such as *Equus* (*Plesippus*) *sim- plicidens* (see next sec- tion). Furthermore, Downs and Miller's measurements of the mandible, and the ratios derived from them, were basically only estimates, because the specimen is incomplete. They were understandably cautious in their taxonomic assignment of the fossil mandible, and the present review emphasizes its uncertain status. More detailed analyses, currently under- way, should help resolve the identification of this important specimen. This is no trivial matter – this particular fossil has been interpreted to represent possibly the youngest occurrence of *Dinohippus* in North America (Downs and Miller, 1994; Cassiliano, 1994, 1999; Bell et al., 2004).

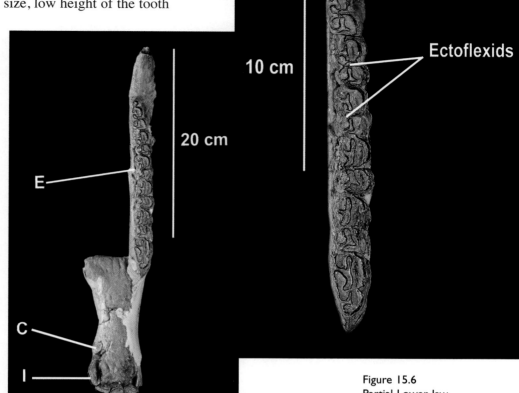

Figure 15.6
Partial Lower Jaw
Assigned to cf. *Dinohippus*.
The front of the mouth is towards the bottom of the image occlusal view.
"I" = incisor;
"C" = canine;
"E" = molar ectoflexid
(ABDSP[IVCM]537/V1873).
(Photographs by Barbara Marrs)

Another fossil, a partial maxilla (upper jaw) with associated teeth, has also been referred tentatively to *Dinohippus,* but has not yet been formally described. This specimen was recovered from the middle to later Pliocene Arroyo Diablo Formation, dating to 3.8–3.6 million years ago. It has features reminiscent of *Dinohippus,* such as short, curved cheek teeth, short, rounded protocones, and a slight depression in the facial bone just in front of the orbit (eye socket), termed the preorbital facial fossa. However, the maxilla is from an older male individual, so the shortness of the teeth and the preorbital depression may be effects of old age rather than phylogenetic hallmarks (features associat- ed with evolution of the taxon as distinct from "ontogenetic" development of the individual). The observed morphology resembles *Equus simplicidens* as well as *Dinohippus,* and a definite assignment cannot yet be made.

Equus (Plesippus) sp. cf. *E. (P.) simplicidens* (American zebra-like equid)

The species *Equus simplicidens* (Cope, 1892) was first named by famed paleontologist Edward Drinker Cope, based upon fossils from Mount Blanco, Texas. The animal represented by those fossils was relatively large, not as sizeable as modern domestic horses, but larger than *Dinohippus* and present-day burros (*Equus asinus*). In many ways, *E. simplicidens* was more primitive than horses as we know them today. For example, the species retained vestiges of the preorbital facial fossa seen in *Dinohippus* and other primitive equids, which is not present in most later Pleistocene and extant horses. In addition, the deeply penetrating ectoflexids of the lower molars present in *Dinohippus* were retained by *E. simplicidens* (and indeed are still exhibited by present-day zebras). However, in other ways, *E. simplicidens* was more advanced than earlier equids. For instance, in addition to details of tooth morphology and placement, *E. simplicidens* had straighter, taller, more hypsodont cheek teeth than *Dinohippus*. As stated previously, hypsodont cheek teeth are an evolutionary adaptation to grazing tough and gritty plant matter.

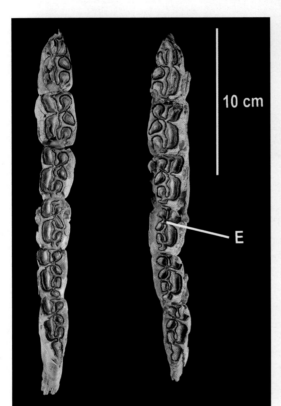

Figure 15.7
Cheek Teeth of *Equus simplicidens*. The front of the mouth is towards the top of the image occlusal view. "E" = molar ectoflexid (ABDSP[IVCM] 790/V2672, ABDSP[IVCM] 790/V2673). (Photograph by Barbara Marrs)

A feature of particular interest is the intermediate tubercle (a bony knob) near the articulation of the humerus with the scapula, or shoulder blade, of *Equus simplicidens*. This tubercle, also present in living equids, reinforces the tendon of the *biceps brachii* muscle and prevents extreme flexion, or bending, of the shoulder joint during stance. Thus the "stay" apparatus of modern horses – the complex arrangement of ligaments, muscles (including the *biceps brachii*), and deep fascia that enables horses to stand for long periods of time, even while dozing – was already fully developed in *E. simplicidens* approximately 3.5 million years ago (Hermanson and MacFadden, 1992).

Three fossils resembling *Equus simplicidens* have been identified from Anza-Borrego. The partial mandible assigned to cf. *Dinohippus* sp. by Downs and Miller (1994) and described above, could equally reasonably be referred to *Equus* sp. cf. *E. simplicidens*, and is roughly contemporaneous with other fossils of *Equus* sp. cf. *E. simplicidens* from Anza-Borrego. Another specimen, an excellently preserved skull with associated lower cheek teeth (Figures 15.7, 15.8), is from the lower Hueso Formation, and is late Pliocene in age. Additionally, a partial skull, probably female, also assigned to *Equus* sp. cf. *E. simplicidens,* was recovered from the Arroyo Diablo Formation, and is mid-Pliocene in age.

The anatomical features employed by Downs and Miller (1994) to determine the specific affinities of *Equus simplicidens* included numerous discrete characters of the skull, mandible, and teeth, as well as ratios derived from measurements of these elements. The cranial specimens resemble *Equus simplicidens* in having relatively elongate palates. Additionally, the skull from the

Hueso Formation exhibits the following features observed in
Equus simplicidens:

1.   the overall size of the specimen

2.   the breadth of the frontal bone relative to skull length

3.   the strong flexion of the basicranium (the base of the skull below
the braincase)

4.   the apparent presence of the preorbital facial fossa

5.   the tightly-folded appearance of the mastoid (behind the ear) and
paramastoid (next to the mastoid) bones when viewed from the side
of the skull

6.   the posterior position of the orbit relative to the last molar in the
upper cheek tooth row. (Downs and Miller, 1994).

Figure 15.8
Cranium of *Equus simplicidens*.
The front of the skull is
to the right, lateral view
above, occlusal view below
(ABDSP[IVCM]790/V2673.05).
(Photographs by Barbara Marrs)

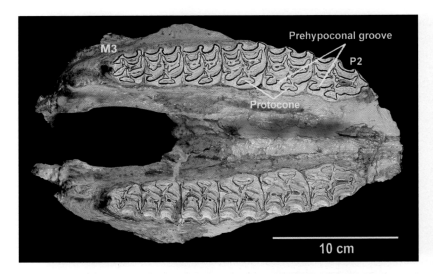

Figure 15.9
Holotype Upper Jaw
of *Equus enormis*.
The front of the jaw is towards the right of the picture occlusal view. "P" = protocone. ABDSP[IVCM]15/V32.01). (Photograph by Barbara Marrs)

Figure 15.10
Holotype Lower Jaw
of *Equus enormis*.
The front of the jaw is to the right lateral view. (ABDSP[IVCM] 15/V32.02). (Photograph by Barbara Marrs)

But the fossils from the Hueso Formation differ from *Equus simplicidens* in several ways, such as the narrowness of the basicranial bones and the shortness of the rostrum. Of particular interest is the shortness of the ectoflexids of the lower molars, because long molar ectoflexids, that fully penetrate the molar isthmus, are a hallmark of plesippine horses including *Equus simplicidens* and extant zebras (Gazin, 1936; Skinner, 1972; Dalquest, 1975, 1978, 1988; Azzaroli and Voorhies, 1993). The absence of this key feature in the lower Hueso Formation fossil cannot be attributed to ontogenetic age-based wear, as the specimen does not show an advanced degree of maturity and the enamel patterns of the molars have not degraded. It might be ascribed to individual variation in ectoflexid morphology, but that is conjectural at present.

Equus enormis (giant zebra-like equid)

A most significant find from Anza-Borrego, *Equus enormis* (Downs and Miller, 1994), was initially considered to be plesippine (zebra-like) in nature. The holotype specimen, a partial cranium with associated right and left dentaries and assorted postcranial material, was recovered from the late Pliocene part of the Hueso Formation. The new species was named for its large size, and indeed, the cheek teeth of this fossil are massive (Figures 15.9, 15.10). However, metapodial measurements showed that *E. enormis* was really no larger than (and in some cases smaller than) other extinct horse species from North America. Other fossils assigned to the new species, including skulls, teeth and jaws and a pelvis, were found in both the Hueso Formation and the Ocotillo Conglomerate, implying an age for *E. enormis* of between 2.1 million and approximately 500,000 years.

In spite of its early assessment as a plesippine, close examination of *Equus enormis* fossils shows some features more derived than those of plesippines, particularly in the holotype specimen. The protocones of its upper premolars closely resemble those of *E. scotti* (see next section) and other Pleistocene horses. The degree of ectoflexid penetration in the lower molars resembles some specimens of the more advanced *E. scotti* rather than those of plesippines such as *E. simplicidens*. In the holotype of *E. enormis,* the ectoflexids barely enter the molar isthmus at all – a distinctly un-plesippine condition – and this cannot be attributed to the effects of variable wear on the teeth. Only one of the fossils referred by Downs and Miller (1994) to *E. enormis* actually exhibits plesippine dental features. The point is not to challenge the taxonomic validity of *E. enormis,* but to consider whether the species is truly plesippine.

The assignment of some fossil specimens from Anza-Borrego Desert State Park to this species is based on the large size, shared dental characteristics, and other features such as the relative length and breadth of the rostrum. However, some question remains as to whether all specimens are correctly assigned. The species, *Equus idahoensis* Merriam, 1918, approaches *E. enormis* in size and exhibits both V-shaped linguaflexids and molar ectoflexids that penetrate the isthmus – characteristic plesippine dental features. Although *E. idahoensis* is not included in the Anza-Borrego assemblages, two specimens recovered locally, from the Indio Hills at the north end of the Salton Trough, are so identified (Azzaroli and Voorhees, 1993). These finds are of particular significance because they occur in depositionally similar sediments of the same age as the Ocotillo Conglomerate.

Equus scotti (extinct large horse, previously reported as *Equus bautistensis*)

Paleontologist J. W. Gidley initially named *Equus scotti* in 1900, based upon fossils from Rock Creek, Texas. *E. scotti*, larger and more robust than *E. simplicidens*, but somewhat smaller than *E. enormis,* was in life approximately the size of the smaller breeds of living domestic horses. Like *E. simplicidens, E. enormis,* and living zebras, *E. scotti* possessed a large head relative to its overall body size. Some morphological features of *E. scotti* are similar to those of *E. simplicidens,* while others more closely resemble *E. enormis*. The lower third incisors of *E. scotti* are open posteriorly, as in both *E. simplicidens* and *E. enormis,* but the preorbital facial fossa of the plesippine equids is generally lost in *E. scotti*. Further, the protocones of its upper cheek teeth are elongate, while the ectoflexids of the lower molars do not fully penetrate the isthmus – features similar to those observed in *E. enormis*. The linguaflexids are variable, often broadly U-shaped, but occasionally V-shaped (Dalquest and Schultz, 1992).

Technically, *Equus scotti* has not been reported from Anza-Borrego Desert State Park prior to this review, although Winans (1989) referred all Vallecito Creek equid fossils to an *E. scotti* "species group" – an amalgamation of fossils of large, stout-limbed North American equids dating from the latest Pliocene through the early Pleistocene Epochs. Since this grouping potentially encompassed more than one species, this cannot be interpreted as an assignment to *E. scotti* in the strict sense.

Scott (1998) reported on equid fossils from late Pliocene – early/middle Pleistocene sediments in Murrieta and Temecula, in Riverside County, to the northwest of Anza-Borrego, which resembled the holotype specimens of *E. bautistensis* as defined by Frick (1921). These fossils were both more numerous and more complete anatomically than the original holotype specimens from Bautista Creek, permitting a more detailed description of *E. bautistensis*. As noted by Scott (1998), this more complete diagnosis very closely resembled published descriptions of the Plio-Pleistocene species *E. scotti*. On this basis, Scott (1998) placed *E. bautistensis* under *E. scotti* as a junior synonym.

A skull and dentary of an elderly individual of large horse, assigned by Downs and Miller (1994) to *"Equus (Equus)* species B,"* shares several characteristics with fossils of *Equus scotti*, including large body size, a short, broad, rostrum, and elongate protocones in the upper cheek teeth (although these latter are very worn). The dentary also shares several characters with *Equus scotti*, including a horizontal ramus that is deepest below the lower fourth premolar, broad, U-shaped linguaflexids in the lower cheek teeth, and molar ectoflexids that enter but do not fully penetrate the molar isthmus. However, given the advanced ontogenetic age of this individual and the lack of diagnostic incisors, more detailed study is needed before this skull and dentary can be confidently assigned to *Equus scotti*.

Whether other fossils recovered from Anza-Borrego may represent *E. scotti* is a subject for future investigations. Cassiliano (1999) documented the presence of numerous large equid fossil teeth with U-shaped linguaflexids at Anza-Borrego. Following Forsten (1986), he assigned these specimens to the subgenus *Equus (Equus)*. While this determination should remain tentative, because other subgenera than *Equus (Equus)* also exhibit similar dental features, the presence of equid teeth with this morphology suggests that *E. scotti* may have been more common in Anza-Borrego than was previously thought.

cf. *Hippidion* sp.

In 1979, paleontologists Bruce MacFadden and Morris Skinner reported on remains of a fossil jaw fragment from the Hueso Formation in Anza-Borrego, dating to the latest Pliocene or early Pleistocene, that closely resembled a very rare equid, *Hippidion*. Prior to their report, *Hippidion* was considered to have been an exclusively South American equid genus. The description of this specimen, an incomplete left ramus with fragments of three cheek teeth, along with the identification of a late Miocene specimen of *Hippidion* from Texas and the description of a related form, *Onohippidium*, from the early Pliocene of Arizona, confirmed that these presumed South American "endemics" actually diversified in North America prior to their dispersal into South America (MacFadden and Skinner, 1979). The jaw fragment from Anza-Borrego has thus played a critical role in shaping our understanding of equid evolution in the Americas.

Initially, the three fossil teeth were presumed to be the third and fourth premolars and the first molar (MacFadden and Skinner, 1979), but closer examination of the specimen revealed them to be the second, third, and fourth premolars. This is an important point, as the key feature used to refer the fossil to cf. *Hippidion* sp. was the depth of penetration exhibited by the ectoflexids of the premolars (MacFadden and Skinner, 1979). In plesippine equids and their descendants, it is the ectoflexids of the molar teeth, not the premolars that are used for identification. But in *Hippidion*, the ectoflexids of the premolars as well as the molars penetrate the isthmuses. The presence of this feature in all three premolars of the fossil jaw, together with the less hypsodont (lower crowned) nature of the cheek teeth, confirms that this specimen closely resembles *Hippidion*.

Hippidion and its close relative *Onohippidium* were one-toed equids, or "hippidiforms," that shared an unusual feature: the deep retraction of the nasal notches. In *Equus* and *Dinohippus,* the nasal notch usually extends to a point above the second or third upper premolar. In *Hippidion* and *Onohippidium,* however, the nasal notch extends much further back, often as deep as the second or even the third molar. Greatly retracted nasal notches do occur in other mammals, such as tapirs (see below), and in these animals the retraction is an evolutionary adaptation to the presence of a proboscis. In such mammals, however, the nasal bones are also retracted towards the back of the skull, while in the hippidiforms, the retracted nasal notch is accompanied by long, forward projecting, unreduced nasal "splints." The anterior position of the delicate nasal splints in *Hippidion* and *Onohippidium* would in fact have restricted the movement of a proboscis. It thus seems unlikely that either *Hippidion* or *Onohippidium* had a tapir-like proboscis, although the function of this unusual morphology cannot easily be explained (MacFadden and Skinner, 1979). Maybe it just had a big fat nose (G.T. Jefferson, personal communication, 2004).

The presence of both *Hippidion* and *Onohippidium* in late Miocene rocks of Texas and Arizona, respectively (MacFadden and Skinner, 1979), suggest that the hippidiforms had diverged by then from the line leading to present day equids, probably evolving from an advanced species of *Dinohippus* (MacFadden, 1992). Between 4 and 3 million years ago, North America connected via the Panamanian isthmus to South America, and the latter continent ceased to be isolated. Great numbers of North American mammals, including equids, flooded into South America, while South American endemics such as opossums and ground sloths headed northwards along the same narrow corridor in what has been termed the "Great American Biotic Interchange" (see McDonald, this volume, *Anza-Borrego and the Great American Biotic Interchange*). The hippidiform equids were evidently part of this interchange. They diversified quickly upon reaching the southern continent, and, within about 2 million years, eight named species had evolved. True *Equus* appears to have arrived in South America some time later (MacFadden, 1992), but hippidiforms found the widespread South American grasslands well suited to their needs, and they prospered there.

While the hippidiforms enjoyed a small but successful adaptive radiation in South America, they disappeared from the fossil record of North America altogether, with one exception – the jaw fragment from Anza-Borrego. Given the absence of North American fossil hippidiform remains between the end of the Miocene and the early Pleistocene, the fossil of cf. *Hippidion* sp. from Anza-Borrego has been interpreted to represent evidence of a "back-dispersal" from South America back into North America (MacFadden, 1992). Additional, more complete remains of hippidiform equids, if such can be discovered in the Anza-Borrego Desert region, would be enormously helpful in shining new light on this subject.

Tapirs: "Evolution Inaction"

Unlike horses, tapirs are evolutionarily very conservative; they can be considered effectively "living fossils." True tapirs – members of the family Tapiridae – first appeared in the fossil record in either the middle Eocene Epoch (Hanson, 1996; Colbert and Schoch, 1998) or in the early Oligocene Epoch (Prothero and Schoch, 2002), depending upon how one defines an early tapir-like animal known as *Protapirus*. This genus is known from early Oligocene rocks in South Dakota, as well as later Oligocene and early Miocene sediments in Oregon; middle Eocene fossils from Texas and Oregon may also fall in this taxon (Hanson, 1996; Colbert and Schoch, 1998). Yet, although *Protapirus* is tens of millions of years old, this animal would – aside from its small size – appear indistinguishable from living tapirs to a casual observer (Prothero and Schoch, 2002).

The cheek teeth of tapirs have undergone only a very few slight changes in anatomy through the millennia, with the premolar teeth becoming increasingly molar-like. By and large, however, tapir teeth have retained their primitive characteristics. Other evolutionary changes include a general increase in size and some refinement of the proboscis (Colbert and Schoch, 1998). There is so little difference between Pleistocene North American tapirs and their extant cousins (living in Central and South America as well as southeastern Asia and its adjoining islands) that they all belong to the same genus, *Tapirus*.

15.11
Tapirus merriami.
Merriam's tapir and her foal feed along the shore of ancient Lake Borrego. (Picture by John Francis)

Pleistocene North American tapirs come in two sizes, large and small. In the eastern U.S., the larger tapirs are usually assigned to the species *Tapirus haysii,* while the smaller forms are referred to the species *T. veroensis.* In the southwestern U.S., the larger forms are commonly assigned to the species *T. merriami* while the smaller animals are often referred to *T. californicus.* Of these latter species, the most complete specimen yet known of *T. merriami* is from Anza-Borrego (see Paleolandscape 3, Figure 15.11). This specimen consists of an incomplete, partially crushed skull with teeth, a partial mandible with teeth, and the partially articulated axial hindquarters of a juvenile tapir (Jefferson, 1989). The size of the teeth indicates that this is a large tapir, as large as or larger than *Tapirus haysii*. In fact, these two species are so closely approximate in size that some scientists have proposed that *T. merriami* was a junior synonym of *T. haysii* (Ray and Sanders, 1984). But there are differences between these two taxa (Jefferson, 1989), albeit these are not pronounced.

Of particular interest is the degree of molarization of the anterior premolar teeth. In *T. merriami*, the anterior second premolar exhibits a protoloph (a transverse ridge across the tooth) that is not as fully developed as in specimens of *T. haysii* (including the holotype of the nominal species *T. copei*, (Jefferson, 1989). In *T. haysii*, the protoloph connects with another tooth cusp, termed the paracone, causing the tooth to "square up" when viewed from the chewing surface to more closely resemble the molars.

Other tapir fossils are also known from Anza-Borrego, but they have yet to be fully described and diagnosed. Fossils of *Tapirus merriami* and the smaller *T. californicus*, have been found in other late Pliocene and early Pleistocene localities throughout southern California (Jefferson, 1989).

Significance and Final Thoughts

The preceding review presented descriptions and interpretations of only a small fraction of the equid and tapir fossils recovered from Anza-Borrego. Many more fossils of these animals remain in the Park's collections, awaiting detailed examination and analysis. Yet even from just this brief overview, it is clear that these few fossils have important stories to tell. The equid remains, in particular, reveal clues about the evolution of these animals in western North America. From small zebra-like ancestors (*Dinohippus*), there emerged larger zebras (*Equus simplicidens*) and giant zebras (*Equus enormis* and possibly *Equus idahoensis*) as well as caballine horses (*Equus scotti*) to roam the plains and hillsides of inland southern California during the Ice Ages. Other equids including *Equus* (*Hemionus*) and *Equus* (*Asinus*) may have been present as well. Clearly, the region provided abundant open space, forage, and water in order for these animals to survive, thrive, and evolve. Nor had the North American equids cornered the resource market, as the presence of the South American hippidiform suggests. The fact that equids appear to have been regionally abundant and widespread for millions of years suggests either that environmental conditions were remarkably stable, or that ancient equids were quite adaptable, or both. Although the Pleistocene Epoch is well documented to have been a period of dramatic changes in climate, the way in which these changes were manifested locally remains to be fully elucidated (see Sussman et al., this volume, *Paleoclimates and Environmental Change in the Anza-Borrego Desert Region*).

The picture of perissodactyl evolution painted by Anza-Borrego's fossils is not unique. Other southern California localities, including the inland valleys, have yielded similar fossils from the same time period. For example, the San Timoteo Badlands in Riverside County have yielded later Pliocene to middle Pleistocene vertebrate fossils including remains of both *Equus idahoensis* and *E. scotti* (Albright, 2000; listed as *"Plesippus" idahoensis* and *E. bautistensis*, respectively) as well as *Equus* (*Plesippus*) *francescana*, a zebra-like horse similar in many ways to *E. simplicidens* (Frick, 1921; Albright, 2000). Nearby, middle

Pleistocene sediments, the Bautista beds along Bautista Creek, produced fossil remains of both tapirs and horses; the large tapir *Tapirus merriami* was initially named from this region, as was *E. bautistensis* (Frick, 1921). Similarly, late Pliocene through middle Pleistocene deposits in Murrieta and Temecula in Riverside County, have yielded remains of *E. scotti* (as *E. bautistensis*), in association with the small tapir species *T. californicus* (Pajak, 1993; Pajak and others, 1996; Scott, 1998, 1999). Outside of southern California, richly fossiliferous badlands near the small fishing village of Golfo de Santa Clara (nicknamed "El Golfo") in northwestern Sonora, Mexico, have also produced tapir and horse remains. These fossils, currently under study, include both large and small tapirs, similar in size to *T. merriami* and *T. californicus,* respectively (Shaw, 1981; Scott and Shaw, 2000). Equids from this region include a large zebra-like form as well as an animal resembling *E. scotti* and a smaller, less diagnostic species (Scott and Shaw, 2000).

These various sites have yet to match Anza-Borrego in overall scientific significance. Exposures in the San Timoteo Badlands and along Bautista Creek are stingier in yielding complete vertebrate fossils than the fossil-bearing beds of Anza-Borrego, while the sediments of Murrieta and Temecula have been largely paved over. The badlands of "El Golfo" continue to yield beautiful vertebrate remains, but the deposits appear to have a more restricted temporal extent. Only Anza-Borrego still boasts vast expanses of accessible, relatively rich fossiliferous sediments dating from the later Miocene through the late Pleistocene, and so retains immense scientific importance for the richness, diversity, and deep temporal span of its fossil record.

Despite the wealth of equid fossils from Anza-Borrego, however, there remains much that we do not know, and may never know, about these ancient ungulates. Were *Equus simplicidens* and *E. enormis* actually zebras in appearance as well as in their bones and teeth? If so, did they have narrow stripes like living Grevy's zebras (*E. grevyi*), or perhaps broader stripes like those of living Burchell's zebras (*E. burchelli*) . . . or even partial striping, like that of the extinct quagga (*E. quagga*)? We have no way to tell. Did *E. scotti* have a long flowing mane and tail, or the shorter, coarse mane seen in today's wild Przewalski's horse (*E. przewalskii*)? Again, we cannot say. Artists may help us envision these extinct forms, but their reconstructions are necessarily equal parts art, science, and conjecture.

Perhaps more interestingly from a scientific perspective, how did these various species live? Modern equids show two distinct forms of social organization, termed Types I and II (Klingel, 1975, 1979; Janis, 1982; MacFadden, 1992). Type I bands (= category D1 of Janis, 1982) are "characterized by non-territorial coherent family units and stallion groups" (Klingel 1979:24). The mountain zebra (*Equus zebra*), the Burchell's zebra or plains zebra (*E. burchelli*) and the horse (*E. caballus*) all exhibit this type of social system. Type II bands (= category D2 of Janis, 1982), characterized by the territorial behavior of mating stallions, has been observed in the Grevy's zebra (*E. grevyi*), the

African wild ass (*E. asinus*) and the Asiatic hemionine (*E. hemionus*) (Klingel, 1979). Type II organization in extant *Equus* is generally considered to be an adaptation to either xeric, dry environments or constantly-varying ecologic conditions (Klingel, 1979; Rubenstein, 1986; Ginsberg, 1989), so determining the social structure of extinct equids is of interest from both a biological perspective and from a paleoenvironmental standpoint. Yet despite the abundance of equid fossils from Anza-Borrego, it is not yet possible to determine whether the various equid species that once dwelled here were Type I or II equids – or whether these animals exhibited an altogether different social structure.

Another imponderable is how the various equid species that coexisted throughout the southwest, including what is now Anza-Borrego, partitioned their environment (see Webb et al., this volume). Modern wild equids are geographically sparsely distributed, and various species generally do not coexist with one another. Grevy's zebras and Burchell's zebras do share one thin region of overlap in Kenya, but that is the exception to the rule. Yet the story told by the Plio-Pleistocene fossil record of North America is consistently one of multiple species of equids living together in a single region. Understanding how these animals coexisted, sharing some resources and using others more exclusively, will require both more fossils and a more thorough understanding of those fossils already recovered.

16

The Smaller Artiodactyls: Peccaries, Oxen, Deer, and Pronghorns

*Time crumbles things;
everything grows old under the
power of Time and is forgotten
through the lapse of Time.*

Aristotle

Lyndon K. Murray

The Smaller Artiodactyls:
Peccaries, Oxen, Deer, and Pronghorns

View of Borrego Valley from Crawford Overlook on Montezuma Grade. (Photograph courtesy of Colorado Desert District Stout Research Center)

Introduction

The abundant remains of fossil artiodactyls, two-toed ungulates, recovered from the thick stratigraphic sections of Anza-Borrego Desert State Park (ABDSP) belong to five modern families: Tayassuidae (peccaries or javelinas); Camelidae (camels and llamas); Cervidae (elk, deer, moose, caribou); Bovidae (sheep, goats, bison, oxen); and Antilocapridae (pronghorns, or prongbuck). Artiodactyls, as a group, share many skeletal characteristics, such as fused lower limb bones (metapodials), an even number of toes on each foot, lack of upper incisors (except in Tayassuidae and some Camelidae), and three of the five North American groups have bony antlers or horns growing from the frontal bone of the skull. Skeletal similarities among the artiodactyls also manifest themselves in the way bones are preserved. Foot bones and the ends of limb bones constitute a large percentage of artiodactyl specimens recovered from ABDSP. This is due primarily to the density of those osteologic elements (resistance to crushing by trampling) and the compact shape of the foot bones (the small surface area reduces the effects of weathering and minimizes breakage).

The Camelidae are discussed in a separate chapter (see Webb et al., this volume, *Extinct Camels and Llamas of Anza-Borrego*). Also, see discussions of the disproportion of carnivores to herbivores in the fossil record of ABDSP (see Murray, this volume, *The Small Carnivorans*) and the methods used for dating the fossil specimens (see Cassiliano, *Mammalian Biostratigraphy in the Vallecito Creek and Fish Creek Basin;* also Murray; also Remeika, *Dating, Ashes and Magnetics,* this volume).

Peccaries and Javelinas: Family Tayassuidae

The peccary family (Tayassuidae) is closely related to Old World pigs (Family Suidae), having evolved almost entirely in North America. There have never been any indigenous suids in the New World, and peccaries have not been found in the Old World since the Pliocene, about 1.77 Ma (million years ago) of Asia and Africa, and the Miocene, about 5.3 Ma, of Europe (Nowak, 1991; see discussion of Old World Tayassuidae in Janis et al., 1998; and Wright, 1998). There are three extant species of peccaries, the Chacoan peccary (*Catagonus wagneri*) of Paraguay and the collared and white-lipped peccaries (*Tayassu tajacu* and *T. pecari*) found from the southwestern U.S. and Central America (respectively) to southern South America (Mayer and Wetzel, 1986; 1987; Nowak, 1991). Other authors placed the extant peccaries in three monospecific (single species) genera, *Tayassu pecari, Dicotyles tajacu,* and *Catagonus wagneri* (Woodburne, 1968; Wright, 1993). *Catagonus* was long considered an extinct taxon, known only from Pleistocene fossils. In the early 1970s, living animals were observed, studied, and reported (Wetzel et al., 1975). With the exception of ?*Tayassu edensis* from the late Hemphillian Eden beds of southern California (Wright, 1998), fossils of the extant peccary genera have not been reported from North America. However, two other genera of peccaries lived in North America during the Pleistocene and earlier times, *Mylohyus,* found from central Texas to the east coast, and *Platygonus,* found throughout North America, south of the polar glaciers.

Figure 16.1
Platygonus.
Illustrated is a large peccary based on the remains of the Rancholabrean form *Platygonus compressus.* (Drawing by Pat Ortega)

Platygonus

Platygonus was among the first fossils reported from North America in the early 1800s. In the two centuries since that find, more than a dozen species of *Platygonus* were described from fossil sites. Most of these species were ultimately synonymized (names combined) into one of four species by Kurtén and Anderson (1980): Pearce's peccary (*P. pearcei*); Cope's peccary

(*P. bicalcaratus*); Leidy's peccary (*P. vetus*); and the flat-headed peccary (*P. compressus*) (Figure 16.1). *Platygonus pearcei,* a strictly Blancan species may have given rise to *P. vetus,* known from Irvingtonian through early Rancholabrean localities, and *P. bicalcaratus,* a Blancan and Irvingtonian species, may have been the lineal ancestor to *P. compressus* from late Rancholabrean localities (Woodburne, 1968; Kurtén and Anderson, 1980). An early Pliocene, unnamed species from Edson Quarry, Kansas, was described by Wright (1993; 1998).

About 40 specimens in the ABDSP collection are currently recognized as *Platygonus.* The known chronostratigraphic range in the Hueso Formation for *Platygonus* begins shortly after 3.04 Ma and ends not long after 1.77 Ma, implying that the genus was only in the Anza-Borrego Desert area during the middle and late Blancan, possibly extending into the early Irvingtonian (see Bell et al., 2004; Cassiliano, this volume). This locally restricted stratigraphic range is consistent with a recent review of the genus *Platygonus* from Idaho, comparing fossil records west of the continental divide to those in the eastern U.S. (McDonald, 2002). The analysis showed a reduction throughout the Pleistocene in the percentage of western fossil vertebrate localities that yielded peccaries.

The fossil beds of the Anza-Borrego Desert produced *Platygonus* specimens, mostly of a large size, although in very fragmentary condition. The earlier specimens probably belong to *P. pearcei* or *P. bicalcaratus.* Specimens from the latest strata may be *P. vetus,* the largest of the *Platygonus* species.

Artiodactyls as a group are generally herbivorous, with the noted exception of the peccaries. Like their pig relatives, peccaries are known to consume both plant and animal matter. The primary food of extant peccaries is vegetation, including seeds, cactus fruits, berries, leaves, roots and tubers, while they also eat mushrooms, worms, larval and adult insects, small vertebrates, and turtle and perhaps bird eggs. *Catagonus* also eats carrion and small mammals (Mayer and Wetzel, 1986; Nowak, 1991).

Today, peccaries are found throughout the western hemisphere in a variety of habitats from tropical rainforest to savannah, desert, and thorn forests. Peccaries of the North American southwest are found in desert scrub and arid woodlands, using heavy vegetation and boulders for shelter from predators and sun while frequenting local water holes. While peccaries in general remain near open water sources, they are capable of surviving strictly on water obtained from vegetation, especially succulents such as cacti (Mayer and Wetzel, 1987; Nowak, 1991).

The records of several North American fossil sites show *Platygonus* succumbing to natural disasters, such as mudslides or floods, in large groups of individuals, indicating they were a gregarious animal (Finch et al., 1972). Similarly, modern peccaries travel in herds of 2 to over 200 individuals, depending on species and local environmental conditions. Extant peccary species possess 4 prominent canine teeth, or upper and lower "tusks", each measuring between 50 and 75 mm (about 2 to 3 in) in length (Meyer and Wetzel, 1986; 1987). Most peccaries are non-aggressive toward other large animal species

(Hoffmeister, 1986; Davis and Schmidly, 1994), however when individuals are wounded or pursued, the whole herd may counterattack (Nowak, 1991), effectively fending off predators as large as *Felis onca* (jaguar) (Mayer and Wetzel, 1987). This combination of "tusks" and "temperament" makes them a formidable item of prey. In South America only the large felids *F. concolor* (cougar) and jaguar, and the large snake *Boa murina* (boa constrictor) were reported to prey on peccaries (Meyer and Wetzel, 1986; 1987). North American peccary predators include *F. rufus* (bobcat) and *Canis latrans* (coyote) (Hoffmeister, 1986). Blancan age predators of the peccaries from Anza-Borrego may have included *Smilodon* (sabertooth cat), *F. onca* (jaguar), *F. rexroadensis* (the large bobcat-like Rexroad cat), and canids such as *Borophagus diversidens* (bone-eating dog), *Canis armbrusteri* (Armbruster's wolf), *C. lepophagus* (Johnston's coyote) and *C. edwardii* (wolf coyote), as well as the giant carnivorous bird *Aiolornis incredibilis.*

Modern peccaries range in size from the smaller *Tayassu* at 44 to 57.5 cm (17 to 23 in) height at the shoulder and 14 to 30 kg (31 to 66 lb) to the larger *Catagonus* 52 to 69 cm (21 to 27 in) shoulder height and 29.5 to 40 kg (65 to 88 lb) (Nowak, 1991). Mounted skeletons of Rancholabrean age *Platygonus compressus* are around 60 cm (24 in) tall at the shoulder (Peterson, 1914) (Figure 16.1). This is within the size range of the modern Chacoan peccary *Catagonus*. The ABDSP peccaries were generally larger than *P. compressus.*

Sheep, Goats, and Oxen: Family Bovidae

Bovids are extremely rare in the ABDSP fossil beds. The bighorn sheep (*Ovis canadensis*) that are a major attraction for visitors to the Park and are responsible for part of the Park's name ("Borrego" is a Spanish word meaning "sheep"), are not found as fossils in the Park. The oldest reported North American *Ovis* records are from mid-Irvingtonian age localities at Porcupine Cave, Colorado (Barnosky and Rasmussen, 1988; Bell et al., 2004), Lake Manix in the Mojave Desert (Jefferson, 2004), and El Golfo, Sonora, Mexico (Shaw, 1981; Lindsay, 1984). Sheep fossils may yet be discovered in the uppermost mid-Irvingtonian fossiliferous strata of the Park. Similarly, the North American mountain goat *Oreamnos* was also a later arrival to the continent, and currently has no representatives in the ABDSP fossil collection.

Only a few definite ox bones have been recovered from the ABDSP fossil beds; a partial horn core and several lower limb bones. Datable ox fossils were recovered from deposits that range from later than 1.77 Ma to slightly younger than 0.76 Ma (Gensler, 2002). The only known North American bovids that occurred in this time range were the shrub-ox (*Euceratherium*) and Soergel's ox (*Soergelia*). Both genera first appeared in North America around the beginning of the Irvingtonian, 1.3 Ma. *Euceratherium* survived to the end of the Pleistocene, about 11,500 BP (before present) while *Soergelia* died out during the early Rancholabrean (Kurtén and Anderson, 1980). Little is known about *Euceratherium* and *Soergelia,* and although rare, fossils of the two taxa are known throughout the western U.S. and Canada.

The shrub ox, *Euceratherium,* likely inhabited low hills, grazing in a manner similar to modern sheep and goats, but unlike some sheep and goats, did not live high in mountains. The skull construction indicates the male shrub ox used head-butting, probably to establish territoriality or dominance as in modern rams and bison bulls (Kurtén and Anderson, 1980).

Soergelia was about the size of a modern steer (*Bos taurus*), and slightly larger than *Euceratherium,* the latter attaining about four-fifths the size of modern bison (*Bison bison*). Both of the extinct taxa were larger than the living musk ox (*Ovibos moschatus*) (Kurtén and Anderson, 1980) which is around 120 to 150 cm (47 to 59 in) tall at the shoulder and weighs 200 to 410 kg (441 to 904 lb) (Nowak, 1991).

Deer and Their Relatives: Family Cervidae

Figure 16.2
Odocoileus.
Most fossil specimens in the collection are close in size and shape to modern deer.
(Photograph courtesy of Photos.com)

At least two species of deer are present in the fossil collection from ABDSP. Most of the cervid material is very close in size and morphology to modern black-tailed deer or mule deer (Figure 16.2) (*Odocoileus hemionus*), distributed throughout western North America today (Anderson and Wallmo, 1984). There are also specimens of very large deer from Anza-Borrego. These are not nearly large enough to be considered elk (*Cervus*), but are much larger than any other modern North American *Odocoileus.* Some of the first vertebrate fossils described from Anza-Borrego were deer, *O. cascensis* (Frick, 1937). This species was subsequently synonymized with *O. hemionus* (Kurtén and Anderson, 1980). The North American fossil record shows the genus *Odocoileus* present by the early Blancan (Webb, 1998).

ABDSP paleontologist George Miller felt that at least some of the cervid fossils were attributable to *Navahoceros,* the extinct mountain deer (Kurtén and Anderson, 1980). *Navahoceros* has skeletal elements that are intermediate in size between *Odocoileus* and *Cervus,* although it has not been reported from sediments older than about middle Rancholabrean age (Kurtén and Anderson, 1980). All of the datable cervid fossils from Anza-Borrego were collected from localities that range in age from shortly after 3.04 Ma until at least 0.78 Ma (see Cassiliano, this volume). It is possible that an ancestor of this extinct Rancholabrean mountain deer lived in the ABDSP area during Blancan through Irvingtonian time.

Odocoileus

Odocoileus inhabit every North American biome except tundra and southwestern deserts such as the central Mojave Desert (Anderson and Wallmo, 1984). Modern North American deer (*Odocoileus hemionus* and the white-tailed deer *O. virginianus*) are found most often in areas with sufficiently tall vegetation for concealment but open enough for swift maneuvering, such as an open forest or tall discontinuous brushland. They are browsers (Figure 16.3). Extant *Odocoileus* males are generally larger than females and attain a maximum height at the shoulder of about 100 cm (40 in). Mule deer males range in weight from 70 to 150 kg (154 to 331 lb) (Anderson and Wallmo, 1984), while male white-tailed deer range from 90 to 135 kg (198 to 298 lb) (Smith, 1991).

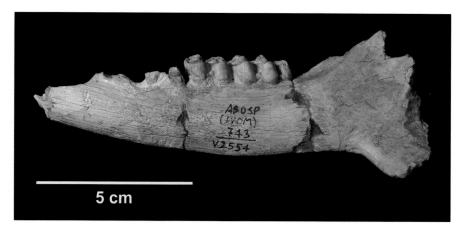

Figure 16.3
Odocoileus.
Deer are browsers with low crowned cheek teeth. Note that the roots are visible at the base of the teeth in this left dentary (ABDSP[IVCM]743/V2554). (Photograph by Barbara Marrs)

One quarter of the 450 catalogued cervid fossils collected from the Anza-Borrego Desert are fragmentary antlers. Mature male deer have branched antlers that are shed and re-grown annually. Antlers are composed of a fast growing bone that arises from the pedicel (a permanent protrusion of the frontal bone, the part of the skull between the eyes). The antlers commence growth between April and May, reach full size by August to September, then become detached from the skull between January and March (Nowak, 1991). Male deer that survive multiple winters may be over-represented in the fossil record due to yearly antler production and shedding. Modern *Odocoileus hemionus* live between 10 and 20 years in the wild (Anderson and Wallmo, 1984; Nowak, 1991). A male deer that dies at 12 years of age may deposit 10 or 11 pairs of antlers throughout its life, in addition to the rest of its bones at death.

Isolated antler fragments can provide important clues to the life history of the individuals that deposited them. Fossil antler material is a positive indicator that the sex of the animal is male. If the base of the antler is present, the season of the year when the antler was deposited may be determined. If the antler base is still firmly fused to the pedicel, the antler was deposited due to the death of the individual. Such an event occurred at some time during the late spring through early fall when the antler was growing and being used for display or battle. If the base is not attached to the pedicel, the antler was shed and therefore was deposited in late winter. If the pedicel is present without the antler attached, the animal probably died in winter. Some features of antlers, such as number of tines or branches may indicate the approximate age of the animal.

Deer predators today include mountain lion, bobcat, coyote, wolf, black bear, and golden eagle (Anderson and Wallmo, 1984). The Pliocene and Pleistocene predators of ABDSP fossil deer may have included the cats *Smilodon, Felis onca,* or *F. rexroadensis,* the dogs *Canis lepophagus, C. edwardii, C. armbrusteri,* or *Borophagus diversidens,* the bears *Tremarctos or Arctodus,* the golden eagle *Aquila chrysaetos,* or the giant carnivorous bird *Aiolornis incredibilis* (see Appendix, Table 3).

Figure 16.4
Capromeryx.
This pair of small antilocaprids forages along a Borrego Valley stream bank, one million years ago. (Picture by John Francis)

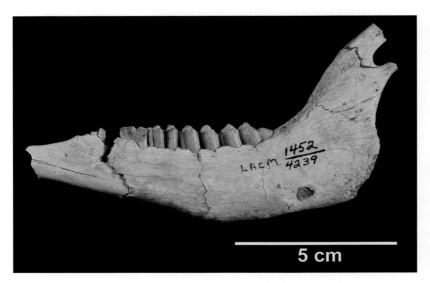

5 cm

Figure 16.5
Tetrameryx.
The antilocaprid *Tetrameryx* has high crowned cheek teeth. Note that the crowns of the teeth continue into the jaw, no roots are visible in this left dentary (see Figure 16.3) (ABDSP[LACM]1452/V4239). (Photograph by Barbara Marrs)

Antilocaprids (Figure 16.4) are an unusual group of artiodactyls, unique to North America. As their family name, Antilocapridae, implies, they appear to be half antelope (*Antilope*) and half goat (*Capra*). When Lewis and Clark noted these animals in their expedition journals (1804-1806), they referred to them variously as "antelopes," "antilopes," and "goats" (Thwaites, 1904). In 1815 they were formally described and named *Antilope americana* Ord 1815, and placed in their own genus three years later as *Antilocapra americana* Ord 1818. The branched horn that gives them the common names "pronghorn" or "prongbuck" is usually found only on the male, while the female typically has a smaller, unbranched horn. The horn core or bony center of the horn is permanent, similar to bovids (bison, sheep, goats) and unlike the deciduous antlers of male cervids (deer, elk, moose, caribou), which are shed and re-grown annually.

Pronghorns do shed and re-grow the keratinous horn sheath annually, a phenomenon not known to occur in any North American cervid or bovid. The *Antilocapra* prong is not bone; it is a keratinous outgrowth produced during the yearly replacement of the horn sheath.

All five subfamilies of bovids have some non-North American members that shed at least part of their horn sheath annually (O'Gara and Matson, 1975). For this reason, and because of similarities in horn structure and development, as well as biochemical studies (Curtain and Fudenberg, 1973), antilocaprids were considered by some authors as a subfamily of Bovidae (O'Gara, 1978). Although subsequent authors maintained that the antilocaprid horn core is more closely related to the cervid pedicel (antler growth platform) than to bovid horn cores, concluding that antilocaprids should remain separate from Bovidae (Solounias, 1988).

The Anza-Borrego fossil pronghorns are represented by at least two, and as many as five genera among: *Capromeryx, Ceratomeryx, Tetrameryx* (Figure 16.5), *Stockoceros* (Figure 16.6), and *Antilocapra.* These taxa are (in part) differentiated by the form of the horn cores. Although the extant Antilocapra horn core consists of a single laterally flattened tine on each side of the skull, most fossil antilocaprids have forked horn cores with at least two or more tines. In *Capromeryx* the forked horn cores of each side of the skull are parallel and project straight up above the orbit (eye socket), with the anterior tine smaller than the posterior (Figure16.8). In *Ceratomeryx, Tetrameryx,* and *Stockoceros* the tines of the horn cores diverge, forming a V-shaped horn on each side of the skull, oriented with one tine pointing more or less anterior and the other more or less posterior. *Ceratomeryx* has a longer anterior than posterior tine, while *Tetrameryx* has a longer posterior than anterior tine, and *Stockoceros* has equal length tines (Figure 16.7). *Capromeryx* species range from about 85% the size of modern *Antilocapra* and smaller. *Tetrameryx* species are known from the size of modern *Antilocapra* up to 20% larger. *Ceratomeryx* is smaller than *Antilocapra,* and *Stockoceros* specimens fall between the size of *Antilocapra* and *Capromeryx* (Kurtén and Anderson, 1980; Janis and Manning, 1998).

The ABDSP pronghorn material ranges in size from about one-half or one-third smaller to about one third larger than equivalent skeletal elements of the modern *Antilocapra americana* (Figure 16.9). Many of these specimens are indistinguishable in size from the modern species. The variable sizes appear consistently throughout the stratigraphic range of the group. This may reflect sexual dimorphism, ontogenetic variation (young and adult), and/or the co-occurrence of multiple taxa.

Ceratomeryx is known only from the Pliocene beds of Hagerman, Idaho. *Capromeryx* appears from middle Blancan through the end of the Pleistocene, from California to Florida (Kurten and Anderson, 1980). *Tetrameryx* has a well documented stratigraphic record as old as the Blancan-Irvingtonian boundary (Kurten and Anderson, 1980), and was reported from other Pliocene localities in California and Arizona (Janis and Manning, 1998). Published records of *Stockoceros* show it appearing in strata of late Irvingtonian age, probably less than 1.0 Ma. All three of these genera became globally extinct at the end of the Pleistocene. *Antilocapra* first appeared in middle Rancholabrean fossil sites, probably less than 0.2 Ma (Kurten and Anderson, 1980), although a Pliocene record of *Antilocapra (Subantilocapra) garciae* was recorded from Florida (Webb, 1973). The latter species was included in the genus *Sphenophalos* by Janis and Manning (1998).

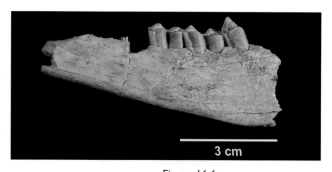

Figure 16.6
Stockoceros.
The antilocaprid, *Stockoceros* has high crowned cheek teeth. Note that the crowns of the teeth continue into the jaw, no roots are visible in this left dentary (ABDSP[IVCM]282/V1047). (Photograph by Barbara Marrs)

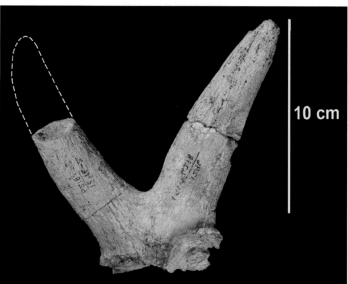

Figure 16.7
Stockoceros or Tetrameryx.
A portion of the orbit is seen below the front margin of the anterior tine of this left horn core, forward is to the right (image reversed for comparison) (ABDSP[LACM]1246/V6218). Although reconstructed, dashed line is approximate, the two tines may have been different lengths and may represent either *Stockoceros* or *Tetrameryx* (see Figures 16.8 and 16.9). (Photograph by Barbara Marrs)

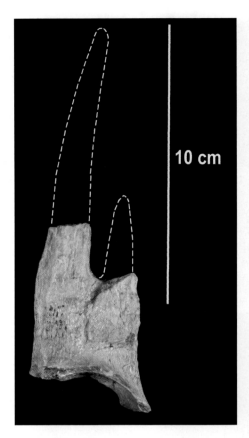

10 cm

Figure 16.8
Capromeryx.
The upper margin of the orbit is at the base of this left horn core, forward is to the right (image reversed for comparison) (ABDSP[LACM]66157/V17615). Dashed lines are an approximate reconstruction (see Figures 16.7 and 16.9). (Photograph by Barbara Marrs)

As with much of the ABDSP fossil material, the antilocaprid specimens consist mostly of isolated, fragmentary osteologic elements. These are very difficult to identify to genus level, much less to a species. Many elements are very close in size and morphology to deer. The earliest datable ABDSP antilocaprids are from sediments deposited just prior to 3.33 Ma and appear to be about the same size as *Antilocapra americana.* The earliest small antilocaprids, noticeably smaller than *Antilocapra,* appeared in sediments between 2.58 and 2.14 Ma. Based on known ages of antilocaprid taxa and general size ranges, the early ABDSP antilocaprids probably include a species of *Tetrameryx* and a large *Capromeryx,* such as *C. arizonensis* (Skinner's pronghorn) or *C. tauntonensis* or *Ceratomeryx.* Later specimens may include *Tetrameryx, Stockoceros,* and smaller *Capromeryx* species, such as *C. furcifer.* A partial mandible from the ABDSP Coyote Badlands (between 0.76 and 0.74 Ma) was identified as cf. *Stockoceros* sp. (Gensler, 2002).

Capromeryx

Capromeryx (Figures 16.8, 16.10) were very small artiodactyls, the smallest species standing less than 60 cm (24 in) at the shoulder, and weighing about 10 kg (22 lb). Throughout the Pliocene and Pleistocene *Capromeryx* evolved into a smaller and smaller animal, culminating in the tiny *Capromeryx minor* known from Rancho La Brea (Janis and Manning, 1998). Between the sizes of living African gazelles (*Gazella*) and the tiny dik-diks (*Madoqua*), *Capromeryx* can be regarded similarly as "everybody's lunch." In order to survive among a multitude of carnivores they may have moved in small groups and hidden within the shelter of trees or other tall vegetation where they foraged for food (see Paleolandscape 4), similar to the small African antelopes (Nowak, 1991). Although large populations of *Capromeryx* at other localities in North America may be indicative of large herds similar to gazelles (Stock and Harris, 1992). The different sized species of *Capromeryx* may have employed different methods of hiding and herding. The ancient habitats of these different species were described variously as plains with copses of trees and shrubs to open, grassy uplands (Kurtén and Anderson, 1980).

Many of the fossil antilocaprid bones and fragments in the ABDSP collection are almost indistinguishable from the complete bones of the modern pronghorn. Antilocapra americana is the fleetest terrestrial mammal in the New World (Nowak, 1991), with estimated top speeds of nearly 87 kilometers per hour (kph) or 54 miles per hour (mph) (Kitchen, 1974). This is rather puzzling from an evolutionary view, since there are no living or recent indigenous carnivorans in North America that can even approach such speeds. The fastest extant New World terrestrial predator is the gray wolf (*Canis lupus*) capable of speeds up to 70 kph (about 43.5 mph) (Nowak, 1991). Assuming the ability to run very fast developed as a response to being chased, rather than from some other unknown cause, the difference in top speeds between prey and possible top predator should probably be very close. The prey only needs to outrun its pursuer by a small amount to be able to reproduce and pass on its genes, while the predator needs to overtake the prey in a reasonably short distance to avoid exhaustion and starvation. Any additional speed for either animal would be unnecessary and an energy drain on its systems. The pronghorn may have developed its speed in response to the presence of the American cheetah-like cat (*Miracinonyx*) during the Pliocene and Pleistocene. A few fossil elements of *Miracinonyx* have been recovered from the Anza-Borrego Desert (see Shaw and Cox, this volume, *The Large Carnivorans*). In many ways *Antilocapra* superficially resembles the Thomson's gazelle (*Gazella thomsoni*) of the African savannah, one of the favorite prey items of African cheetah (*Acinonyx jubatus*). The top speed of *G. thomsoni* is 80 kph (49.7 mph) while reported maximum speeds of *A. jubatus* range from 80 to 112 kph (49.7 to 69.6 mph) (Nowak, 1991). There is currently no way of knowing the top speed of either the fossil antilocaprids or their fastest predators, most likely the American cheetah-like cat, but they were undoubtedly very fast animals.

Antilocapra

Antilocapra (Figure 16.9) also has the ability to maintain high speeds (56 to 72 kph or 35 to 45 mph) over several kilometers without reaching exhaustion, whereas its African relative, the Thomson's gazelle, becomes fully exhausted traveling at 48 to 64 kph (30 to 40 mph) over a similar distance (Kitchen, 1974). The African cheetah can maintain its top speed for only a few hundred meters (Nowak, 1991). Endurance may be an evolutionary response of the North American pronghorn to the extended pursuit by wolves which are able to maintain a very fast pace for at least 20 minutes, or continue fast pursuit over long distances (as much as 20 km or about 12.5 mi) (Nowak, 1991). *Antilocapra* is prepared genetically to escape at blazing speed from an extinct sprinting predator (probably *Miracinonyx*) as well as maintain a high rate of speed over a long distance to escape a slower but more determined extant predator (gray wolf).

Historical records indicate that the southwestern boundary of the range of *Antilocapra* once included the Colorado Desert of southern California. At least four isolated elements of modern *Antilocapra americana* were recovered as surface finds in the Anza-Borrego Desert. These were probably deposited within the last century. The extensive modern North American distribution of *Antilocapra*, throughout the western and plains states, emphasizes its ability to survive in variable climatic conditions. Nevertheless, most of its preferred habitats show a broad physical similarity, primarily open, relatively dry grasslands and brushlands, as well as mesquite and desert environments (O'Gara, 1978). Records of *Antilocapra* in Kansas and Texas show a preference for cactus, wheat grass, forbs, and browse. A study in western Utah showed that *Antilocapra* does not drink water when the moisture content of the available plant material is above 75% (Beale and Smith, 1970).

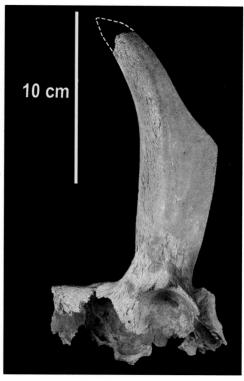

10 cm

Figure 16.9
Antilocapra americana.
The round orbit is immediately below the base of this right horn core, forward is to the right (ABDSP Z622-1-79). Dashed line is an approximate reconstruction (see Figures 16.7 and 16.8). (Photograph by Barbara Marrs)

Figure 16.10
Capromeryx.
This drawing is based on the smallest of the antilocaprids, *Capromeryx minor* from Rancho La Brea.
(Drawing by Pat Ortega)

Plio-Pleistocene antilocaprids had lower-crowned cheek teeth (the chewing and grinding teeth) than modern *Antilocapra*. A high tooth crown (that portion of the tooth above the gum line) is most often seen in grazers, animals that eat grasses and low-lying vegetation. This is, in part, an evolutionary response to the silica phytoliths that form as structural support in grasses. The silica of phytoliths is the same substance as in the quartz grains of beach sand (SiO_2). It is very hard and abrasive. Herbivores that eat grass continuously (grazers) grind their teeth down with this silica plus any of the dirt clinging to roots they may inadvertently ingest. Higher-crowned cheek teeth allow them to live longer before their teeth are too worn to eat. Other large herbivorous mammals are browsers, those animals that primarily consume the green shoots and flowers of bushes and low trees, lacking in silicates and soil. Their teeth therefore do not wear down as fast and are often low crowned as seen in the Plio-Pleistocene antilocaprids. *Antilocapra* is found today in small herds up to about 50 individuals. Based on fossil discoveries of large groups of apparently associated antilocaprids elsewhere, most of the larger ABDSP antilocaprids may have been herding animals.

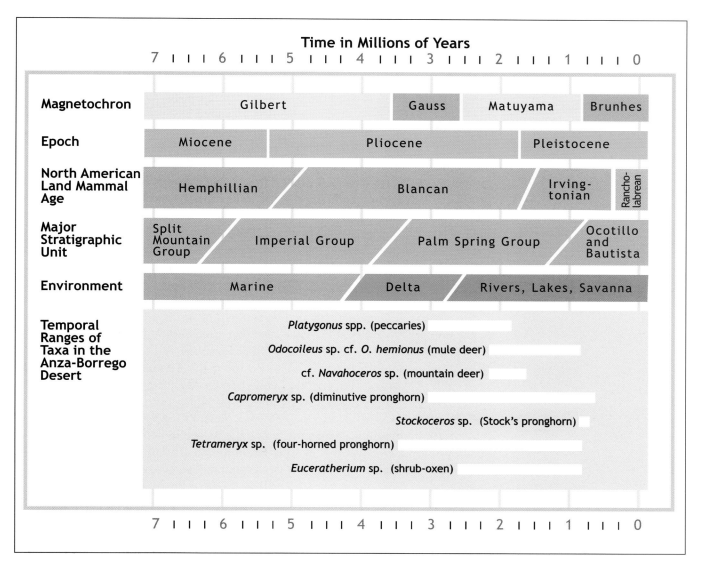

Figure 16.11
The Smaller Artiodactyls
of Anza-Borrego.
Although some of these seven
small artiodactyl taxa exhibit
relatively long temporal ranges,
several of these ranges,
Euceratherium for example, are
based on very few Anza-Borrego
specimens.

History of the Artiodactyls

Primitive forms of Artiodactyla first appeared in North America in the central Great Plains around the beginning of the Eocene Epoch, after about 55.5 Ma (Janis et al., 1998). Tayassuidae first appeared as a recognizable group distinct from all other artiodactyls near the Eocene-Oligocene epochal boundary (around 33.4 Ma) (Wright, 1998). Camelidae first appeared in Middle Eocene sediments (around 45.9 Ma) (Honey et al., 1998). The latter 2 families were indigenous to North America, arising from earlier groups of North American artiodactyls. Although antilocaprids are indigenous to and known only from North America, their immediate ancestors may have emigrated from Asia prior to about 18.8 Ma (Janis and Manning, 1998). Genera of Bovidae and Cervidae first emigrated from Asia in the late Miocene and early Pliocene, after about 8.8 and 5.2 Ma, respectively (Janis et al., 1998; Webb, 1998).

Between their respective first appearances and the late Miocene the North American tayassuids, camelids, and antilocaprids underwent major expansion and diversification, each reaching a maximum during the middle Miocene (roughly between 15 and 10 Ma). From the end of the Miocene (beginning of the Pliocene at 5.2 Ma) the diversity of these 3 groups diminished noticeably. Tayassuids grew from 2 genera in the Oligocene to maintain a minimum of 4 distinct genera at any given time throughout most of the Miocene (Wright, 1998). Camelids expanded from 2 genera to at least 13 contemporaneous genera in middle Miocene and ended the Miocene with at least 4 genera (Honey et al., 1998; Webb et al., this volume). The 10 million years centered on the middle Miocene saw 7 distinct contemporaneous genera of early antilocaprids while most of the final 3.5 million years of the Miocene witnessed a completely different 7 contemporaneous genera (Janis and Manning, 1998). Between the end of the Miocene and the end of the Pleistocene, around 10,500 years ago; tayassuids reduced from at least 4 genera to 2 or 3; camelids went from a minimum of 4 genera to continental extinction in North America; and antilocaprids dropped from at least 6 genera to one genus with only one species (Wright, 1998; Honey et al., 1998; Janis and Manning, 1998). Cervids expanded from one or two Pliocene genera to at least 6 by the late Pleistocene. Most of the latter appeared through multiple immigration events (probably from Asia via the Bering Land Bridge). The modern endemic cervids include 4 genera. North American bovids expanded from one poorly known late Miocene genus to as many as 12 during the late Pleistocene. Most of these arrived via separate immigration events, several never venturing further south than eastern Beringia (modern Alaska and Yukon). Of the Pleistocene North American bovids, 4 genera are extant (Kurtén and Anderson, 1980).

Database records and dating protocols show camelids as the earliest artiodactyls in the ABDSP fossil record, appearing at least as early as 5 Ma. About two million years later antilocaprids showed up, followed soon by tayassuids and cervids. Nearly a million and a half years after that bovids appeared.

Environmental Inferences

With the exception of late immigrants from Asia and South America, most genera of mammals living in North America at the time of European contact were present by the beginning of the Pleistocene. Aside from some overall body size differences, most of the modern genera are indistinguishable from their early Pleistocene ancestors. In addition, many Pliocene genera, although some are distinctly different from the later Pleistocene forms, are recognizable as close relatives to their successors. Because of the similarities between living and fossil skeletal specimens, paleontologists make an educated assumption that the ancestral animals had habitat and dietary requirements very comparable to their modern descendants. This allows us to propose general reconstructions of ancient habitats based on the plants and animals found together as fossils in the ABDSP sediments.

Three of the five families of artiodactyls represented in the Anza-Borrego Desert fossil collection are adapted to environments lacking free-standing water at least part of the year; camels, peccaries, and antilocaprids. The presence of the smaller ABDSP artiodactyls imply the availability of multiple local habitats including: open countryside with sufficient succulent plants for a source of water and occasional low widely spaced shrubs providing visibility and clear routes for swift, high-speed escape from predators, useful for larger, fast antilocaprids (see Paleolandscape 5); abundant, well spaced, tall shrubs or open stands of trees, probably near mountains or canyons, suitable sources of food and shelter for deer and large peccaries; and relatively continuous, thick patches of foliage for hiding and foraging, used by *Platygonus* and the small species of *Capromeryx* (see Paleolandscape 4).

The timing of arrival of the different groups into the Anza-Borrego Desert indicates environmental changes within the area of fossil deposits. During the early tenure of the camels (see Webb et al., this volume) the landscape was probably covered with low dry grasses, taller plants being restricted to stream channels and lakeshores. Camels were not the only large grazing ungulates in the area, as horses have an equally long history in the ABDSP fossil beds (see Scott, this volume, *Extinct Horses and Their Relatives*). Tayassuids, cervids, and antilocaprids arrived when the environment offered more green plants and browse away from lakes and streams to provide the water necessary for their survival in savannah and desert-like environments. Also, brush and scattered trees were abundant enough to provide shelter from predators. The first bovids arrived at ABDSP soon after their first appearance in North America through Beringia, indicating they were able to travel readily to this area, finding sufficient food and water along their route. While the local landscape underwent significant change, it was not drastic enough to eliminate the camels and horses that require large open spaces with ample grasses.

With the influx of the new artiodactyls into ABDSP came the large predators: cats, wolves, and bears (see Shaw and Cox, this volume). None of the large bodied carnivorans (greater than 40 kg, or about 88 lbs) have yet been found in the local record prior to the appearance of tayassuids, cervids, and antilocaprids.

Paleolandscape 5: A Late Spring Mid-Morning 1 Million Years Ago

INDEX

A. Location of Landscape #3

B. Location of Landscape #4

1. *Buteo lineatus* (red-shouldered hawk)
2. *Mammuthus meridionalis* (southern mammoth, extinct)
3. *Mammuthus columbi* (Columbian mammoth, extinct)
4. *Hemiauchenia* sp. (llama, extinct)
5. *Camelops hesternus* (yesterday's camel, extinct)
6. *Canis priscolatrans* (wolf coyote)
7. *Masticophis flagellum* (coachwhip snake)
8. *Callipepla californica* (California quail)
9. *Crotalus* sp. (rattlesnake)
10. *Spermophilus* sp. (ground squirrel)
11. *Equus scotti* (Scott's horse, extinct)
12. *Gambelia corona* (crowned leopard lizard)
13. *Yucca schidigera* (Mojave yucca)

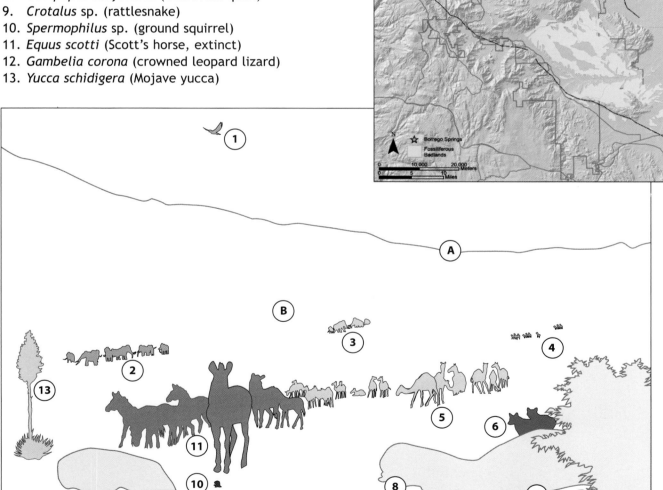

Ancient Borrego Valley supports a brushy savanna, and is flanked by alluvial fans and low granitic mountains to the north. Riparian forests (Landscape 4) follow a braided stream that flows east into Lake Borrego (Landscape 3). *Equus scotti* (Scott's horse) are a common inhabitant of the open grasslands. The view today is to the east from the Visitor Center overlook. Landscape 4 (B) is in the mid distance along the stream, and Landscape 3 (A) is at the lake margin in the distance. (Picture by John Francis)

Extinct Camels and Llamas of Anza-Borrego

17

"In many respects the horses and camels followed parallel courses of evolution, keeping remarkably even pace with each other, and eventually vanishing from this continent about the same time."

William Berryman Scott (1937)

Professor W. B. Scott of Princeton University was one of the world's leading vertebrate paleontologists at the end of the nineteenth and during the first half of the twentieth century. His great monographic series on the White River Fauna of North America and on the Santa Cruz Fauna of South America remain classics to this day. Late in life, he wrote the book cited here which places the entire history of land mammals of the Western Hemisphere in grand perspective.

S. David Webb
Kesler Randall
George T. Jefferson

293

Extinct Camels and Llamas of Anza-Borrego

Carrizo Overlook. (Photograph by Lowell Lindsay)

Introduction

The family Camelidae had its origins in North America during the middle Eocene, about 44 million years ago. Camels went extinct about 11,000 years ago in North America, but are survived by the modern llamas in South America, and camels in Africa and Asia. In the Anza-Borrego paleontological record, camels are the second-most common group, second only to the Family Equidae, horses. But their diversity is greater than that exhibited by the Equidae, with at least eight different forms of camels known. These various members of the family range from the oldest to the youngest Anza-Borrego deposits, from greater than 5 to less than 0.5 million years.

The richest assemblages of large herbivores ever known in North America lived during the mid-Miocene, about 15 to 10 million years ago. Gradually as climates turned more extreme and annual rainfall decreased, especially in rain-shadow regions of the southwest, the diversity of large mammals trailed off, particularly among the browsers, those species that fed on soft leafy vegetation. In general, species that grazed coarse gritty vegetation such as grass were less severely impacted. Indeed some groups of grazers, notably those in branches of the Equidae and Camelidae that developed high-crowned cheek teeth, became more diverse and more abundant during the Pliocene, 5.2 to 2.8 million years ago. Such browsing groups as tapirs, peccaries, and deer became more narrowly confined to stream systems and forests, and eventually many surviving representatives of these groups retreated into subtropical latitudes where climates remained more moderate and green vegetation was available on a year-round basis. The large fossil herbivores of Anza-Borrego record critical phases in the Pliocene and Pleistocene selection for progressive grazers and the decline of browsers, culminating in the late Pleistocene extinction of a majority of the largest mammals.

The modern Anza-Borrego Desert is a treasure trove of camelid fossils and is quite possibly the greatest for any site in North America. Although other Pliocene and Pleistocene localities in California, such as Bautista Creek, Clark Park, Irvington, Lake Manix, Rancho La Brea, San Timoteo, and Tehama yield fossil camelid remains, none has the diversity of Anza-Borrego. Here, the paleofaunal assemblages open a window into Plio-Pleistocene evolution of camelids in North America. Camelid finds include both Tribes (subdivisions of Families) of modern camelids: the Tribe Lamini, represented today by llamas of South America, and the Tribe Camelini, today's camels of the Old World. Together these encompass five genera, including four Lamini and one Camelini. Taxonomically, this constitutes the richest assemblage of Plio-Pleistocene camelids known anywhere, and these finds have helped shape our knowledge of camelid emergence and evolution in North America.

Figure 17.1
Camelus.
Modern camels are very similar to some of the camels and llamas that roamed Anza-Borrego millions of years ago. (Photograph courtesy of Photos.com)

Family Tree

Modern camelids, Tribe Camelini, have been traced to North American stocks that date back some 10 million years into the Clarendonian NALMA (Honey et al., 1998). They evolved from late Eocene ancestral ruminants in North America (Carroll, 1988). Camelids thrived and radiated into 7 subfamilies and some 36 genera. They attained their greatest diversity in the middle Miocene, when as many as 13 contemporaneous genera are known to have lived (Honey et al., 1998). This diversification and radiation was related to the emergence of open grasslands. Camels' highly cursorial or walking behavior and generally grazing nature provided the foundation for them to flourish during the late Cenozoic. The ancestor of modern camels is believed to have reached Asia in the late Miocene, and today is represented by the largely domesticated Bactrian and dromedary camels of Asia and Africa. At least three genera of llama migrated to South America in the early Pleistocene, and survive today as the llamas, vicuñas, alpacas, and guanacos (Honey et al., 1998).

Unique Feet Distinguish Early Camelids

Camelids were among the earliest large herbivores to develop adaptations for living in open country, which helps explain their subsequent success as dry land dwellers. The first evidence of such adaptations are the elongation and consolidation of their feet (Figure 17.2). Camelids evolved with such distinct foot bones that when fossils are discovered they often can be identified to the generic level by a single foot bone.

Four toes on each foot were reduced to two, those articulated to the two most central long bones of the foot, called metapodials III and IV. This increased the weight-bearing properties per bone. These metapodials eventually fused into a single structure called the cannon bone. This change occurred independently in several subfamilies of camelids (Harrison, 1979), as it also did in other ruminants such as cervids and bovids. The reduction of foot bones and elongation of the distal limbs is an adaptive trait for most large cursorial herbivores.

Secondarily, in the camelid stock that led to both Tribes Camelini and Lamini, the foot dropped down onto padded toes, as digits III and IV splayed apart to better support the weight of the animal. Splaying of the toes and a broad pad provided the adaptive ability to walk on soft sediment, like desert sand, and stability on rugged slopes. Camelids have adopted this limb architecture so well that they have some of the longest foot bones in any animal. Some camelids have metapodials so long that they are nearly equal to the length of their upper hind leg bone (femur). This elongation allows for a longer gait and enables them to traverse great distances for food and water.

In addition to distal limb elongation, the pacing gait of camelids has allowed them to be highly successful by increasing their stride length. Most animals exhibit movement of contralateral limb pairs. This is where the diagonally opposite limbs, for example the left fore and right hind limb, move parallel with each other. As the animal walks, on a given side, the forelimb and hind limb come together and then move apart. Camelids utilize ipsilateral limb pairs by

Figure 17.2
Modern Camel Foot.
Note the splayed toes and wide pads for support on soft sandy ground. (Photograph courtesy of Photos.com)

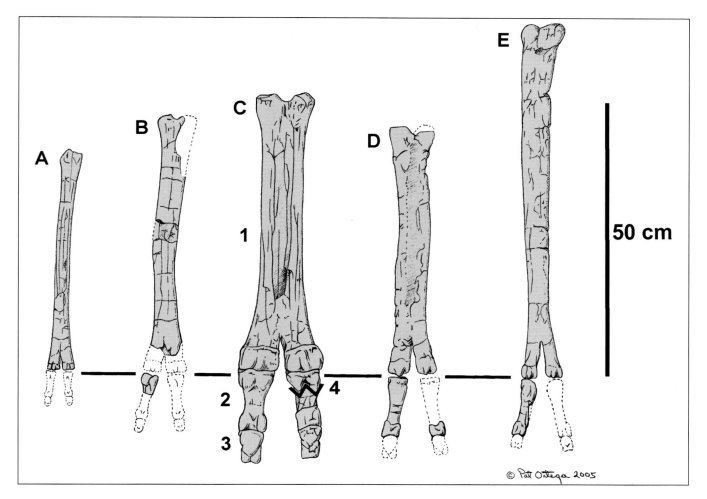

Figure 17.3
Comparative Foot Structure in Extinct Camels.
Note the relative proportions, length versus mid-shaft width, of the different metacarpals. These dimensions reflect adaptions to different terrains. Explanation: A. small species of *Hemiauchenia* sp. (ABDSP[LACM]1599/V6438); B. large species of *Hemiauchenia* sp. (ABDSP[LACM]3842/V64658) (ABDSP[IVCM]282/V1262.03); C. *Gigantocamelus spatula* (ABDSP[IVCM]786/V2657); D. *Camelops* sp. (ABDSP[LACM]1656/V16829) (ABDSP[IVCM]463/6623.01); E. *Blancocamelus meadi* (ABDSP[LACM]1125/V1548); 1. metacarpal; 2. first phalanx; 3. second phalanx; 4. W-shaped ligament attachment. (Drawing by Pat Ortega)

having the fore and hind limbs on the same side move forward and back at the same time. With this arrangement, the front and hind limbs cannot interfere with each other during walking or running and, through natural selection, the limbs can become longer over time (Webb, 1972; Janis et al., 2002). Evolution of the pacing gait has increased the length of the limb elements, most notably the metapodials. This special gait is characteristic of all nearly modern camelids, although there is considerable diversity in limb proportions.

Tribe Lamini

This Tribe has its roots in the late Miocene of North America. In the Pliocene, it is represented by three genera, namely *Alforias, Hemiauchenia,* and *Blancocamelus.* By the early Pleistocene, lamine genera in North America included *Hemiauchenia, Camelops,* and *Paleolama.* Central America may have been inhabited by early representatives of genera that migrated into South America, including *Eulamaops, Vicugna,* and *Lama,* but the fossil record there is poorly known at present.

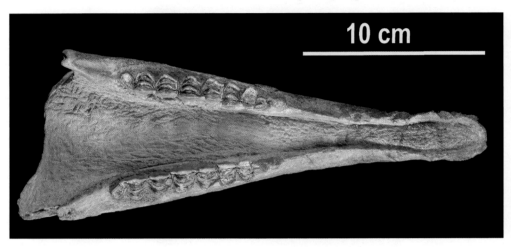

Figure 17.4 A
Hemiauchenia vera.
This small llama mandible ABDSP2340/V6616, (above) occlusal view and (below) right side, was recovered from five million year old Imperial Group near shore marine deposits. (Photograph by Barbara Marrs)

Hemiauchenia

Hemiauchenia is the typical North American llama and probably was ancestral to all other llamas in both North and South America. It most likely migrated to South America and gave rise to the modern llamas, guanacos, alpacas, and vicuñas (Webb, 1974). As indicated by its elongated and slender limbs, *Hemiauchenia* was highly cursorial, with a long stride, and thus could easily travel considerable distances. Like the modern llama, it is believed that *Hemiauchenia* was predominately an open plains animal that fed mostly on grasses.

The genus can be distinguished from other camelids by its metapodial proportions. The foot bones are elongate and narrow, much like in *Blancocamelus.* The suspensory ligament scar, a roughly textured bone surface for the attachment of ligaments that is present on the posterior surface of the proximal end of the proximal phalanx bone, is useful for generic identification, for not only this camelid but also others in the family. This ligament scar is shaped like a "W" and does not extend onto the shaft of the proximal phalanx, but is restricted to the edge of the proximal articular surface of the phalanx (Figure 17.3).

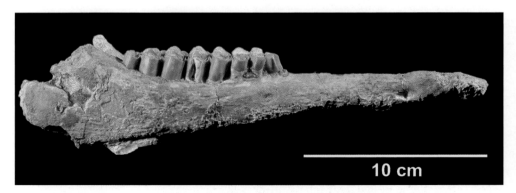

Figure 17.4 B
Hemiauchenia vera.
(Photograph by Barbara Marrs)

Three North American species are known: *Hemiauchenia blancoensis,*
H. macrocephala (Figure 17.5), and *H. vera. Hemiauchenia vera*, is temporally
restricted to the Hemphillian NALMA. It is represented by only a few Anza-
Borrego specimens recovered from the 5 million-year-old shallow marine sandstones of the Latrania Formation, including a complete lower jaw (Figure 17.4).

The presence of *Hemiauchenia* in the Anza-Borrego assemblages from
the Hueso Formation is based on long metapodials (Figure 17.3B) and dental
features. Unfortunately, the diagnostic element that would most distinguish
H. blancoensis from *H. macrocephala,* the lower fourth premolar tooth, has not
yet been found in Anza-Borrego. Both *H. blancoensis* and *H. macrocephala,*
owing to their known geographic and temporal ranges, most likely occur in
Anza-Borrego. However, to further substantiate this, the discovery of more
diagnostic material is needed.

A fourth, yet undescribed, form of *Hemiauchenia* was recently recognized in the Anza-Borrego Desert collection. This new animal is represented
by a very slender metapodial (Figure 17.3A) and small phalanges, smaller than
either *H. macrocephala* or *H. blancoensis.* Features in the lower dentition,
although only slightly smaller than *H. macrocephala*, also distinguish it from
H. blancoensis, H. macrocephala, and *H. vera.*

Camelops

Camelops is a large camelid, about 2.2 m (7 ft) tall at the shoulder in the largest species; some species are approximately 20% larger than the extant dromedary camel. Although it was probably ecologically similar to the modern genus *Camelus*, it is more closely related to the llamas than to the camels (Figure 17.1). However, it differs from other llamas in the increased crown height of the teeth and generally larger size (Webb, 1965; Honey et al., 1998; Harrison, 1979).

Because of its similarity to the modern camel, it is believed that *Camelops* was highly cursorial and a grazer that also occasionally browsed. Like the modern camel, it is adapted for traveling long distances and eating a variety of foods based on their availability. It is uncertain if *Camelops* had the water retention abilities for survival in extremely arid climates similar to that in modern camels. This may have been a trait that developed only later in the camels that migrated to the Old World.

The genus can also be distinguished, like other camelids, by the proportions of its metapodials and ligament scar morphology on the proximal phalanx bone. *Camelops* has metapodials that are stockier than llamas, and the ligament scar on the posterior surface of the proximal phalanx is not W-shaped, and extends 1/3 to 1/2 of the way down the proximal phalanx shaft (Figure 17.3D).

The taxonomy of *Camelops* is complicated, as six species have been named in North America, and a revision of the genus is recommended (Kurtén and Anderson, 1980; Dalquest, 1992). Of the six species known, only two have been reported in California: *C. minidokae* and *C. hesternus*. *Camelops minidokae* is a smaller animal than *C. hesternus*. *Camelops minidokae* is the predominant Irvingtonian NALMA species on the west coast and is present in the Anza-Borrego paleofauna. Other North American species include *C. sulcatus* and *C. traviswhitei*, small and large forms from Texas, and *C. huerfanensis*, a large form, from the western Great Plains.

Camelops hesternus is the best-known species of *Camelops*. It occurs widely throughout central and western North America and is probably most famous for its occurrence at Rancho La Brea, California. Specimens of a large species of *Camelops* (Figure 17.6) have been recovered from the Irvingtonian NALMA sediments of Anza-Borrego. These materials differ from specimens of *C. hesternus* from Rancho La Brea and may represent either a variety of *C. hesternus* or the species *C. huerfanensis*. Further studies are needed to resolve identification of this material.

Camelops minidokae and the large species of *Camelops* most likely occupied different ecological niches as suggested by their size difference. They occur together at Anza-Borrego and in both Irvingtonian and Rancholabrean

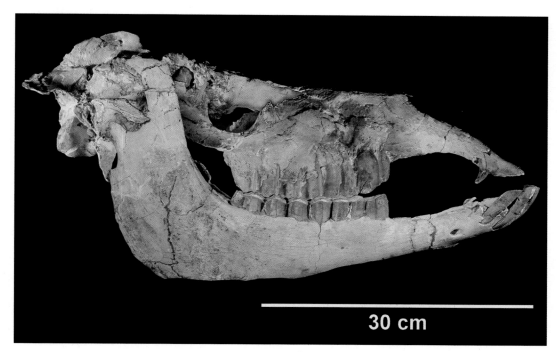

Figure 17.6
Camelops sp.
Skull and mandible of
ABDSP[IVCM]667/V403.
(Photograph by Barbara
Marrs)

NALMA sites in the Mojave Desert, California.
Blancocamelus

Blancocamelus is a monotypic genus, with the single recognized species. It is described from the Blanco local fauna of Texas and is only known from postcranial material. Based on limb proportions and morphology, *Blancocamelus* was probably very similar in general appearance to the llama *Hemiauchenia*, but was approximately 1.5 times larger (Honey et al., 1998). It has extremely long limb bones, and was probably taller than the giant camelids such as *Titanotylopus* and *Gigantocamelus,* with shoulders about 2.3 m (7.5 ft) above the ground.

Blancocamelus had a very long stride and was highly cursorial. Like *Hemiauchenia,* it may have inhabited plains and eaten grasses. *Blancocamelus* from Anza-Borrego is represented by metapodials and associated phalanges (Figure 17.3E) recovered from Blancan NALMA sediments that date between 2.4 and 2.1 million years old. Its affinities to the Texas animal are supported by nearly identical proportions in the metacarpals and metatarsals and the fact that no other

Figure 17.7
Paleolama sp.
The ancient llama.
(Drawing by
Pat Ortega)

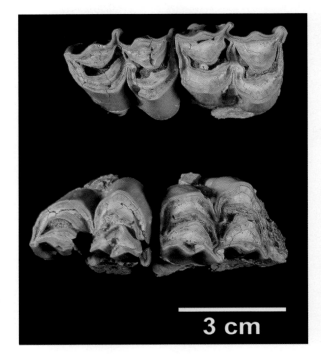

3 cm

Figure 17.8
Paleolama.
Partial upper dentition of ABDSP2204/V6305, occlusal view of right and left
second and third molars. (Photograph by Barbara Marrs)

camelid has limbs this elongate and slender.

Paleolama

Paleolama is unique among camelids, distinguished by
its low crowned teeth and stocky limbs (Figure 17.7). This
animal probably was a browser and ate the leaves of trees
and bushes (Kurtén and Anderson, 1980). The stocky limbs
suggest it was well adapted for rugged terrain (Webb, 1974).
Three species of *Paleolama* are presently recognized: *P. mirifica* from North America, and *P. weddelli* (larger) and *P.
aequatoralis* (smaller) from South America. *Paleolama* from
Anza-Borrego (Figure 17.8) is larger than the known North
American species *P. mirifica*. It is similar in size to the larger

South American species, *P. weddelli,* but could represent a new species. The Anza-Borrego material is from late Blancan and Irvingtonian NALMA deposits that predate the oldest Pleistocene North and South American records (Webb, 1974). Prior to the recent discovery of the Anza-Borrego specimens, it was thought that *Paleolama* and the modern llama originated in the Andes of South America from an ancestral *Hemiauchenia* type animal. Then *Paleolama* presumably migrated to North America during the Irvingtonian NALMA. However, it now appears more likely that this genus may have originated in western North America.

Tribe Camelini

The Camelini originated in the late Miocene of North America. Most students of fossil Camelidae consider *Procamelus* to be the Miocene genus with closest affinities to modern camels. Taxa assigned to this Tribe are generally very large and tend to have longer rostra, Greek for snout or nose, and flatter skulls than the Lamini. The four currently recognized genera, *Megacamelus, Megatylopus, Titanotylopus*, and *Gigantocamelus*, are well known from the Great Plains states, but also occur in the western U.S. Only the *Gigantocamelus*

Figure 17.9
Gigantocameulus spatula.
The extinct giant camel stood over 3m (10 ft) at the shoulder. (Drawing by Pat Ortega)

occurs in Anza-Borrego.
Gigantocamelus

The genus is monotypic, with the single recognized species, *Gigantocamelus spatula* (Figure 17.9). Its presence in the Blancan NALMA aged sediments of Anza-Borrego is a new record for the genus and extends its range into California. The taxon is restricted to the Blancan and has been reported from Colorado, Idaho, Kansas, Nebraska, New Mexico, Oklahoma, South Dakota, and Texas.

Gigantocamelus is most closely related to *Megacamelus,* and these two genera form a group that is related to the modern camel, *Camelus* (Honey et al., 1998). Based on this relationship, it is logical to assume that *Gigantocamelus* may have been biologically similar to the modern camel. Like *Camelus,* it most likely was adapted for open areas and was highly cursorial. The desert adaptive traits of *C. dromedarius* and *C. bactrianus* include slit-like nostrils for keeping out sand, and the ability to eat almost any vegetation available, to drink brackish and salty water, and to drink up to 57 liters (15 gallons) at a time to restore lost body water (Walker, 1968). *Camelus* is believed to have evolved from the common ancestor of *Gigantocamelus* and *Megacamelus* and migrated to Asia in the Miocene. From there it moved through the deserts and steppes of Eurasia and during the middle Pleistocene migrated into north Africa (Gauthier-Pilters and Dagg, 1981).

Gigantocamelus is described as having spatulate incisors and high-crowned molars (Barbour and Schultz, 1939; Harrison, 1985), with metapodials that are long and stocky, and more massive than in *Camelops*. The proximal phalanges are also very stocky and have a suspensory ligament scar that is W-shaped and covers approximately the upper 1/3 of the posterior surface of the phalanx shaft (Voorhies and Corner, 1986; Corner and Voorhies, 1998). This ligament scar shape is very diagnostic (Figure 17.3C) and allows a positive identification of *Gigantocamelus* in the Blancan-aged sediments of Anza-Borrego.

The taxonomy of this Blancan and Irvingtonian NALMA camel has also had a complicated history. Because of many skeletal similarities, including size, for many years *Gigantocamelus* was considered a junior synonym of *Titanotylopus*. Only recently it was resurrected as a valid genus (Harrison, 1985; Corner and Voorhies, 1998). Thus, descriptions in the literature for *Titanotylopus* may include specimens of *Gigantocamelus*. Additionally, the species *Megacamelus merriami* was previously named *Titanotylopus merriami,* so again older generic descriptions, as well as geographic and temporal ranges, may have included *T. merriami*. Because of these taxonomic problems, specimens of giant camelids in many collections simply were referred to as *Titanotylopus* (Kurtén and Anderson, 1980), creating a "garbage bin" generic

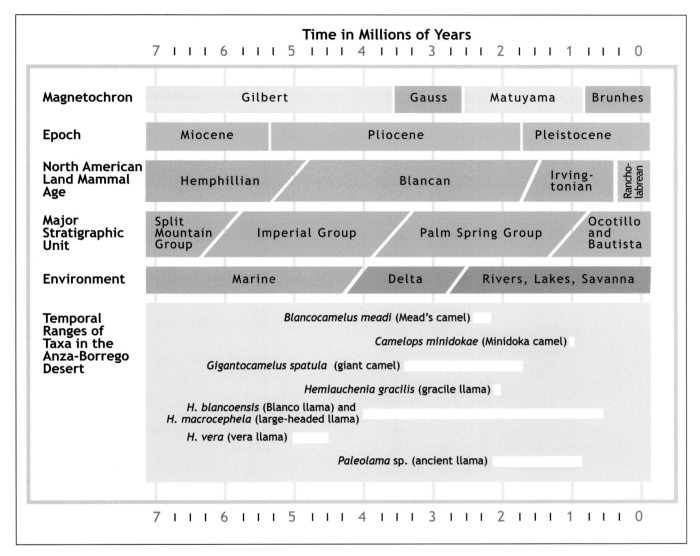

Time in Millions of Years

Magnetochron	Gilbert			Gauss	Matuyama	Brunhes
Epoch	Miocene	Pliocene		Pleistocene
North American Land Mammal Age	Hemphillian	Blancan	Irving-tonian	Rancho-labrean
Major Stratigraphic Unit	Split Mountain Group	Imperial Group	Palm Spring Group	Ocotillo and Bautista
Environment	Marine	Delta	Rivers, Lakes, Savanna

Temporal Ranges of Taxa in the Anza-Borrego Desert

Blancocamelus meadi (Mead's camel)

Camelops minidokae (Minidoka camel)

Gigantocamelus spatula (giant camel)

Hemiauchenia gracilis (gracile llama)

H. blancoensis (Blanco llama) and *H. macrocephela* (large-headed llama)

H. vera (vera llama)

Paleolama sp. (ancient llama)

How Could So Many Similar Types of Camels Live Together?

name. *Titanotylopus* currently is not recognized from Anza-Borrego.

The Anza-Borrego Camelidae include at least eight species (Appendix, Table 3), the richest assemblage of nearly modern camelids known anywhere. The presence of a wide array of camels and llamas also is confirmed by their extensive trackways in muddy deposits of the Anza-Borrego geological section (Remeika, 2001, and this volume, *Fossil Footprints of Anza-Borrego*). This rich and diverse record raises the question of how so many different types of camels and llamas could have lived together within one area. A partial answer is, of course, that these animals did not all live at exactly the same time. Not more than four or five species lived within the same NALMA (see Cassiliano, this volume, *Mammalian Biostratigraphy in the Vallecito Creek-Fish Creek Basin*).

Figure 17.10
Extinct Camels and Llamas of Anza-Borrego.
Continued research will undoubtely add much to our knowledge of the time ranges of the eight identified taxa of extinct Anza-Borrego camels and llamas. Large species of *Camelops* are not included in this chart and two species of *Hemiauchenia* remain combined.

Even the co-occurrence of four species of camelids seems quite remarkable. One is led to postulate that these animals exhibited a number of divergent biological adaptations that allowed them to divide or partition the available food resources. Differences in size and limb proportions probably indicate that they inhabited different ecological habitats within the region.

The means by which numerous sympatric (Latin, literally "together in the same country," or coexisting) herbivore species partition their resources usually involves different feeding modes or strategies. For example, the many kinds of bovids or antelopes in the east African Serengeti ecosystem are readily divided into browsers, grazers, and mixed grazer/browser types of feeders. Furthermore, these animals have specific habitat preferences, such as marshlands, scrublands, grasslands, forested, or open uplands. In addition to obvious differences in body size, there may be morphological clues that signify different feeding and chewing methods. For example the presence of high-crowned, hypsodont cheek teeth is often considered a hallmark of grazing ungulates, although in reality there are many exceptions in which non-grass diets prevail at least seasonally.

Figure 17.11
Modern *Camelus*.
Today's camels have forward-projecting lower incisors, much like those in extinct *Camelops*. (Photograph courtesy of Photos.com)

Most members of the Lamini had short rostra or noses, and rather procumbent, forward-projecting lower incisor teeth (Figure 17.11). These characters suggest that they could regularly feed upon coarse vegetation close to the ground. This does not mean that lamines were strictly "grazers," as they may well have taken in more shrubs than grasses. Among the different taxa of Lamini, *Camelops* and the younger species of *Hemiauchenia* are notable for their larger size and very high-crowned cheek teeth with abundant cementum (cementum, a tough mineral-like substance, is deposited on the outer enamel surfaces and inhibits tooth wear). Such nearly modern features in these two genera suggest that they were among the coarsest feeders in the Anza-Borrego landscape. *Paleolama* most likely falls near the browsing end of the dietary spectrum with its relatively delicate incisors and less-hypsodont cheek teeth.

Members of the Tribe Camelini generally have longer rostra and less procumbent lower incisors than Lamini. They also are large and have extremely high-crowned cheek teeth with heavy cementum. Thus, *Gigantocamelus* and *Titanotylopus* might be seen as near the coarse-feeder end of the dietary spectrum, and potential competitors of *Camelops* and the large species of *Hemiauchenia*. Their very spatulate lower incisors and usually short metapodials are evidently adaptations for coarse grazing and rough or steep terrain respectively.

Two additional methods for determining feeding adaptations, quite independent of morphology, have been applied to some of the advanced camelids. One is simply to examine chewed plant material preserved in the fossettes and fossettids of herbivore cheek teeth. In this manner, Akersten et al. (1988) showed that *Camelops hesternus* from the Rancho La Brea did very little true grazing. It was a mixed feeder, eating mostly shrubs from the southern California coastal landscape. Another independent approach to the diets of extinct herbivores is to analyze the stable carbon isotopes deposited in tooth enamel during growth (see Sussman, et al, this volume, *Paleoclimates and Environmental Change in the Anza-Borrego Desert Region*). The ratio of the carbon isotopes ^{13}C to ^{12}C is directly correlated with the percentage of tropical grasses in the diet of an herbivore. Considerable empirical evidence shows that this method works except at high elevations or latitudes where tropical grasses do not grow. Several such studies have included the extinct *Hemiauchenia*, and generally show that it was not primarily a grazer, but rather a mixed feeder (Feranec, 2003). On the other hand, there is a trend for the larger later species of *Hemiauchenia*, such as *H. macrocephala,* to incorporate more grass in their diet. The same is generally true of modern llamas living on the altiplano of Peru and Bolivia, where other shrubs far outweigh grasses in their diet.

An additional line of evidence regarding the possible dietary spectrum in the diverse Anza-Borrego camelid assemblage is drawn simply from the adaptations of their nearest living relatives. As noted above, living species of the genus *Lama* are generally mixed feeders. This is not strictly a contradiction of their progressive "grazing" morphology with procumbent lower incisors and hypsodont cheek teeth. The living *Vicugna* is the only true grazer, this characteristic being reflected in its extremely hypsodont cheek teeth and ever-growing lower incisors. These are adaptations of a grazing llama, and even then, vicuñas consume much low shrubbery along with the tropical grasses that grow in the Andean altiplano. The North American lamines did not evolve the extreme morphological features that characterize the grazer end of the dietary spectrum seen in *Vicugna*.

The diet of living *Camelus* also sheds light on the diet of the extinct giant camelids, *Gigantocamelus* and *Titanotylopus*. Modern camels, in the arid regions of the Asiatic steppe and the north African desert, do not consume primarily grasses, but a diversity of low shrubs including sagebrush. Presumably, the giant camelids of North America were adapted to similar habitats. Their masticatory system, dentition, and musculature suggest consumption of a similar mixture of coarse low shrubs and grasses. Moreover, as noted above, their relatively short metapodials with broadly splayed toes would have allowed them to pace in sandy and/or rough terrain.

Why Did Camelids Become Extinct
in North America?

One of the most difficult questions for paleontologists to answer concerns the causes of large-scale extinctions. The loss of camelid diversity in southern California, and for that matter in all of North America, is simply shocking. The problem is much the same for the Equidae as for the Camelidae, since both vanished together at the end of the Pleistocene. If biologists had lived in the late Pliocene or early Pleistocene, it is doubtful that they would have predicted the mass extinction of the North American megafauna that occurred about 11,000 years ago. The fact that camels, and also horses, were very successful when reintroduced on the continent for labor and transportation (Figure 17.12) and thrive in feral populations, makes their extinction even more difficult to understand.

Two general causes are widely offered as explanations for these extinctions. One cause is the impact of Paleoindian peoples as they colonized the New World, both directly by hunting and indirectly by environmental modification. The other cause is the disruption of environments produced by rapid climatic shifts at the end of the Ice Ages. One can cite evidence to support each of these cascades of causal events. Certainly, one can draw scenarios of increasing disruption of the elaborate environmental systems that are required to support large and diverse herbivore communities. We are all too familiar with present examples of diminishing populations of endangered large animal species all over the World. A few bad years of climatic crises, multiplied by encroachment and poaching by human populations, are well known to close the noose around wild populations that have already dwindled below their critical numbers. Quite possibly similar events, involving both human impacts and climate changes, had already taken their toll by 11,000 years ago.

Figure 17.12
Camels Thrived When Reintroduced to America's Southwest.

Extinct in North America for 11,000 years, camels were reintroduced during the nineteenth century for a variety of projects based on their famous ability to serve as beasts of burden, battle, and transport in desert regions. Most famous was the establishment in 1856 of the "American Camel Corps" unit of the U.S. Army, a pet project of Jefferson Davis, Secretary of War. They were also used in mining operations in British Columbia, in surveying expeditions in the southwestern U.S., as ranch and farm labor, and in mail delivery. The above illustration, originally published in *Harper's Weekly* in 1877, shows *Camelus dromedarius* in Nevada delivering military supplies from San Antonio to Los Angeles in 1857. The only documented combat operation using camels, however, was a mounted charge on hostile Mojave Indians by a group of threatened civilian laborers (it was successful).

In spite of the creature's extraordinary capacity to work and sustain itself in harsh, arid conditions, none of the camel projects was entirely successful. Some observers noted physical problems, such as their tender feet, well-suited to sand but less adapted for the rocky trails of the West. More cited difficulties managing the animals, who became sulky when separated from the main and could hold a grudge against an insensitive handler. The mere presence of camels occasionally caused horses and mules to stampede. By Civil War times, they were seen as more problematic than useful. As they were decommissioned, shifted around and ultimately sold at auction, many camels wandered off or were released into the desert, establishing feral herds. Occasional sightings of the animals, usually by hunters, were reported through at least the 1930s. Their second extinction on this continent was clearly the result of human activity.

Much of what we know about ancient camels we see not only in skeletal remains but also in their footprints (Figure 17.13; Remeika, this volume). Individual footprints showing the size and proportions of feet can often allow scientists to identify the animals that made them. In addition, a trackway, a series of footprints left by one individual, can give clues to the limb length and gait. Webb (1972) was able to show, based on a set of parallel trackways from the Barstow region in the Mojave Desert, that certain kinds of camelids had already developed the pacing gait during the mid-Miocene. This may have been the beginning of the characteristic gait that carries camels efficiently over soft desert substrates. The next chapter of the Anza-Borrego story shows us in more detail how much we can learn from footprints (ichnites) and what they tell us about the fossils we find.

Figure 17.13
Camelid Footprints.
Two prints pointing in opposite directions are clearly visible in the center. Others, fainter and on the left, point to the top. (Photograph by Paul Remeika)

Fossil Footprints of Anza-Borrego

*A thousand pyramids had
mouldered down
Since on this rock thy footprints
were impressed;
Yet here it stands, though since then
Earth's crust has been upheaved
and fractured oft.*

**Edward Hitchcock
"The Sandstone Bird," 1836**

Paul Remeika

Fossil Footprints of Anza-Borrego

Track-bearing Vertical Rock Strata, Truckhaven Rocks Area, Coachwhip Canyon. (Photograph by Paul Remeika)

Introduction

During a rainy season some three million years ago, on low-lying grass-lands of rolling hills, dales, and marshes to the south of where the Vallecito Mountains now stand, a large group of llamas made their way across damp

Figure 18.1
Ceiling Casts of the Large Llama
Megalamaichnum albus
at Hanging Tracks.
(Photography by Paul Remeika)

mudflats to drink. The green country they roamed is now called the Vallecito-Fish Creek Badlands or Basin. Three million years ago, it was similar to the present-day valley of Lake Henshaw in the mountains some 50km (30 miles) west of Borrego Valley. At that time, streams and rivers of a huge, progressively developing delta, being laid down across the Salton Trough by the ancestral Colorado River, interlaced this basin. The delta sands would eventually become the sandstones of the Palm Spring Formation (Woodring, 1931, see Appendix, Table 4). We do not know whether the llamas drank from a river, a lake, or simply from a shallow depression in the floodplain. We do know they were there together, because they left a hoofprint record of their

Figure 18.2
Pointing Out Llama Footprints
at Camel Ridge.
(Photograph by Lowell Lindsay)

Figure 18.3
Ranger Paul Remeika Interprets Exposed Track-bearing Horizon in Fish Creek Canyon. (Photograph by Lowell Lindsay)

journey in the mud. Most of these hoof-prints belong to two distinguishable footprint species, the Borrego small llama (*Lamaichnum borregoensis*) and White's large llama (*Megalamaichnum albus*). Be aware that the names of these footprint taxa are similar to, but not the same as, the names of the track-makers (see Appendix, Table 3, last page for footprint taxa).

Today that mudflat, hardened into a sedimentary ledge that seems almost turned on its edge, is a showcase for layer upon layer of mud cracks, ripple marks, raindrops, and footprints. Named by the author as "Camel Ridge," it is the first place where researchers documented vertebrate fossil footprints in Anza-Borrego (Collins, 1981; Everett, 1981). Other tracks and trackways in the same late Pliocene horizon were also discovered, including those of the sandpiper bird (*Gruipeda diabloensis*), cheetah-like cat (*Pumaeichnum stouti*), river otter (*Mustelidichnum vallecitoensis*), deer (*Odocoilichnum* sp.), and possibly a peccary (*Tayassuichnum* sp.).

Not far away, but lower in the stratigraphic section and one million years older, in a unique locality in the Fish Creek Badlands called "Hanging Tracks" (Stout and Remeika, 1991), we find abundant hoofprints of the large llama (*Megalamaichnum albus*). These are the oldest known vertebrate tracks in Anza-Borrego. Here they are preserved in a different setting, in marine-deltaic clays littered with fossils of ancient sea creatures – broken sand dollars (*Encope* sp.) and high-spiraled gastropods (*Turritella imperialis*) – along with what appears to have been an early Pliocene (Blancan NALMA) beach of the northern Gulf of California. The soft, moist sands tracked by these robust llamas as they followed the shoreline at low tide are now part of the Yuha Formation of the Imperial Group (Remeika, 1998, see Appendix, Table 4).

Between the Camel Ridge and Hanging Tracks sites, an abundance of well-preserved footprints testifies to the presence of a wide variety of extinct track-makers (Stout et al., 1987; Remeika, 1999a, 2001a, 2001b). At Carnivore

Figure 18.4
Equid Manus (front) Hoofprint. Undistorted well-preserved forefoot mold impression of *Hippipeda downsi* (ABDSP V6283) in clayey siltstone from Fish Creek Canyon. 6" ruler for scale. (Photograph by Paul Remeika)

Figure 18.5
Posterior Oblique View of the
ABDSP 1732 Ceiling Panel.
Note set of canoid tracks
of *Celipus therates* with the
trackway overprint by the
large elephantoid track of
Stegomastadonichnum garbanii
(ABDSP V6292). Also note faint
presence of an isolated, partially-
preserved feloid track of
Pumaeichnum milleri to the left of
the canoid impressions.
(Photograph by Paul Remeika)

Ridge are extinct dog (*Chelipus therates*) and various-sized cat tracks (*Pumaeichnum milleri* and *Pumaeichnum* sp.), preserved in exquisite detail, alongside bird tracks (*Gruipeda diabloensis*), llama hoofprints (*Lamaichnum borregoensis*), and a large gomphothere footprint called *Stegomastodonichnum garbanii*. Another nearby locality, Scavenger Ridge, features a sandstone esplanade replete with the paw prints of this same extinct dog. South of Scavenger Ridge is "The Corral," a site containing a trampled horizon of more than one hundred beautifully preserved single-toed horse hoofprints of *Hippipeda downsi*.

The picture of a group of animals that once roamed the floodplains of Pliocene Anza-Borrego emerges from these many fossil tracks – the Fish Creek Canyon Ichnofauna (Remeika, 1999a) of the Vallecito-Fish Creek Badlands. Similarly recognizable by the footprints in the younger Pleistocene horizons of the more northern Borrego Badlands is the San Felipe Ichnofauna, a younger-track fossil assemblage. Llama, camel, elephant, horse, and dog tracks mark these sediments.

So, what is an "ichnofauna"?

Paleoichnology

The scientific study of fossil footprints and other fossilized traces of animal activity is called paleoichnology, from the Greek words *palaios*, meaning ancient, and *ichnos*, meaning trace. Besides footprints and trace fossils (also known as ichnofossils, or ichnites), included in this study are burrows, nests, tooth and claw marks, trails, gastroliths (gizzard stones), coprolites (fossilized excrement), and skin impressions. A footprint or track, by definition, is the indentation or outline left by a foot in sediment. It is preserved in sedimentary rock in two ways: as a negative relief or mold of the original foot; and as a positive relief or cast if overlying sediments have filled the original impression.

An animal leaves behind only one skeleton when it dies, but most likely will have made hundreds of thousands of footprints during its lifetime. However, not every locale, habitat, climate, and walking surface is equally receptive to footprints, and geologic processes destroy footprints more often than they protect and reinforce them. Only a very small fraction of the number of footprints actually made remain preserved as ichnites in the fossil record, making fossil footprints incredibly rare. Each one can yield significant insight into the behavior of an extinct animal, and give us clues about its size, shape, and movement style, dynamic information that may not be available from bones alone.

The distribution and density of tracks at a site not only indicate how many animals may have inhabited the area, but also tell us about their relative abundance, social behavior, and direction of travel. Features of the sedimentary rock in which we find footprints are clues to the paleoenvironment that these track-makers inhabited. Although bone evidence provides much data, we cannot be sure that animals actually lived and died in the same place where their body fossils are found. Oftentimes, a bone is buried after the body has been scattered by scavengers or redeposited by fluvial processes; but a footprint cannot be reworked and still be preserved. *A footprint found intact, therefore, is an undistorted window into the past. It is direct* in situ *evidence that an animal walked on that very spot.* If the depositional horizon is dated by stratigraphic, isotopic, paleomagnetic, or other methods (see Cassiliano, this volume, *Mammalian Biostratigraphy in the Vallecito Creek-Fish Creek Basin*, also Remeika, *Dating, Ashes and Magnetics*), then the track-maker is dated as well.

While no two footprints are identical, features of shape and size, track morphology, length of stride, and pattern may make it easy to assign the footprint to a given family. A well-preserved footprint may show enough detail to identify the track-maker down to the genus or species level. However, associating this identification with a particular animal species known from its bony fossils is extremely difficult. Therefore, footprints and other ichnofossils are used as the foundation for a classification system based on the preserved traces of their activity instead of by their bony remains. Genus and species when so defined are called "ichnogenera" and "ichnospecies" (see Appendix, Table 3). The entire animal population of a given area, environment, formation, or time span *as revealed by its ichnofossils* is known as an "ichnofauna." Some track-makers can be correlated with known body fossil taxa (generally skeletal data) with reasonable confidence based on coeval (similar period) track data. However, in many cases, there is no supporting skeletal evidence. This is the case *for most of the track-makers in Anza-Borrego, and we know them only by their footprints.*

What Do Footprints Tell Us?

Nearly 150 years ago, the environmentalist John Muir, founder of the Sierra Club, hiked along a sandy roadway in the Sierra Nevada foothills. He looked down at the beveled furrows created by passing wagon wheels and noted, *"A tiny lizard darted into the stubble ahead of me, and I carefully examined the track he made. I was excited with delight in seeing an exquisitely beautiful strip of embroidery about five eighths of an inch wide, drawn out in flowing curves behind him as from a loom."* (Muir, 1911) Soon Muir discovered another track pattern in the roadway. He was able to trace it to a grasshopper. Again he noted, *"I glowed with wild joy as if I had found a new glacier; copied specimens of the previous fabric into my notebook and strode away with my own feet sinking a dull crunch, crunch, crunch in the hot gray sand."* (Muir, 1911)

The pleasure John Muir took in tracking comes clearly across the decades as we read his notes. Observation of wildlife is a great thrill, but direct observation of wildlife is not always possible. Many fascinating track-makers, such as coyote, bobcat, deer, mountain lion, ringtail cat, and raccoon are nocturnal or crepuscular (twilight) in their habits, and they usually avoid direct interaction with humans. Much of what we learn about them is based on the study of their traces, primarily footprints left behind in soft muds, sands, or soils associated with cienegas, arroyos, or dry-lake shorelines where animals may leave marks of their presence or passing. Footprints made on a pebbly alluvium, of course, are not preserved with the same kind of detail as those planted in soft mud. The discovery and documentation of a good-quality footprint or set of footprints in a trackway can be just as exciting and informative as observing the track-maker in its natural habitat.

Direct observation of *extinct* wildlife, of course, is never possible. Therefore, we use the same process that we would use for the close observation and systematic analysis of an unseen modern animal: reconstructing their behavior and establishing possible identification by evaluating fossil footprints and trackways.

Since different types of animals produce distinctive footprints that can be attributed to general classifications, such as dog, cat, or ungulate, it is important to pay close attention to the main structural elements of a footprint. Some of the most important features to observe are the presence, number, and characteristics of toe impressions. Are claws present? What shape are the toes? Are they elongated? Do they splay or diverge outward or converge inward? How are the digital pads shaped? Are there interdigital crest lines between the toes, or webbing? Modern equines, such as horses and zebras, make one-toed (monodactyl) footprints. Multi- and even-toed ungulates such as llamas, camels, and deer make two-toed (bidactyl) impressions. Cats and dogs make four-toed (tetradactyl) prints. Elephants produce five-toed (pentadactyl) footprints, but not all five toes are seen in all the impressions.

Other diagnostic characteristics of footprints include the shape and lobe pattern of the interdigital pad, a round or cloven hoof, and distinctive heel marks. Two or more of these morphologic features in a single footprint are often sufficient to assign the track-maker to a specific ichnogenus. And far more useful than an isolated footprint is a series of footprints that makes up a "trackway," defined as two or more consecutive tracks made by the same animal. We can get an idea of the size, weight, speed, and posture of an animal, and tell whether it walked on two feet (bipedal) or four (quadrupedal). Impressions made by forefoot (manus) and hind foot (pes) may be distinguished in a quadruped trackway, since the forefoot is usually the larger, load-bearing foot. This helps us to analyze the gait pattern, or "locomotor cycle" of the track-maker.

Most quadrupedal animals are diagonal-sequence walkers; they advance the right forefoot and the left hind foot at about the same time, alternating with the left forefoot and right hind foot. This pattern, called a "contralateral gait," is evident in most trackways. But llamas and camels, very important in Anza-Borrego paleontology (see Webb et al., this volume, *Extinct Camels and Llamas of Anza-Borrego*), use a different locomotor cycle, a syncopated "ipsilateral gait" in which both legs on one side move forward at the same time.

The Fish Creek Canyon Ichnofauna

The group of animals that left footprints in the Vallecito-Fish Creek Badlands during the Pliocene epoch is called the Fish Creek Canyon Ichnofauna. It contains at least five distinct taxa of track-makers (Remeika, 2001a, 2001b) including birds, carnivorans, equids, artiodactyls (two-toed ungulates), and proboscideans (elephants and their relatives). For many of the tracks described below, some to the ichnogenus and ichnospecies level, we have identified good candidate track-makers.

Birds

The smallest tracks belong to the ichnospecies *Gruipeda diabloensis* (Figure 18.6). Specimens of *G. diabloensis* consist of small bipedal bird footprints showing four slender, well-defined digits, measuring 1.3 to 1.6 cm (1/2 to 1/3 in), one of which is directed forward, two directed outward, and the fourth pointed or directed to the rear with its axis coinciding with, or at an angle to, that of the middle digit. This fourth digit is shorter than the others are. Digits may be united or separated and taper to a pointed claw tip. Webbing is not evident. The footprints are faintly impressed on what was a moist, subaerially exposed (terrestrial) clay surface. The footprints form a narrow trackway with a short stride, and are deep enough to leave behind a clear outline, retaining significant characteristics representative of the majority of *G. diabloensis* impressions. The uniformity of preservation indicates that the footprints were impressed within a short period. Since there are no mud cracks, the claystone was moist enough and was not exposed to the drying effects of air for a prolonged period prior to burial (Bucheim, et al., 1999). The tracks appear to be made by several individuals of the same species walking in various directions. The size and morphology of *G. diabloensis* closely resemble footprints of the modern least sandpiper (*Calidris minutilla*), which today may be spotted feeding in and around the agricultural wetlands of the Imperial Valley (see Jefferson, this volume, *The Fossil Birds of Anza-Borrego*).

Figure 18.6
Gruipeda diabloensis.
Well-preserved avian natural mold trackway impression in hardened claystone from Camel Ridge (ABDSP V6107). (Photograph by Paul Remeika)

Carnivorans

Well-preserved footprints suggest the presence of at least five carnivorans, including river otter, dog, and three cats of various sizes (Remeika, 2001a). The river otter footprint (*Mustelidichnum vallecitoensis*) features an isolated and undistorted left hind paw impression in what was a moist, subaerially exposed, ripple-bedded clay surface (Figure 18.7). The paw print exhibits five well-developed digits, each with a spheroidal to ovoidal digital pad. The three central digits are parallel to each other; the outer digits are offset and angle slightly outward. Arrangement of the digits forms a single semicircular arc in

Figure 18.7
Mustelidichnum vallecitoenis.
An isolated natural mold impression of a left pes (hindfoot) paw print in hardened claystone from Camel Ridge (ABDSP V6280). (Photograph by Paul Remeika)

front of the interdigital pad. Digital pad impressions are of equal or similar size, with the first digit subordinate to the central digits. Each digit possesses a nonretractile claw that may or may not "register" (make an imprint). There is no separation between claw and digit, and the claw marks are indistinct, giving each digit an elongated, pointed appearance. Webbing is present, impressed between the digits. The interdigital pad is large, deeply lobed, and asymmetric, with an elongated heel. The anterior of the interdigital pad is chevron shaped, longer than wide. The morphological attributes of the track not only resemble those made by an aquatic mustelid, but also represent the first unequivocal musteloid track reported from California. Although any identification is somewhat speculative, the only likely track-maker candidate recognized from the Vallecito-Fish Creek Badlands fossil record is the small-clawed, fish-eating river otter (*Satherium piscinarium*). This otter is well represented in North America during the Pliocene period (see Murray, this volume, *The Small Carnivorans*).

Figure 18.8
Chelipus therates.
Well-preserved ceiling cast set of canoid tracks in sandstone from Fish Creek Canyon. Note larger load-bearing left manus, fore footprint (ABDSP V6276), and its relationship to the smaller left pes (hind) footprint (ABDSP V6277). (Photograph by Paul Remeika)

Footprint structure that exhibits four digits, each with a spheroidal to ovoidal digital pad and the registration of sharp, permanently extended claw marks and a triangular-shaped, deeply lobed interdigital pad, are diagnostic characteristics of the dog family, and distinguish them from cat footprints. Footprints assigned to the ichnospecies *Chelipus therates* are gracile (slender, slight) in size and morphology, barely within the lower limits of the modern juvenile wolf (*Canis lupus*) track dimensions, and differ in no significant way from the modern footprints of a moderate-sized coyote (*Canis latrans*). The fossil footprint impressions at Carnivore Ridge represent an undistorted set of paw prints made by an intermediate-sized dog, and are preserved as ceiling casts on the underside of a massive channel sandstone ridge bedding plane (Figures 18.8, 18.9). Other footprints are also on strike (along a single stratigraphic layer), one of which, a forefoot print, exhibits four claw marks clearly registered in the impressio. On the load-bearing foot, paw prints exhibit four well-developed digits, each with an elongate spheroidal to ovoidal digital pad. Central digits are thick, exhibit bilateral symmetry, and give rise to a prominent nonretractile claw, fully extended. Outer digits angle slightly outward, are also thick, and possess nonretractile claws that may or may not register.

The hind foot is typically narrow or ovoid, with a lack of space between toes and anterior edge. The leading edge tends to be convex. The overall shape of the forefoot and hind foot is rectangular, longer than wide. Footprints are definitely doglike in structure, with fore and hind footprints identical in size and morphology to those of a typical contemporary coyote. Therefore, the candidate track-maker of *Chelipus therates* is a coyote-sized dog. Since the morphological attributes of the fossil footprints are indistinguishable from the extant *Canis latrans*, a plausible candidate track-maker species is the coyote-sized extinct rabbit-eating dog (*C. lepophagus*) (see Murray, this volume). Remains of *C. lepophagus* are reported from the Vallecito-Fish Creek Badlands (Remeika et al., 1995), and this dog is well represented throughout western North America during the Pliocene Epoch. It is the direct ancestor of Pleistocene to Holocene (recent) coyotes (Kurtén and Anderson, 1980).

Figure 18.9
Chelipus therates.
Ceiling cast impression of an isolated track in claystone from Fish Creek Canyon. Note triangular-shaped interdigital pad configuration and presence of claw marks. (Photograph by Paul Remeika)

Three types of cat footprints are represented: a small bobcat-sized footprint (Figure 18.10) (*Pumaeichnum* sp.) discovered at Scavenger Ridge, an intermediate lynx-sized cat (*P. milleri*) at Carnivore Ridge, and a mountain-lion-sized, cheetah-like cat (*P. stouti*) (Remeika, 2001a) at Camel Ridge. Diagnostic feloid (cat-form) footprint characteristics, such as arrangement of the digits and pad impressions, deeply bi-lobed anterior/tri-lobed posterior interdigital heel-pad configuration, and impressions with claws normally retracted and claw tips usually absent except in the case of canid-like feloid tracks (cheetah-like cat), distinguish footprints assigned to the ichnogenus *Pumaeichnum* from typical dog footprints.

Footprint impressions of *Pumaeichnum milleri* occur as an undistorted set of tracks on the underside of a massive channel sandstone strike-ridge bedding plane at Carnivore Ridge. The tracks are preserved in convex hyporelief (underhanging, in raised relief), and represent the right fore and left hind paw prints of a lynx-sized cat. The forefoot is noticeably larger than the hind foot, with an overall forefoot print length of 6.1 cm (2.4 in) and print breadth of 6.3 cm (2.5 in), compared with print length of 4.5 cm (1.8 in) and print breadth of 5.5 (2.2 in) cm for the hind foot. Paw prints exhibit four well-developed digits, each with a spheroidal to ovoidal digital pad. Central digits are parallel and asymmetric; outer digits angle slightly outward. Arrangement of the digits forms a single semicircular arc in front of the interdigital heel pad. Digital pad impressions are of equal or similar size, almost wider than long. Impressions of claw tips are absent. The interdigital pad is large, deeply lobed, and equilateral, especially in the hind foot, with two lobes found anteriorly (toward the front) and three lobes found posteriorly (toward the rear). The overall shape of the fore and hind paw prints is round, or wider than long. Digits of the manus (forefoot) are spread more than in the pes (hind foot); front digits are imprinted more deeply than the small digit, indicating a right foot with principal load-bearing digits. The left hind foot tends to be wider than long. This recognition is significant and strongly suggests a feloid (cat-form) track-maker.

The fossil footprint impressions of *Pumaeichnum milleri* are larger than tracks assigned to *Pumaeichnum* sp. (bobcat sized) and more robust and larger than modern tracks of the bobcat (*Lynx rufus*), yet smaller than the tracks of *P. stouti* (Stout's cheetah-like cat, see below) and modern tracks of the mountain lion (*Felis concolor*). *Pumaeichnum milleri* is intermediate in size between the genera *Lynx* and *Felis*. Though distinctive, the tracks share a resemblance to both lynxes and mountain lions. This is compelling evidence that *F. rexroadensis* is the plausible candidate track-maker. The intermediate lynx-sized *F. rexroadensis* (Stephens, 1959) is well represented throughout western North America during the Pliocene Epoch (Werdelin, 1985), and cranial and postcranial elements of *F. rexroadensis* are reported from the Vallecito Creek Local Fauna (Remeika et al., 1995).

Figure 18.10
Pumaeichnum stouti.
Feloid right manus (front foot) natural mold impression (ABDSP V6285) in hardened claystone from Camel Ridge. Rock hammer for scale. (Photograph by Paul Remeika)

Cat tracks of *Pumaeichnum stouti* (Figure 18.10) are preserved in concave epirelief (impressed in the strata in normal relief), and represent paw prints of a mountain-lion-sized, cheetah-like cat track-maker. The right front paw print is the best preserved, and exhibits four well-developed digits, each with a spheroidal to ovoidal digital pad. Central digits are parallel and asymmetric; outer digits angle slightly outward. Arrangement of the digits forms a single semicircular arc in front of the interdigital heel pad. Digital pad impressions are of equal or similar size, longer than wide, ranging from 1.4 cm (0.6 in) to 2.8 cm (1.1 in) in length and 1.4 cm (0.6 in) to 1.8 cm (0.7 in) in breadth. Claw tips register in the footprint impression, directly joined to the anterior end of each digital pad. There is no separation between claw and pad. The interdigital pad is large, equilateral, and lobate, with two lobes found anteriorly and three lobes posteriorly. The overall shape of the hind foot is round, or wider than long, with an overall print length of 6.8 cm (2.7 in) and overall print breadth of 7.0 cm (2.8 in). The impression indicates a right front foot, similar to *P. milleri* above.

The skeletal fossil record of the area includes at least three large cats (see Shaw and Cox, this volume, *The Large Carnivorans*): The jaguar (*Panthera onca*), a cheetah-like cat (*Miracinonyx inexpectatus*), and a gracile sabertooth cat (*Smilodon gracilis*). However, *Panthera* or *Smilodon* could not have produced the *Pumaeichnum stouti* paw prints because tracks from both the former taxa differ from the *P. stouti* tracks in size, morphology, and diagnostic criteria, as described above. This leaves only the extinct cheetah-like cat as the likely candidate track-maker for *P. stouti*. Extinct mountain-lion-sized, cheetah-like cats inhabited North America during the Pliocene period (Jefferson and Tejada-Flores, 1995). The occurrence at Camel Ridge from 3.4 Ma represents the earliest record of cheetah-like cats in the Pliocene of California. Overall, this is clear evidence for the existence of a new, third feloid ichnospecies for Anza-Borrego (see Appendix, Table 3).

Equines

Even though remains of horses are common throughout the fossil vertebrate assemblages of Anza-Borrego, evidence of their footprints had been noticeably absent. This was problematic because horse remains are the most abundant fossils in the Vallecito-Fish Creek Badlands. Horse (equoid) footprints should be more common than those of llamas (llamoid), which is not the case. Recently, however, we found the first equoid track site in the Colorado Desert at The Corral. (Figure 18.11) The abundance of equoid footprints at this locale was at first glance difficult to grasp. The track formation originally appeared, from a distance, to be physically induced soft-sediment deformation structures, super-ficially resembling a ripple-bedded mudflat deposit. However, closer scrutiny revealed the structures to be a large bedding-plane exposure of closely spaced monodactyl (one-toed) equoid footprints, confirming a biogenic (life-generated) origin. The abundant footprints range in size from juvenile to adult. The distribution and size of footprints throughout this thoroughly trampled stratigraphic horizon is a major resource, yet untapped, that should eventually provide significant sedimentological and paleoecological information about the age, structure, and size distribution of the animals. Adjacent beds, above and below, lack equoid footprints.

Figure 18.11
Equoid Tracks at the Corral.
The prevalence, size, and morphology of the footprint impressions strongly suggest that equids are the responsible track makers at The Corral; 6"ruler for scale.
(Photograph by Paul Remeika)

These *Hippipeda downsi* tracks were preserved as natural mold impressions in relatively moist but compactable sands prior to hardening. They are ovoid to rectangular in shape, distinguished by a single central digital pad impression. There is no evidence or impressions of lateral digits (side toes). The digital pad (sole) is bordered anteriorly by a strongly curved digit and hoof wall. The sole is well pronounced, concaved downward to the bar (the support structure of the hoof wall that extends from the heel and ends at the point of the frog), defining the posterior ends of the pad. The pad surrounds a nondescript V-shaped frog (a rubbery cushionlike structure in the rear of the underside of the hoof) that is either subtly expressed or absent. The overall shape of the manus (forefoot) print is, in most examples, wider than long. Observations of similar-sized living domestic horses, leaving behind unshod imprints in moist sand, are nearly identical in size and morphology. Overall length of adult footprints is 6.2 cm (2.4 in); overall print breadth is 12.0 cm (4.7 in).

Although many of the *Hippipeda downsi* footprints are extraordinarily clear, the taxonomic and systematic classification of North American Plio-Pleistocene horses based on the fossil record is not clear (see Scott, this volume, *Extinct Horses and Their Relatives*). This makes a confident assessment of candidate track-makers for The Corral quite difficult. The Plio-Pleistocene assemblage contains several fossil equid taxa (see Appendix, Table 3), of which two could be considered valid candidate track-makers: *Dinohippus* sp. and *Equus* (*Plesippus*) *simplicidens* (see Scott, this volume).

The structurally primitive genus *Dinohippus,* the Pliocene horse, known primarily from the Miocene (Hemphillian NALMA), was around too early to trample the younger Pliocene sands of The Corral. However, in a recent revision of equids from Anza-Borrego, Downs and Miller (1994) extended its temporal range upward (younger) into the late Blancan NALMA, based on diagnostic craniodental elements found stratigraphically below and above The Corral site. With some hesitation, this occurrence of *Dinohippus* sp., if appropriately diagnosed, may qualify it as the likely candidate track-maker. On the other hand, *Equus* (*P.*) *simplicidens*, although exclusively a Blancan taxon, is represented by cranial material in the Vallecito Creek-Fish Creek Badlands, also below and above The Corral site (see Scott, this volume). Its candidacy as a plausible track-maker is also possible.

Artiodactyls

Abundant bidactyl (two-toed or cloven-hoofed) footprints suggest the presence of at least five artiodactyls, including deer, possibly peccary, camel, and large and small llamas. All these taxa, except camel, occur at Camel Ridge.

The deer (*Odocoilichnum* sp.), as represented by an isolated footprint impression, remains to be studied. It is unknown which deer species made the footprint (both *Odocoileus* and *Navahoceros* are known from Anza-Borrego). Based on a measured hoof length of almost 9 cm (3.5 in), the most likely candidate track-makers are the Blancan NALMA recent white-tailed deer (*Odocoileus virginianus*) or the mule deer (*O. hemionus*) (Kurten and Anderson, 1980). The extinct mountain deer (*Navahoceros* sp.) might be considered, but it has not been reported from sediments as old as Camel Ridge (see Murray, this volume, *The Smaller Artiodactyls*).

The peccary-like footprint impression (*Tayassuichnum* sp.) is yet to be fully studied. However, a candidate track-maker is the extinct peccary *Platygonus* (see Murray, this volume).

The large camel track, from the Pleistocene of the Borrego Badlands, measuring 27 cm by 24 cm (10.6 in by 9.4 in), is the only definitive cameloid footprint (*Camelopichnum* sp.) discovered in the Colorado Desert. Unfortunately, these tracks have been vandalized and may not be salvageable. As of yet, no cameloid footprint impressions have been identified in the Vallecito-Fish Creek

Badlands, though *Gigantocamelus spatula* occurs in Anza-Borrego (see Webb et al., this volume, *Extinct Camels and Llamas of Anza-Borrego*).

There are two distinct populations of llama footprints, *Lamaichnum borregoensis* from a moderate-sized animal and *Megalamaichnum albus* from a larger form. These two ichnotaxa recur in fossil track assemblages throughout the Vallecito-Fish Creek Badlands stratigraphic section.

Lamaichnum borregoensis (Borrego small llama track) is represented by trackways with gracile, small- to moderate-sized footprints that average 12 cm (4.7 in) in length. (Figure 18.13) The footprints show digital cushions (pads) below the phalanges (toe bones), and have an oval to rounded-rectangular shape, with two distinct weight-bearing hooves (pads plus toe-nail) in each foot of both manus (forefoot) and pes (hind foot). The medial (toward the center) and lateral (toward the outside) hooves of each foot are nearly identical. The axes of these hooves are parallel or splayed outward to the front, separated by a linear interdigital (between the toes) crest line. The claw-like toes point forward. Each hoof is broadest at the heel, which is round to sharply parabolic in shape with apexes (points) rounded in the manus and pointed to round in the pes.

The earliest known occurrence of *Lamaichnum borregoensis* is from the transition zone between the marine-deltaic sediments of the Yuha Formation and the delta-plain sediments of the Palm Spring Formation in the Vallecito-Fish Creek Badlands. The footprint morphology is similar to llamoid impressions discovered at Camel Ridge (Figure 18.12) and along Arroyo Tapiado and Arroyo Hueso, as well as in the Borrego Badlands. *Lamaichnum borregoensis* is the most common ichnotaxon from Anza-Borrego. Footprint size and morphology clearly indicate that a llama with an ipsilateral gait produced the tracks, which do not vary significantly from contemporary llama (*Lama glama*) footprint impressions.

The morphology of the manus and pes footprints generally fits the skeletal parameters of the metacarpus and metatarsus (front and back foot bones respectively) and phalanges (toe bones) of an extinct *Hemiauchenia* candidate track-maker (Stout and Remeika, 1991). Four species of *Hemiauchenia* have been identified in the Vallecito-Fish Creek Badlands. However, these tracks are larger than the feet of the two smaller species of *Hemiauchenia*, *H. vera* and *Hemiauchenia* sp. Furthermore, *H. vera* is no younger than Hemphillian in age (too old for the track localities). Unfortunately, the two larger species of *Hemiauchenia*, *H. blancoensis* and *H. macrocephala*, cannot be identified from

Figure 18.12
Ranger Paul Remeika Interpretes Camel Ridge Depositional History.
Members of the Society of Vertebrate Paleonology listen on a 1991 field trip to Anza-Borrego. (Photograph by Lowell Lindsay)

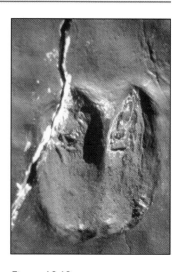

Figure 18.13
Lamaichnum borregoenis.
Right manus (front foot) footprint natural mold impression (ABDSP V4945) in hardened claystone from Camel Ridge. The fracture to the left of the footprint impression is partially filled with a white-colored liquid plaster. (Photograph by Paul Remeika)

postcranial (below the skull) remains, including foot bones (see Webb et al., this volume). Larger *Hemiauchenia* body fossils are common in the Arroyo Seco and Vallecito Creek Local Faunas of the Vallecito-Fish Creek Badlands (Remeika et al., 1995) and in the Borrego Local Fauna of the Borrego Badlands (Remeika and Jefferson, 1993), but definitive identification of the *Lamaichnum borregoensis* track-maker in the fossil record awaits further discovery and analysis.

Megalamaichnum albus (White's large llama track) is represented by robust moderate- to large-sized (overall length 18.5 cm by greatest breadth 14.5 cm, 7.3 by 5.7 in) llamoid footprints and trackways. The medial and lateral hooves, two per foot, are mirror images in outline separated by a narrow linear interdigital crest line. Hoofprints are widest near the heel, and taper forward to an apex at the toe; the rounded heel forms a broad parabolic curve. This ichnogenus differs in its larger size and morphology from the smaller *Lamaichnum* discussed above.

The late Blancan-Irvingtonian NALMA body fossil record in the Vallecito Creek Badlands includes several large llamine taxa (see Appendix, Table 3). *Megalamaichnum* seems to be about the size of *Blancocamelus* or possibly *Camelops minidokae*, a smaller species of *Camelops*. *Paleolama* sp. may also be a candidate track-maker, and should be considered even though only two specimens have been identified in the Vallecito-Fish Creek area. The relative abundance of these taxa has not been fully determined.

Proboscideans

Large, deeply impressed, plantigrade elephantoid trackways and footprints are known from the greater Borrego Badlands, and at Carnivore Ridge in the Vallecito-Fish Creek Badlands. One trackway is located in medial alluvial fan sandstones that are uplifted to view in the Truckhaven Rocks of the Santa Rosa Badlands. Another trackway is preserved in the playa margin deposits of the Borrego Badlands. This particular trackway, assigned to the ichnogenus *Mammuthichnum* sp., is significant because it occurs in *Mammuthus*-bearing strata of middle Pleistocene age. In the Fish Creek Badlands, the Carnivore Ridge track-bearing layer yields a large, deeply impressed footprint assigned to the ichnospecies *Stegomastodonichnum garbanii* (Figures 18.5, 18.14). This impression is well preserved, and is the first recognized gomphotherioid footprint discovered in the Colorado Desert. It resides in direct association with a number of other noteworthy footprints, such as *Pumaeichnum milleri, Chelipus therates, Gruipeda diabloensis, Lamaichnum borregoensis,* and *Megalamaichnum albus*.

The *Stegomastodonichnum* footprint is preserved in convex hyporelief, featuring a broad oblong to oval-shaped impression, length 37.5 cm (14.8 in) and breadth 27.5 cm (10.8 in), apex forward with three pronounced digits oriented anteriorly. Tridactyl (three-toed) functional morphology in a pentadactyl (five-toed) foot is consistent with the anatomy of extant elephantoid pes (rear) foot impressions and is not unusual. Digital pads are equal in size and character, and nearly parallel in orientation. Digits are robust and short, with a blunt anterior and a semicircular shape. Lateral digits are indistinct or effaced, reduced as the circumference of the depression shrank when the foot was withdrawn. The metatarsal pad is broad and flat, about 2.5 cm (1 in) thick, with a rounded heel along the posterior margin. The depth of the imprint is 13.5 cm (5.3 in). The track preserves a three-dimensional record of locomotor behavior in this Pliocene proboscidean.

Figure 18.14
Stegomastedonichnum garbanii.
Elephantoid ceiling cast impression (ABDSP V6292) in sandstone from Fish Creek Canyon. Posterior oblique view with three toe impressions pointing downward. (Photograph by Paul Remeika)

The attributes of the footprint impression suggest that it was made by a robust proboscidean. Candidates include the stegomastodon gomphothere (*Stegomastodon mirificus*), the gomphothere (*Gomphotherium* sp.), and the American mastodon (*Mammut americanum*). *Stegomastodon* sp. is primarily a Blancan taxon and is reported from the Vallecito-Fish Creek Badlands (Remeika et al., 1995; McDaniel and Jefferson, 2002). *Stegomastodon mirificus* has been recovered from similar-aged sediments in central Arizona (Agenbroad et al., 1998). The recent recovery of a fragmentary dentary of *Gomphotherium* sp. from deltaic deposits on strike with this track site (Jefferson and McDaniel, 2002) is significant and greatly supports the identification of the footprint. Lastly, *M. americanum* ranges from Blancan to Rancholabrean in age. Though its occurrence cannot be ruled out, no specimens of *Mammut* have been positively identified from the Salton Trough (McDaniel and Jefferson, 1998).

The San Felipe Ichnofauna

By exhaustively examining different footprint types, paleoichnologists can identify distinctive tracks that belong to general classifications of animal groups, and can estimate their relative abundance. It is usually easy to identify a track-maker at the family level, and often, if enough detail has been preserved in the ichnite, its ichnogenus and ichnospecies can be identified as well. If the fossil record contains skeletal evidence of one or more animals compatible in size and morphology with preserved footprints, and the skeleton and the footprints are known to have been present at the same time and place, then the animal that left the fossil remains may have produced the footprints. Such "body fossil" taxa are considered "candidate track-makers."

Middle and late Pleistocene ichnofossil assemblages make up the San Felipe ichnofauna of the Borrego Badlands. The oldest assemblage is about 850,000 years of age (Remeika and Beske-Diehl, 1996; Remeika, 1998a; Remeika and Liddicoat, in prep). It occurs along shorelines of presumably pluvial (glacial episode) freshwater lakes fed by the ancestral Colorado River, which flowed into the Salton Trough prior to the formation of Lake Cahuilla.

Two track-bearing sites exclusively yield robust llamoid hoofprints as long as 18 cm (7 in), identified as *Megalamaichnum albus*. At least half a dozen track-bearing sites contain gracile llamoid hoofprints identified as *Lamaichnum borregoensis* with lengths of 10 to 13 cm (3.9 to 5.1 in). Most of these tracks are preserved in fine-grained claystones and siltstones in the upper part of the Bautista beds (Frick, 1921; Remeika and Beske-Diehl, 1996) and are of Irvingtonian age (Remeika and Beske-Diehl, 1996). One notable site has a *L. borregoensis* trackway in direct association with *Hippipeda downsi,* as well as the first documented footprints of *Camelopichnum* sp. in the Anza-Borrego area. *Camelopichnum* tracks are a rare addition to the fossil record. The candidate track-maker may be either *Camelops hesternus* and/or *C. huerfanensis,* both of which may have occurred here concurrently during the early Pleistocene (see Webb et al., this volume).

Later Pleistocene layers higher in section yield a *Mammuthichnum* sp. trackway preserved in fluvial deposits and a *Chelipus* sp. trackway. The paw prints in the *Chelipus* sp. trackway are identical to, but larger than, those made by modern coyotes. Because the footprints are relatively young, in this case the most reasonable candidate track-maker is believed to be the coyote *Canis latrans,* not extinct *Canis lepophagus.* The elephantoid trackway is stratigraphically younger than the Bishop Tuff (see Remeika, this volume), which has been isotopically dated at 758,000 years before the present (Sarna-Wojcicki and Pringle, 1992). This confirms its middle Pleistocene (Irvingtonian NALMA) age. The candidate track-maker is believed to be the Columbian mammoth (*Mammuthus columbi*).

Conclusions

The Fish Creek Canyon and San Felipe Ichnofaunas are widespread and abundant avian and mammal hoofprint and footprint track assemblages are recognized in the Vallecito-Fish Creek and Borrego Badlands. Nine morphofamilies are now established, represented by fourteen ichnotaxa, making these ichnofaunas some of the most varied yet discovered. Indeed, only the ichnofauna from the Hemphillian of Death Valley National Park (Scrivner and Bottjer, 1986; Nyborg and Santucci, 1999) shows a comparable diversity of morphotypes. In summary, these two vertebrate ichnofaunas include the following:

- Footprints and trackways of *Gruipeda diabloensis* that closely resemble those made by the modern least sandpiper, *Calidris minutilla,*
- An aquatic musteloid, *Mustelidichnum vallecitoensis,* the first unequivocal footprint evidence of an extinct river otter reported from the Pliocene of California, that resembles the fish-eating river otter, *Satherium piscinarium,*
- Footprints of typical canoid, *Chelipus therates* and *Chelipus* sp., similar to those of modern coyotes, *Canis latrans,*
- Footprints of *Pumaeichnum milleri,* intermediate in size between the tracks made by modern species of the genera *Lynx* (bobcat) and *Felis concolor* (mountain lion),

- A robust footprint impression of a right manus (forefoot), representing cheetah-like cat, *Pumaeichnum stouti,* the first occurrence of an extinct cheetah-like cat footprint and the earliest record of cheetah-like cat in the Pliocene of California,

- Abundant footprints of *Hippipeda downsi,* which may represent the candidate track-makers *Dinohippus* sp. or *Equus simplicidens,*

- A single footprint of a deer, *Odocoilichnum* sp., which may represent *Odocoileus hemionus* or *O. virginianus,*

- Abundant footprints and trackways of *Lamaichnum borregoensis* which occur throughout the strata and whose morphology generally matches the foot osteology of the extinct candidate track-maker *Hemiauchenia* sp.,

- Abundant footprints and trackways of *Megalamaichnum albus,* clearly indicating that a larger llamoid ichnospecies was common through the Anza-Borrego stratigraphic record,

- Llamoids, the most common track-makers, with *Lamaichnum borregoensis* ranking first and *Megalamaichnum albus* ranking second in relative abundance, and the tracks of *Hippipeda downsi* under-represented relative to body fossil evidence,

- A large, deeply impressed footprint, *Stegomastodonichnum garbanii*, the first recoginized gomphotherioid or mastodontoid footprint discovered in the Colorado Desert; possible candidates include *Stegomastodon mirificus, Gomphotherium* sp., and *Mammut americanum,*

- And a *Mammuthichnum* sp. trackway in fluvial deposits of middle Pleistocene age thought to represent the candidate track-maker *Mammuthus columbi.*

The presence and abundance of large mammal tracks throughout multiple stratigraphic levels in the Vallecito-Fish Creek and Borrego Badlands spans from early Pliocene (Blancan NALMA) to late Pleistocene time. This treasure of well-preserved footprints occurs in a vertically continuous sequence of basin-margin strata deposited in many environments, as fresh and saltwater flowed through and in and out of Anza-Borrego Desert State Park. Layer after layer of marine-delta, delta-plain, fluvial-alluvial fan, and lake-margin sediments contributes to the amazing collection of mammalian-dominated ichnites that testify to the life that once thrived and evolved on these lands.

Because of the nature of the sediments and of the sedimentary processes in Anza-Borrego, the quality of fossil footprint preservation is often extraordinary, whether as positive ceiling casts on the undersides of thick overhanging sandstone ledges, or as negative floor impressions. This clarity and detail makes it possible to learn enough from the tracks to make associations with the bony fossil record and, one toe at a time, to see the story of past life a little more clearly. The osteologic remains of a majority of the older candidate track-makers have not been recovered, but the younger candidate track-makers are represented nicely in the late Blancan through Irvingtonian age deposits of Anza-Borrego Desert State Park.

Anza-Borrego and the Great American Biotic Interchange

*Darwin's impression can be ascribed
to a canny intuition of what we
now call the great interchange.
This impression arises from
the presence in the Pleistocene
on both continents of such striking
creatures as giant ground sloths
and some mastodonts,
now extinct everywhere.*

G.G. Simpson, 1980

H. Gregory McDonald

Anza-Borrego and the
Great American Biotic Interchange

Rainbow Wash from Fonts Point to Borrego Mountain. (Photograph by Paul Remeika)

Throughout the Tertiary Age, the North American mammalian paleo-fauna was enriched by species dispersing from other continents. These intermixed with the local endemic forms to create entirely different new ecological associations. At the same time, many North American species migrated in the opposite directions. The spread of horses from their North American homeland into Eurasia via the Bering Land Bridge of Alaska and eastern Siberia is one of the better-known examples. During most of the Tertiary, the primary source of new taxa entering North America was Eurasia, and as a result, the North American mammalian fauna has a distinct Eurasian flavor. The ancestors of many large mammals in today's fauna – bison, moose, elk, deer, and bighorn sheep, to name a few – originated in Eurasia and subsequently entered North America via the Bering Land Bridge.

Such interchanges of animals and plants between Northern Hemisphere continents date to the early part of the Cenozoic, when the northern Atlantic Ocean was first forming. Some animals were able to cross northern Canada, across Greenland and into northern Europe, until the widening of the Atlantic eventually disrupted this route. For most of the later part of the Cenozoic, the principal route of animal migration between North America and Eurasia was across the Bering Land Bridge (Figure 19.1). This bridge was not a permanent feature, but was only intermittently exposed, appearing and disappearing with changes in sea level. It was most recently above water during the glacial episodes of the Pleistocene, when ocean levels fell as seawater became trapped in continental glaciers.

Beginning in the Cretaceous and throughout most of the Tertiary, South America was an island continent with a mammalian fauna that evolved in essentially total isolation. The endemic paleofauna of South America included groups of mammals every bit as strange as or perhaps stranger than those that live today on the island continent of Australia. During the Tertiary, new animal groups entered South America by dispersal from both Africa and North America, and in the latest Tertiary, South American animals were able to move northward across the Panamanian Land Bridge.

This bridge formed as a result of the movement of continental plates that uplifted land from the ocean floor, in contrast to the Bering Land Bridge, which appeared whenever sea levels dropped low enough to expose the land. As the North and South American

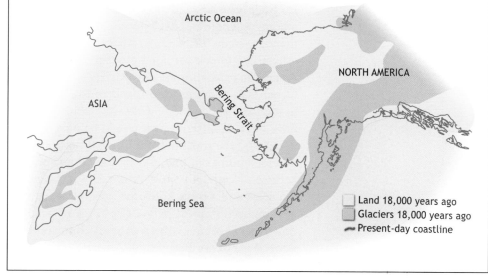

continental plates converged and the Pacific Plate subducted under these plates, mountains such as the Andes were uplifted. In the seaway that separated North and South America, this process led first to the formation of a chain of islands and eventually to what we know today as the Isthmus of Panama. At different stages in the development of the Isthmus, different animals were able to move between the two continents (see Demére and Rugh, this volume, Figure 3.1, *Invertebrates of the Imperial Sea*). Consequently, ever since Darwin's explorations, study of the history of South American mammals has led to the recognition that the modern South American fauna includes three faunal "strata" or dispersal episodes that define its development (Simpson, 1980).

The first and oldest stratum is considered the foundation of the modern South American mammalian fauna and contains the earliest mammals on the continent. These groups originated in the late Mesozoic or earliest Cenozoic and include the living marsupials and xenarthrans (e.g., sloths and armadillos) and the extinct South American ungulates (hoofed animals) including condylarths, litopterns, notoungulates (southern hoofed mammals), pyrotheres (fire beasts), and astrapotheres (lightning beasts). People who have looked only at the North American mammals probably have never heard of most of these groups, but South American researchers are as comfortable with them as their northern counterparts are with bison, moose, elk, bighorn sheep, and horses.

The next two (younger) strata are defined by the appearance of new groups of mammals that dispersed into South America from other continents. The second stratum is marked by the first appearance of primates, still present today and well represented by spider and howler monkeys, and the caviomorph rodents, a diverse group that includes guinea pigs, capybaras, and porcupines. "Caviomorph" refers to cavy (a name for the guinea pig) and "morph" means form. Therefore, the term refers to rodents with features similar to the guinea pig.

These mammals first appear in South America in the Oligocene, when South America was still an island continent. The ancestors of both groups probably traveled across open water barriers on rafts of vegetation. The source for both primates and caviomorphs has long been a subject of disagreement among paleontologists. One school of thought favors an African origin since there are many primates in Africa and the anatomy of the African hystricomorph ("hystrix," Greek for porcupine) rodents is very similar to that of the South American caviomorphs. Supporters of this idea also point out that the Atlantic Ocean was much narrower, and rifting had not yet separated the two continents to the extent seen today. Crossing the ocean on rafts of vegetation would not have been as difficult as it would be now. Another theory derives both the primates and the rodents from tropical portions of North America. Proponents of this idea point out that, during the Oligocene, the width of the water barrier between North and South America was much narrower than between South America and Africa. Unfortunately, we have essentially no fossil record from the southern tropical parts of North America. Only with continued research in this tropical region, which extends from southern Mexico through Central America into northern South America, will it be possible to uncover evidence to support this idea.

The third and youngest stratum of the South American mammalian fauna is marked by the appearance of animals of a decidedly North American/Eurasian flavor: cricetine (e.g., wood rat), heteromyid (e.g, kangaroo rat), and sciurid (e.g., squirrel) rodents, placental carnivores (e.g., dogs, cats, weasels, and bears), insectivores (e.g., shrews), lagomorphs (e.g., rabbits), proboscideans (e.g., elephants), perissodactyls (e.g., horses and tapirs), and artiodactyls (e.g., peccaries, deer, and camels). The descendants of many of these groups are found only in South America today. Many people fail to realize that they originated in North America. Some of the best examples include the llama, tapir, and spectacled bear.

This third stratum appeared as a result of South and North America becoming connected by the final formation of the Isthmus of Panama about 3 million years ago. Prior to that, about 9 million years ago, two South American ground sloths appeared in North America (see McDonald, this volume, *The Ground Sloths*), marking the first appearance of South American forms on the northern continent. At about the same time, the peccary, an animal of North American origin, appears in South America. The Isthmus of Panama was not fully formed at this time and these animals probably "island hopped" across the water barrier between the two continents. By about 3 million years ago, the formation of a complete land connection permitted a greater number of animals to cross between the two continents. It is important to remember that the Panamanian Land Bridge not only permitted the dispersal of many types of mammals of North American origin into South America but also permitted members of South American groups such as xenarthrans (e.g., sloths), marsupials (e.g., opossum), and caviomorph rodents to enter North America. The movement of animals in both directions impacted and changed the faunas on both continents and is referred to as the "Great American Biotic Interchange" (GABI) (Webb, 1985; Stehli and Webb, 1985). In some cases, animals that entered

South America from North America may have originated in Eurasia, and so are present in South America as the result of two dispersal events: Eurasia to North America via the Bering Land Bridge, and North America into South America via the Panamanian Land Bridge.

The sediments at Anza-Borrego span the entire length of time during which the GABI occurred, from the first appearance of North American forms in South America and vice versa. Understanding the GABI is essential to understanding the origin of many taxa present in the Anza-Borrego fossil record. Because of the continuous history preserved at Anza-Borrego, the fossils found here play an important role in our interpretation of this interchange that so profoundly impacted the mammalian faunas of both continents. Many of the fossil mammals found at Anza-Borrego are closely related to groups found only in South America today despite their North American origin (see Table 19.1).

Studies of paleofaunas on both continents, from roughly 9 to 0.5 million years ago, show that the GABI was completed by about half a million years ago, after which time the faunas seem to have become stable with only a few exchanges taking place. However, it is important to remember that the GABI took place in stages and was not a single event. Webb (1976), in his analysis of the dispersal events between North and South America during the Cenozoic, identified four separate periods of faunal interchange. The first event was in the Hemphillian/Huayquerian, a second in the Blancan/Chapadmalalan, a third in the Irvingtonian/Uquian, and a final event in the Irvingtonian/Ensenadan. (The first name in each pair is the North American Land Mammal Age and the second is its South American equivalent; see Cassiliano, this volume, *Mammalian Biostratigraphy of the Vallecito Creek-Fish Creek Basin*, for a discussion of NALMAs). South American species dispersed northward in the first three events but in the last event, only North American forms dispersed southward.

The exchange of mammals between the two continents was not balanced; many more species of northern origin entered South America than entered North America from the south (Webb, 1985). Another disparity is that, while many of the North American groups gave rise to many new genera and species following their entry into South America, the same did not happen to South American forms that entered North America. As a result, the modern South American mammalian fauna is dominated by members of the third stratum (74 genera) of North American origin, followed by members of the earlier second stratum (58 genera) and the earliest first stratum (29 genera) (Simpson, 1980). Given the age of the rocks at Anza-Borrego, all of the animals found here as fossils, which are related to modern South American species, are members of the third faunal stratum.

In any faunal interchange, the appearance of invaders in a new region can have various ecological consequences. The impact may be simply an enrichment, as some invaders fill empty niches and become integrated into the fauna with no disruption of native species. In the GABI, this seems to have been the case for ground sloths, glyptodonts, pampatheres, and armadillos in North

America. But sometimes following an interchange there is a marked decrease in the diversity of native taxa with some native species becoming extinct (Marshall, 1981). This appears to have happened in South America after the GABI. Some of the North American invaders may have had ecological requirements very similar to those of the natives, successfully competed with them for resources, and thus replaced them in the ecosystem. Another possible impact is that of direct predation. Many of the North American species that entered South America were predators; they appear to have been more efficient in preying on the native South American herbivores than were their South American counterparts. These counterparts included marsupial carnivores (borhyaenids) and large, extinct, flightless, predatory birds (phorusrhacids). The appearance of sabertooth cats, *Smilodon*, and short-faced bears, *Arctodus*, along with smaller canids (dogs), felids (cats), and mustelids (e.g., skunk) may have had a significant impact on the survival of native South American herbivores. The first two were prominent predators in North America, well adapted for preying on large herbivores anywhere. Similarly, the smaller carnivores would have had an impact on smaller prey. Many of these predators of North American origin are present in Anza-Borrego fossil assemblages.

The mammalian fauna preserved at Anza-Borrego includes forms ancestral to groups that dispersed into South America, as well as forms that originated in South America and dispersed into North America. Mammals of South American origin found here include three ground sloths, *Megalonyx, Nothrotheriops,* and *Paramylodon,* and a caviomorph rodent, the porcupine, *Coendou* (White, 1968). The absence of some South American forms such as the armored xenarthrans (glyptodonts, pampatheres, and armadillos) in Anza-Borrego reflects the distinctive distribution patterns of these animals. They tend to be restricted to the wetter eastern U.S. Another South American form apparently restricted to wetter areas, and also absent from Anza-Borrego, is the capybara. Fossil capybaras are common in the southeastern U.S., but they are less common in the west. There is a single record of a capybara from San Diego County but it was found on the coast, which was probably wetter than Anza-Borrego. The presence of some forms and absence of others provides insights into the ancient environments of Anza-Borrego and suggests possible barriers – physical, environmental or climatic – that prevented some groups of animals from entering the area and surviving here. For example, faunas in Arizona of the same age as those found at Anza-Borrego contain glyptodonts and capybaras, so we know that these animals were nearby. The Irvingtonian age El Golfo fauna in western Sonora, Mexico, includes capybara and giant anteater, yet neither of these animals has been found in Anza-Borrego. Comparative studies of the paleoecology of Anza-Borrego taxa and of contemporary nearby faunas may eventually provide explanations for the absence of these animals here.

Equally important are those animals found at Anza-Borrego that were ancestral to mammals that dispersed into South America. These include: Carnivora, represented by the families Procyonidae (raccoon), Mustelidae (e.g., skunk), Ursidae (bear), Felidae, and Canidae; Rodentia with the families

Cricetidae (e.g., wood rat, white-footed mouse), Heteromyidae (kangaroo rat), Geomyidae (pocket gopher), and Sciuridae (e.g., ground squirrel); Artiodactyla (even-toed ungulates) including the Camelidae, Tayassuidae (peccary), and Cervidae (e.g., deer); Perissodactyla (odd-toed ungulates) with the Equidae (horse) and Tapiridae (tapir); Lagomorpha (rabbit) with one family the Leporidae; Insectivora also with one family the Soricidae; and finally the Proboscidea represented by the Gomphotheriidae (see Appendix, Table 3 for common names). Except for the Proboscidea and the horses, which are now extinct in the Americas, all of the above taxa still have living representatives in South America. As can be seen in Table 19.1, all of these groups and most of the families are well represented in Anza-Borrego assemblages.

There is still much work to be done to determine the degree of relatedness between some taxa found on the two continents. In some cases, a South American genus is closely related to a different North American genus that may be its direct ancestor. A good example of this is how closely related the extinct otter, *Satherium,* is to the living giant Amazonian river otter, *Pteroneura*, its presumed descendant. *Satherium* is a large otter whose fossil remains have been found at Anza-Borrego as well as at numerous other localities across North America. It became extinct in North America prior to the Irvingtonian. Did *Satherium* disperse into South America and give rise to the giant Amazonian river otter or did the South American form come from another otter lineage? In contrast, the living river otter, *Lontra* (= *Lutra*), is present in both North and South America. *Lontra* was a late immigrant to North America from Eurasia, first appearing here in the Irvingtonian, it quickly dispersed into South America.

Llamas, guanacos, and vicuñas, all members of the camel Family, are often thought of as typical South American mammals. Llamas evolved in North America prior to the formation of the Isthmus of Panama. Several species of these early llamas, in the genus *Hemiauchenia,* are found at Anza-Borrego. It is not clear whether a species of *Hemiauchenia* is the direct ancestor of the South America llamas or merely a close relative. The genus is also known from South America (Webb, 1974). *Hemiauchenia*, along with another extinct genus, *Paleolama*, first appears in South America in the early Pleistocene (see Webb et al, this volume, *The Extinct Camels and Llamas of Anza-Borrego*).

The tapir, with three living species, is another group we commonly associate with South America. Yet, like the llama, its origin and most of its evolution took place in North America, with dispersal into South America a late event in its history. It would appear that tapirs were good dispersers. The only other living species of the genus, besides the South American forms, is found in Southeast Asia, implying that it must also have moved northward, and entered Asia via the Bering Land Bridge. The species found at Anza-Borrego, *Tapirus merriami*, was a large animal, bigger than all of the living forms (see Scott, this volume, *Extinct Horses and Their Relatives*). Tapirs first appear in South America at the same time as the llamas. During the Pleistocene, there were at least four different species of tapirs in North America and almost that many are present today in South America. While we think of tapirs primarily as tropical forest animals, living tapirs are quite flexible in their choice of habitat. One living species, the mountain tapir, *Tapirus pinchaque*, lives at elevations of up to 4,500 meters in the Andes. All living New World tapirs are gap specialists, meaning they utilize available openings in forest environments.

Another distinctive New World group is the peccary or javelina. Once widespread across North America, today our single living species, the collared peccary, is restricted to the southwestern part of the U.S. and south through Mexico and Central America into South America. In South America, there are two additional species, the white-lipped and the Chaco, along with the collared peccary. The Chaco peccary, *Catagonus*, is often referred to as a living fossil and appears to be closely related to *Platygonus*, the extinct genus found at Anza-Borrego (see Murray, this volume, *The Smaller Artiodactyls*). Peccaries, represented by *Selenogonus*, first appeared in South America in the late Miocene, long before the arrival of llamas and tapirs in the Chapadmalalan (Blancan of North America).

Of the carnivores closely associated with South America, perhaps the best known is the spectacled bear, *Tremarctos ornatus*. *Tremarctos* is the sole surviving representative of a subfamily of bears that also includes the extinct giant short-faced bear, *Arctodus*. Both of these bears had a long history in North America after their ancestors entered the continent from Eurasia, and both were among the North American invaders that entered South America. *Arctodus* was widely distributed across North America, from Alaska to Mexico, but *Tremarctos* is found primarily in the southern part of the U.S. and southward. Fossils of both are found at Anza-Borrego (see Shaw and Cox, this volume, *The Large Carnivorans*). Although both genera appear in South America in the early Pleistocene, the circumstances attending their entrance must have been tremendously different. The spectacled bear is one of the most herbivorous of the living bears while its extinct relative, *Arctodus*, is thought to have been extremely carnivorous. Recent studies of the tremarctine bears in South America (Trajano and Ferrarezzi, 1994) do not recognize *Arctodus* on that continent but rather place the large bears resembling *Arctodus* in a distinct genus, *Arctotherium*. Despite the differences in opinion as to what to call these animals, they are closely related, and the South American forms are derived from North American ancestors.

There are many groups of North American mammals found at Anza-Borrego with either fossil or living relatives in South America; the above survey touches on only a few of them. Nevertheless, it illuminates the relevance of the Anza-Borrego fossil record not only to the history of the North American mammalian fauna but also to that of South America, and to one of the grandest biological and geological events that ever occurred on earth, the Great American Biotic Interchange. As observed by George Simpson (1980), "The history of South American mammals can be considered as an experiment without a laboratory, fortuitously provided by nature." The fossil record preserved at Anza-Borrego is, among other things, a notebook of that experiment.

ORDER	FAMILY	ANZA-BORREGO	SOUTH AMERICA
Lagomorpha	Leporidae	*Sylvilagus audubonii* *Sylvilagus hibbardi*	*Sylvilagus brasiliensis*
Rodentia	Cricetidae	*Calomys* (*Bensonomys*) sp. *Sigmodon curtisi* # *Sigmodon hispidus* *Sigmodon lindsayi* # *Sigmodon medius* # *Reithrodontomys* sp.	*Calomys* spp. *Sigmodon* spp. *Reithrodontomys* spp.
Carnivora	Canidae	*Canis dirus* # *Canis edwardii* # *Urocyon* sp.	*Canis dirus* # *Urocyon cineroargenteus*
	Ursidae	*Arctodus simus* + *Tremarctos* cf. *floridanus* #	*Arctodus* (*Arctotherium*) + *Tremarctos ornatus*
	Procyonidae	*Bassariscus* cf. *casei* # *Nasua* sp. *Procyon lotor*	*Bassariscus sumichrasti* *Nasua nasua* *Procyon cancrivorus*
	Mustelidae	*Mustela* cf. *frenata* *Satherium piscinarium* + *Trigonictis* sp. +	*Mustela frenata* *Pteroneura brasiliensis* *Galictis* spp.
	Felidae	*Felis concolor* *Smilodon fatalis* +	*Felis concolor* *Smilodon populator* +
Proboscidea	Gomphotheriidae	*Cuvieronius* or *Stegomastodon* +	*Cuvieronius* sp.+
Perissodactyla	Equidae	*Equus bautistensis* # *Hippidion* sp. +	*Equus* spp. *Hippidion* sp.+
	Tapiridae	*Tapirus merriami* #	*Tapirus* spp.
Artiodactyla	Tayassuidae	*Platygonus vetus* +	*Catagonus wagneri* *Tayassu* spp.
	Camelidae	*Hemiauchenia blancoensis* + *H. macrocephala* +	*Hemiauchenia* sp.+
	Cervidae	*Odocoileus* cf. *virginianus* *Navahoceros* sp. +	*Odocoileus virginianus* *Hippocamelus* spp.

Table 19.1
Anza-Borrego Genera of North American Origin Present in South America.
Genera in bold are either closely related or may be congeneric (same genus.) A + indicates an extinct genus, # an extinct species. Where possible, a species name is provided, but for those genera in South America represented by multiple species, this is indicated by spp. (see Appendix, Table 3 for common names).

Paleoclimates and Environmental Change in the Anza-Borrego Desert Region

*You don't need a weatherman
to know which way
the wind is blowing.*

Bob Dylan

Sharron Sussman
Lowell Lindsay
Howard J. Spero

Paleoclimates and Environmental Change in the Anza-Borrego Desert Region

Sand Verbena Near Coyote Canyon Entrance. (Photograph by Paul Remeika)

Introduction: Climate, Weather, and Gravity

"Climate" is defined as the characteristic weather of a region, averaged over some significant time interval. "Weather" is what we see reported on the evening news: temperature, barometric pressure, and humidity of the air; direction, speed, and motion pattern of the wind and the waves; and the form and quantity of precipitation.

Earth's climate seems to be in a warming phase today, as is evidenced by melting polar icecaps and record high summertime temperatures. Climate changes have occurred both in long-term trends lasting tens of millions of years and in relatively rapid cycles over as little as a few thousand years (such as during some of the glacial periods of the past two million years). Although tectonic events and occasional disastrous impacts by outer space objects may influence our climate, its major determinants are the astronomical characteristics of the solar system. Weather and climate on Earth are probably somewhat affected by tiny variations in solar radiation, such as those associated with the eleven-year sunspot cycle. However, the main factor that dictates how much energy we receive from the Sun is our distance from it.

Our solar system is a spinning, wheeling arrangement of planets and other bodies circling a star – the Sun – at various distances. The planets turn on axes of rotation as they move around the Sun in more or less circular paths called orbits. Several planets have satellites of their own, moons, which orbit around them. Every body in the solar system acts on every other body through the force of gravity, establishing and then subtly perturbing the complex planetary motions. This has profound implications for climate.

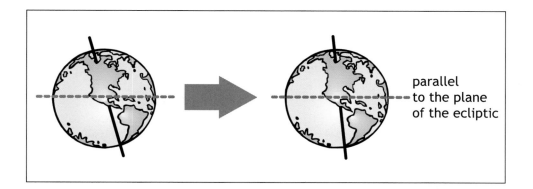

Figure 20.1
The Tilt or Obliquity Cycle.
Over a period of 41,000 years,
the axis of Earth's rotation
changes its orientation to the
orbital plane (the ecliptic). The
difference between maximum tilt
(24.5°) and minimum tilt (22.2°)
is slight.

parallel
to the plane
of the ecliptic

Both the axial orientation and the orbital motion of the Earth affect the climate by bringing the planet's different regions alternately closer to and farther away from the Sun. The axis of Earth's rotation is tilted, not vertical, to the plane of its orbit. As the Earth moves around its orbital path, it passes through a point at which the Northern Hemisphere is maximally tilted toward the Sun. On this date, the Anza-Borrego Desert experiences the longest day of the year, which is the summer solstice. Six months later, or 180° through the orbit, the Northern Hemisphere is maximally tilted away from the Sun and we experience the winter solstice. This cyclic variation in received solar energy is what causes our annual seasonal weather patterns.

Currently, the axis of Earth's rotation is tilted about 23.4° from the vertical, but this is not a fixed angle. Over time, the axis very slowly increases and decreases its tilt from 24.5° to 22.2° (the tilt is slowly decreasing today, Figure 20.1). The tilt cycle requires 41,000 years to complete a full cycle. While the Earth's tilt is slowly changing, the direction the axis points to the heavens slowly shifts

Figure 20.2
The Perihelion or Aphelion
Regression Cycle.
Because the tropical years and
the anomalistic year (see text)
do not match exactly, each year
perihelion and aphelion occur
later with respect to the seasons.
It takes 23,000 years to complete
this cycle around the entire
tropical year. This figure shows
our current condition, with the
summer solstice occurring close
to aphelion (in the lower
diagram).

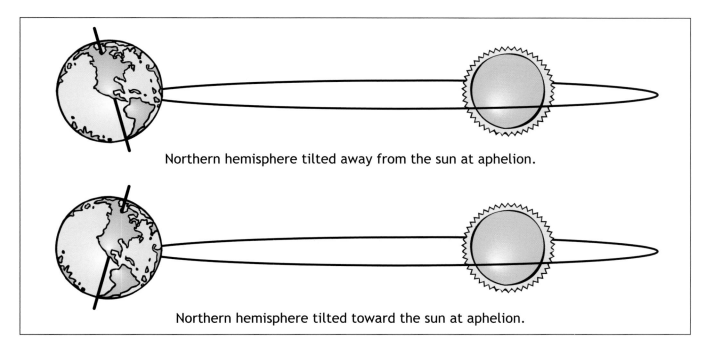

Northern hemisphere tilted away from the sun at aphelion.

Northern hemisphere tilted toward the sun at aphelion.

such that, in the not-too-distant future, Polaris will no longer be the pole star. This axial direction change describes a wobbling motion such that the geographic poles trace circular paths in opposite directions. This effect, called precession, is similar to the motion of a spinning top as it slows. It requires 25,700 years for the wobble to move completely around the cycle.

The time measurement from solstice to solstice, known as the "tropical year," is the basis for our civil calendar. However, there is another way to measure time. As the Earth orbits the Sun, the Earth-to-Sun distance varies because the orbit is slightly elliptical rather than circular. Once each year, the Earth makes its closest approach to the Sun, called perihelion from the Greek for "near the Sun." (Aphelion is the point farthest from the Sun.) The time measurement from perihelion to perihelion, known as the "anomalistic year," is approximately twenty-five minutes longer than the tropical year. This is due to a combination of precession of the axis of rotation, which causes the exact timing of the solstice to vary from year to year, and small orbital perturbations. The date of perihelion thus occurs later each year, regressing with respect to the solstice (Figure 20.2). Perihelion currently occurs only two weeks after the winter solstice. The perihelion regression cycle, combined with the axial wobble, creates a dominant cycle near 23,000 years as the date of perihelion progresses through the entire tropical year.

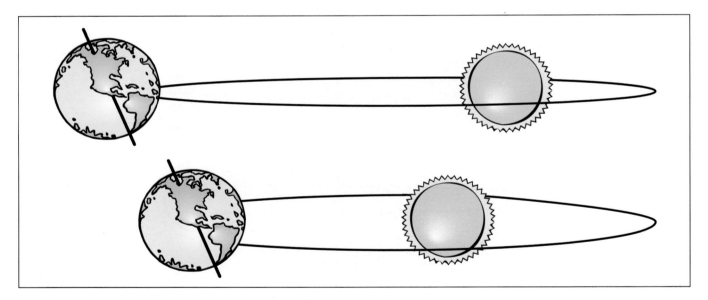

Figure 20.3
The Eccentricity Cycle.
The Earth's orbital path around the sun varies between more and less elliptical over a period of 100,000 years. These diagrams are greatly exaggerated for emphasis — the orbit is never really very far from circular.

A larger orbital fluctuation is the 100,000 year "eccentricity cycle," during which the elliptical shape of Earth's orbit becomes slowly more or less pronounced, varying from nearly circular to about three times present values (Figure 20.3). The difference between aphelion and perihelion distances is actually quite small, currently about 3 percent, not enough to cause a pattern of seasonal variation but capable of affecting its severity. During epochs of large eccentricity, the effect of the perihelion shift cycle is magnified.

The 100,000 year eccentricity cycle, the 41,000 year tilt or obliquity cycle, and the 23,000 year perihelion regression cycle are caused by gravitational interactions not only involving the Sun, Earth, and Moon, but also Jupiter and Venus. Together the cycles are known as Milankovitch cycles, after Milutin Milankovitch, a Serbian engineer who worked out the model and its mathematics during the 1920s and 1930s. Although Milankovitch cycles are well grounded astronomically, the relationship of long-term climatic variations to astronomic causes is not yet completely laid out, most notably lacking a clear mechanism by which the 100,000 year eccentricity cycle exerts a direct effect on climate. However, the potential is there for a perfect storm conjunction of all three cycles to tip us into dramatic climate change.

Other modifiers may play a leading role in amplifying orbital effects. The various ways in which the Earth receives, reflects, stores, and utilizes its solar radiation allotment interact to determine the climate of a given region. The distribution of water, ice, rock, and vegetation on the Earth's surface affects the planet's tendency to reflect rather than absorb sunlight, a phenomenon called albedo. Long before there were power plants and automobiles, volcanoes would periodically spew high loads of particulates and gases into the upper atmosphere, temporarily changing the way solar energy reached the surface of the Earth. They still do, causing major short- and long-term weather changes. Patterns of air and water circulation become established, fixing regional climates for a while. Sea levels rise and fall depending on the amount of water frozen into ice sheets and glaciers. Islands rise and coalesce into land bridges, altering the patterns of ocean currents. Mountain ranges are pushed up to wind-blocking heights by deep tectonic forces, then erode down again. These Earth processes also change the chemistry of the atmosphere, as crystalline rock reacts with carbon dioxide, trapping and burying this greenhouse gas in new sediment, thus cooling the planet.

Paleoclimate Inferences from the Fossil Record

Paleontology opens windows on the past, a major theme of this book. The fossil record offers broad perspectives on ancient climates, environments, and habitats that may be made more definitive by climate proxies, as discussed in the following section.

The climatic and environmental story told by the rocks and fossils of the Park begins about 7 million years ago in the later Miocene with the development of the ancestral Gulf of California. This marine environment was warm and tropical, with affinities to the modern Caribbean Sea. The oldest sandstone strata of the Imperial Group were being laid down around and under the warm, clear waters of the Imperial Sea. Anza-Borrego was part of the Tertiary Caribbean Province, an unbroken tropical region encompassing the Caribbean Sea, the western Gulf of Mexico, and the equatorial eastern Pacific Ocean.

Many of the clams, snails, and corals found fossilized in the Imperial Group sediments have affinities (Figure 20.4), fossil and living Caribbean species (see Deméré, this volume, *The Imperial Sea* and Deméré and Rugh, *Invertebrates of the Imperial Sea*).

This sunny picture began to change around 5 million years ago, at the very end of the Miocene. Locally, the ancestral Colorado River started to build its massive delta at the head of the proto-Gulf of California, rapidly depositing clay-stones and siltstones from its turbid flow on top of the older sandstones. The very rapidity of the erosion process speaks of a period of increased precipitation in southwestern North America. Although the latest Miocene (6.5–5 million years ago) was marked by the cooling at high and middle latitudes that is associated with global marine regression due to Antarctic ice sheet expansion, the earliest Pliocene was a period of warming temperatures (Kennett, in Vrba et al., 1995). By the early to middle Pliocene (4.0–2.6 million years ago), the Anza-Borrego region was a floodplain near sea level, still receiving the sediments of the Colorado River, whose delta had already walled off the upper part of the proto-gulf, present day Coachella and Imperial Valleys.

Figure 20.4
Coquina.
This specimen of sedimentary rock from the southern part of the Anza-Borrego Desert is a sandstone-cemented jumble of whole and fragmentary fossil shells of marine invertebrates that lived in the Imperial Sea. (see Deméré, this volume.) They bear witness to a time when the desert was a tropical ocean. (Photograph by Jim Zuell)

Fossil hardwoods from these deltaic deposits reveal that the river flood-plain was occupied by a temperate woodland community consisting of bay laurels, walnuts, avocados, cotton-woods, willows, ashes, buckeyes, and palms (Figure 20.5) (see Paleolandscape 2). This flora is an excellent indicator of wet soil and permanent water in a moderate but four-season climate with annual rainfall of 38–62 cm (15–25 in) mostly in the winter or spring, and temperatures ranging from -9.4° to 26.6° C (15° to 80° F) (see Remeika, this volume, *Ancestral Woodlands of the Colorado River Delta Plain*).

Figure 20.5
Pliocene Woodland Community.
Artist's conception of the wet, temperate Colorado River delta floodplain, about 3.5 million years ago. (Picture by John Francis)

Ostracodes are microscopic arthropod crustaceans commonly called water fleas or mussel shrimp. Their numbers in freshwater lakes and ponds will bloom under favorable conditions. Fossil ostracode abundances in the 1–2 million-year-old Hueso Formation in the Vallecito Creek-Fish Creek basin indicate significantly higher annual precipitation. Peak abundances appear to fluctuate with a 100,000 year Milankovitch cycle (Cosma et al., 2001).

Figure 20.6
Hesperotestudo.
Artist's conception of the extinct
giant tortoise that roamed Anza-
Borrego during the Pliocene and
Pleistocene, when the climate
was much more wet and
temperate than it is today.
(Picture by John Francis)

The presence of at least five species of fossil tortoises and turtles in the Hueso and Ocotillo deposits suggest that the region had warm winters, cool summers, and enough precipitation to maintain permanent bodies of water. The presence of the giant tortoise suggests that winters never dropped below freezing. The record of neotropical iguanid lizards further supports this picture. However, the appearance in the fossil record of spiny lizards, alligator lizards, skinks, ground lizards, and whiptails at about 2.5 million years ago speaks for the spread of grassland and savanna habitat in the region (see Gensler and others, this volume, *The Fossil Lower Vertebrates*).

The fossil birds of the Park, mostly water birds and perching birds, range from mid-Pliocene to mid-Pleistocene age (3.5–0.5 million years ago) and also indicate a setting of lakes, ponds, and streams during this period (see Jefferson, this volume, *The Fossil Birds of Anza-Borrego*).

Several groups of large animals arrived early and stayed late in the region from late Miocene through the close of the Pleistocene age. Among these were the camelids. This group apparently attained its greatest diversity in response to the emergence of open grasslands. They exhibit a number of adaptations, mostly of the limbs, gait, and teeth, which facilitated a highly cursorial, mostly grazing mode of feeding with variations among coexisting species that would have enabled them to partition resources effectively. The group had prolonged success, surviving and thriving through many climate shifts in what became an increasingly arid environment. They became extinct in North America about 11,000 years ago, along with most of the other large animals.

Another group of large animals that arrived early and stayed late was the giant ground sloths (see McDonald, this volume, *The Ground Sloths*). These xenarthrans originated in South America and entered North America in three stages across the slowly developing Isthmus of Panama beginning in the late

Miocene around 9 Ma. The first group to arrive is the megalonychids. They were browsers more dependent on forest habitat than the other two sloth groups, and were more abundant in the eastern U.S. than the western, where their distribution likely followed watercourses and riparian vegetation. This group is represented by several species of *Megalonyx* from Anza-Borrego. The second group is the mylodonts, which are represented in Anza-Borrego by *Paramylodon harlani*. Members of this group have dental and limb adaptations suitable for grazing or mixed feeding in country that is more open. The third group is the megatheriids, which are represented locally by *Nothrotheriops*, the Shasta ground sloth. These animals appear to be adapted for an eclectic diet and survival in a more arid or even desert habitat.

Each of the three groups had a different preferred habitat. Although they may occur jointly in a North American paleofauna, it is thought that this only occurred at the contact area, or ecotone, between their preferred habitats. A diversity of habitats prevailed here during the Pliocene and much of the Pleistocene, even as the overall climate became more arid with the development of the rain shadow of the Peninsular Ranges after around 1.5 million years ago (see McDonald, this volume).

Anza-Borrego's more than 3,800 identified small mammal fossils include a variety of both warm-adapted taxa such as the cotton rat and cool-adapted animals such as lemmings and voles (see White et al., this volume, *The Small Fossil Mammals*). Through the middle part of the record, these forms are found in the same strata. This makes it difficult to draw conclusions about paleoclimates during this time. Conditions may have varied or shifted back and forth on a scale that is not possible to resolve from the fossil record. However, some trends do emerge. Many of the small mammals found here as fossils had affinities for warm, moist, tropical environments, more so in the older levels of the Hueso Formation. However, a fair number of the taxa represented are thought to have preferred a cooler environment, with seasonal variation even including winter frost and snow. Two cool-adapted forms, *Eutamias* the chipmunk and *Peromyscus* the white-footed mouse, make their first appearance about 3.3 million years ago, suggesting a shift toward a cooler climate at that time (see White et al., this volume).

Vivid as reconstructions from the fossil record may seem to the active imagination, these snapshots and action sequences are only limited views of the paleoclimates of Anza-Borrego, rough guides to their history. They do not tell us how different climatic regimes came to exist in this part of the World or by what means and how rapidly or slowly the region transformed from one kind of climate to another. The history of climate as revealed solely by the evidence of fossil assemblages is analogous to the evolutionary and geologic story told by relative dating methods (see Cassiliano, this volume, *Mammalian Biostratigraphy in the Vallecito Creek-Fish Creek Basin*).

In order to generate a more meaningful narrative, including accurate descriptions of the ancient climates and their meteorologic parameters, it is necessary to devise ways of measuring *indirectly* the important variables that we have no *direct* way to measure. There are a number of such methods, known as "climate proxies," in use and in development (Bradley, 1985). Residual biological and chemical evidence left in the rocks and fossils can be analyzed and measured with modern techniques. The results of these analyses can be precisely correlated with the specific values of some climatic variables. In other words, they are proxies or "stand-ins" for the non-directly-measurable variables. Examples of successful climate proxies include the stable isotopes of oxygen and carbon (Figure 20.7).

^{18}O and ^{16}O are naturally occurring, nonradioactive isotopes of oxygen that occur in nature. These isotopes are found in molecules such as water (H_2O) in temperature-dependent proportions as the molecules undergo the cycle of precipitation and evaporation. Compounds containing oxygen will have an oxygen isotope ratio that reflects the oxygen isotope composition of the water at the time the compounds formed. The ratio of ^{18}O to ^{16}O ratio in meteoric water (e.g., rain), generally presented in 'delta notation' relative to an international standard ($\partial^{18}O$), is very sensitive to mean annual temperature and the amount of evaporation relative to precipitation. Delta values decrease (move in a negative direction) as the climate cools and becomes wetter. These values increase (become more positive) as the climate warms and gets drier. The shells, bones, and teeth of animals, often fossilized, contain abundant oxygen in compounds such as carbonates and phosphates, which are synthesized by the animal's cells from ingested water.

The stable, nonradioactive carbon isotopes ^{13}C and ^{12}C (Figure 20.7) are sorted or fractionated in characteristic patterns by plants using different biochemical pathways of photosynthesis. A stable carbon-containing compound, such as the mineral hydroxyapatite ($Ca_5[PO_4\ CO_3]_3[OH, F, Cl]$) that is contained in the tooth enamel of herbivores, will reflect the carbon isotope patterns of the plants in their diet. This is because the relative proportions of the carbon isotopes from the ingested plants are incorporated directly into the animal's bones and teeth. These body parts are then found as fossils that can be chemically analyzed. The ratio of ^{13}C to ^{12}C ($\partial^{13}C$ when compared to an international standard) serves as a proxy for the relative proportion of the different types of plants in

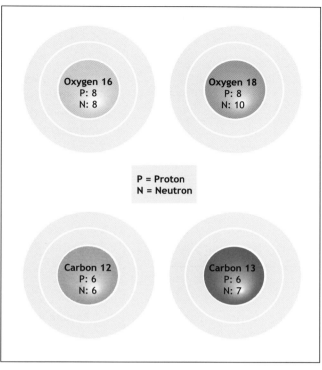

Figure 20.7
Stable Isotopes of Oxygen and Carbon.
Only the atomic nuclei are shown, for simplicity. The number of nuclear protons, which determines the chemical properties of an element, remains the same for all its isotopes. The heavier isotopes are heavier because they contain more neutrons, particles which have mass but no electrical charge, and thus do not change chemical behavior. ^{18}O is heavier than ^{16}O, the most common isotope of oxygen, and similarly, ^{13}C is heavier than the ordinary ^{12}C.

the diet. The $\partial^{13}C$ value in vegetation is lower (i.e., there is less ^{13}C) in C3 plants (the grasses, trees, and shrubs of cooler, wetter climates), which make a three-carbon intermediate molecule during photosynthesis, than in C4 plants (the grasses of more arid climates, which have more ^{13}C) that manufacture a four-carbon intermediate sugar. Since different plants are associated with distinct climatic conditions, the carbon isotope composition of bones and teeth is a proxy indicator of climate.

Determining the mechanisms responsible for variations in proxies like stable oxygen and carbon isotopes deepens our understanding of their actual relationship to the long-vanished climates and biological systems they represent. By applying proxy techniques to modern specimens under known conditions, the paleoclimatologist can achieve increasingly accurate climate-proxy calibration, something akin to the way in which we use the skeleton of a modern animal to help us understand the fossil bones of its remote ancestors. In addition, and most importantly, correlating a good proxy climate measurement with a well-constrained date for a specimen using modern absolute dating methods can help us to reconstruct both the nature and the history of an ancient climate (see Remeika, this volume, *Dating, Ashes, and Magnetics*).

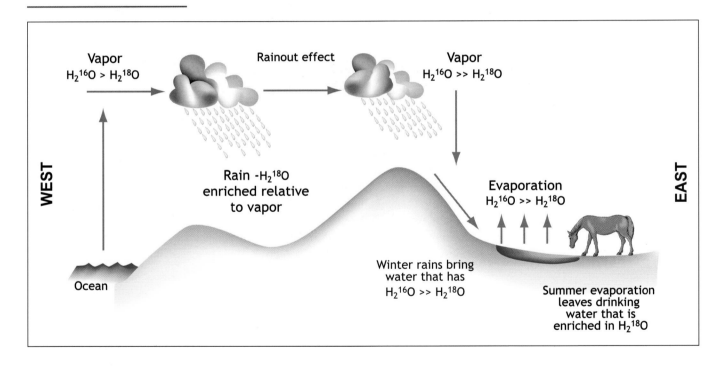

What Do Fossil Horse Teeth Have to Do with Paleoclimates of Anza-Borrego?

The relative proportions of the oxygen isotopes ^{18}O and ^{16}O of the calcium carbonate ($CaCO_3$) shells of fossil microorganisms that make up much of the bulk of Earth's marine sediments has been studied since 1947 (Urey, 1947, 1948) to determine the history of ocean temperatures. Such studies have yielded valuable information about the sequence of glacial periods during the Pleistocene. Recently this method was applied to a collection of well-dated fossil horse teeth from Anza-Borrego, ~2.9 to ~0.8 million years old (Brogenski, 2001). *Equus* remains are the most common and abundant large mammal from the Anza-Borrego Desert (see Scott, this volume, *Extinct Horses and Their Relatives*). Thus, a study of their teeth from the Hueso Formation could yield a continuous proxy record of climate extending over more than two million years, including the Pliocene-Pleistocene boundary.

Fossil teeth are ideal for geochemical analysis for much the same reason that we find so many of them: they are quite resistant to abrasion and weathering. Tooth enamel is composed of an extremely hard, polished, nonporous form of the mineral hydroxyapatite ($Ca_5[PO_4\ CO_3]_3[OH, F, Cl]$), the basic substance of both bones and teeth. The enamel surface has low porosity and large hydroxyapatite crystals, leaving less than 1% of the atoms on the crystal surface of a tooth available for chemical reactions. Thus, the proportion of oxygen isotopes fixed in the phosphate and carbonate groups of the hydroxyapatite when the tooth is formed remains stable over very long time periods, continuing to reflect the geochemistry of the animal's drinking water as it changed with evaporation and precipitation. Similarly, the proportion of different carbon isotopes in the enamel is a stable proxy for the proportions of different types of plants eaten by the animals.

Brogenski (2001) analyzed the oxygen and carbon isotope ratios in a total of 28 fossil horse teeth from the Park, each with well-constrained ages ranging from 2.9–0.8 million years ago (see Cassiliano, and Remeika, this volume). Each specimen was examined both in a serial sampling method, which collected enamel at 6 mm (1/4 in) intervals from root to crown, and in a bulk sampling method, which collected enamel from the entire length of the tooth. The serial samples were used to determine $\partial^{18}O$ and $\partial^{13}C$ in the enamel layers of each specimen, demonstrating seasonal variations within a single tooth. Analysis of the bulk samples provided evidence for long-term climatic change over the 2.1 million years represented by the specimens.

Ninety separate samples of tooth enamel were analyzed by a complex process of extraction, purification, and mass spectrometric determinations of the isotope ratios for both the carbonates and phosphates of the hydroxyapatite. Results for both oxygen and carbon isotopes were plotted against time. Seasonal variations of as much as 2-3% in isotopic values were seen in the serially sampled enamel, with greater variability in the oldest sample at 2.92 million years

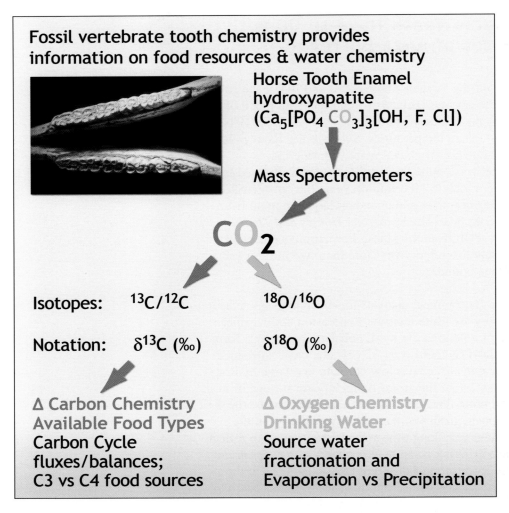

Fossil vertebrate tooth chemistry provides information on food resources & water chemistry

Horse Tooth Enamel
hydroxyapatite
$(Ca_5[PO_4\ CO_3]_3[OH,\ F,\ Cl])$

Mass Spectrometers

CO_2

Isotopes: $^{13}C/^{12}C$ $^{18}O/^{16}O$

Notation: $\delta^{13}C$ (‰) $\delta^{18}O$ (‰)

Δ Carbon Chemistry
Available Food Types
Carbon Cycle
fluxes/balances;
C3 vs C4 food sources

Δ Oxygen Chemistry
Drinking Water
Source water
fractionation and
Evaporation vs Precipitation

Figure 20.9
How Do We Reconstruct Earth's Climate History from Fossil Teeth?
The carbonate portion of the hydroxyapatite molecule is extracted chemically from a tooth enamel sample and analyzed by mass spectrometry into the isotope ratios of its carbon and oxygen atoms. From these ratios we can infer the climatic condition that led to a preponderance of one type of vegetation over the other ($\partial^{13}C$) and the evaporation/precipitation regime that provided the animal with its drinking water ($\partial^{18}O$). (n.b. the "per mill" symbol "‰" means "per thousand," and is not the same as the "percent" symbol "%".)

ago and less seasonality in the samples across the Plio-Pleistocene boundary at 2.18, and 1.06 million years ago. Clear trends were evident in the long-term (bulk samples) data.

High $\partial^{18}O$ values (enriched in ^{18}O) near 2.9 million years ago decreased gradually between 2.9 and 2.0 million years ago, suggesting a change from warm/dry conditions to a cooler/wetter climate during the latest Pliocene. Then, near the Plio-Pleistocene boundary at 1.9 million years ago and into the more recent Pleistocene, the oxygen isotopic values increased to those similar to 2.9 million years ago, suggesting an abrupt return to a warmer/drier climate. $\partial^{13}C$ values, however, showed a distinct shift at the Plio-Pleistocene boundary, indicating a change from C4 (more arid vegetation) to C3 (vegetation requiring wetter conditions) plants, which would suggest a transition to a cooler and/or wetter climate. This apparent paradox is an intriguing mystery that research scientists are currently studying. While we do not have a clear explanation yet, several geological and oceanographic changes were occurring at this time that could provide clues to explain this conundrum.

The late Pliocene and Plio-Pleistocene boundary period between ~2.7 and ~2 million years ago was a time when marine fossil evidence shows that ocean temperatures cooled substantially, and in the Northern Hemisphere, the Greenland ice sheets grew for the first time. It was also a time of major tectonic change, with the final closure of the Isthmus of Panama producing the continental pattern that we see today. In all likelihood, the final closure of Panama during the late Pliocene initiated the transformation of ocean and atmospheric circulation patterns into the patterns that exist in the twenty-first century. The oxygen isotope data obtained by Brogenski (2001) suggests that horses living in the vicinity of Anza-Borrego were drinking water that had experienced elevated evaporation. However, the carbon isotope data suggests that the grasses the

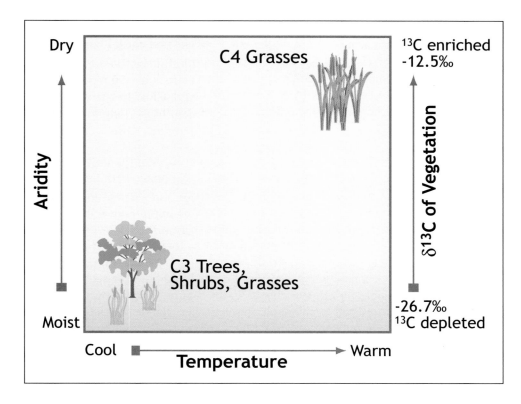

C4 Grasses

¹³C enriched
-12.5‰

C3 Trees,
Shrubs, Grasses

-26.7‰
¹³C depleted

Dry

Aridity

Moist

Cool ▪ → Warm

Temperature

δ¹³C of Vegetation

Figure 20.10
Carbon Isotopes and the Climatic Preference of C3 and C4 Plants.
A high ratio of ¹³C to ¹²C (∂¹³C) in a fossil bone or tooth suggests that the living animal's diet consisted mainly of "C4" plants, grasses that flourish in warm, dry conditions. A low ∂¹³C suggests a diet rich in the "C3" trees, shrubs and grasses that need a cooler, wetter environment. Intermediate values of ∂¹³C, of course, are less useful as indicators of ancient climate conditions.

horses were feeding on required more rainfall to thrive. Could the closure of the Isthmus of Panama, a change in ocean and atmospheric circulation, and the glaciation of Greenland have conspired in some complicated way to create conditions that can solve this contradictory data set? We can only wait and see what answers next come – quite literally from the horse's mouth.

Conclusions

Climate is a complex system, the sum of many mechanisms that interact both in obvious and unobvious ways. Life and its evolution always have and always will depend intimately on climate. Understanding the climate of our planet, how it changes from one state to another or maintains long-term stability, is of crucial importance to our species. The challenging puzzle of how the climate of the Anza-Borrego region evolved to its present state, including an understanding of the stages and transitions it passed through over the eons, requires more than one key for its full solution.

Some of those keys surely reside in cabinets and drawers full of fossil specimens. The fossil collection, neatly prepared, curated, dated, stored, and catalogued, is not merely the raw material for exhibits. It is an enormous natural archive of environmental data containing more information than we currently know how to extract. As Colleen Brogenski says in her 2001 thesis, "The fact that the Anza-Borrego Desert State Park stratigraphic sequence represents the longest continuous Plio-Pleistocene sequence in the United States, and is one of the few North American sections in which late Pliocene and early Pleistocene faunas occur in superposition, makes it an ideal section for the study of biological and ecological change at geologically short intervals." This, combined with abundant evidence of concurrent local tectonic and hydrographic changes, invites the development and confirmation of models for the processes that occurred here.

It is interesting to reflect on the early paleontologists who combed dry washes under a blazing sun to collect many of the fossil horse teeth that are being studied today (see Marrs, this volume, *The History of Fossil Collecting in the Anza-Borrego Desert Region*). At the time, those intrepid rock hounds could not have anticipated the intricacies of stable isotope mass spectrometry nor the wealth of environmental information that was locked within the small fragments of horse teeth they recovered. Even today, it is hard to imagine that the microchemistry of a newly exposed fossil tooth could hold the secret to the mechanisms responsible for climate change in the region two million years ago. Who can say what tools will be available to researchers ten or fifty years in the future, or what new information might be waiting to be deciphered from the bones and teeth of long-dead organisms that roamed this region in the distant past?

Epilogue

*Let us reflect carefully on the lessons to be learned from this fascinating time and place,
as the fate of our own species may very well depend upon
how deeply we understand them.*

Jacques Gauthier, Yale Peabody Museum of Natural History

Lowell Lindsay
George T. Jefferson

At the beginning of this book, Jacques Gauthier offers us a lofty yet humbling reflection on the role of paleontology in Anza-Borrego Desert State Park. We set out at the dawn of this decade to compile and portray a story, which began some seven million years earlier, and ended a few hundred thousand years ago. This desert window on the distant past and those of other sites in southern California, such as Rancho La Brea, bring this story up to the end of the Ice Ages and the appearance of humans in the Americas. Gauthier, and many other scientists, now think that the biosphere is "faced with the most daunting environmental challenges" since the last mass extinctions of the terminal Pleistocene. A paramount task of paleontology is to laboriously and exactingly unearth the fossil record, one tooth and one animal track at a time. Such evidence of prehistory is essential to research and is the basis of conclusions that will help us, as stewards of our environment, to make the best decisions for the future health of the planet. *Fossil Treasures of the Anza-Borrego Desert* is the story of this paleontologic work, performed by hundreds of volunteers and dozens of professional researchers over the last 150 years. It presents the results of this initial phase of discovery and analysis, paves the way for future paleontologic studies, and reveals past patterns of biotic and environmental change crucial to understanding our own impacts on the global ecosystem.

However there is another, possibly more engaging reason to pursue distant prehistory. On the surface, paleontology may seem to have little impact or relevance to our modern digital culture. But we must realize that inherent in the human spirit is a deep and profound curiosity about our past and a compulsive longing to understand our biological origins. In some ways paleontology, although a foggy view into deep time, satisfies these basic needs. And it does so on the same human experiential level as a Bach fugue or a Monet water lily. Why do paleontology? The remains of ancient life help us understand our place in time and space, and allow us to touch the past.

Figure E.1
A faraway group of paleontology volunteers makes a one mile trek up Virgin Wash, following the 1.2 Ma old contact between the fluvial Ocotillo Conglomerate (gray green) and the lacustrine Borrego Formation (pale red), into the distant past. (Photograph by John Strong)

Appendix

TABLE 1
Systematic List of Fossil Plants from Anza-Borrego Desert State Park.

Explanation: †= extinct taxon; sp. = single species; spp. = two or more species present; ? = tentative assignment. Most taxonomic names have been revised to conform with current usage. Higher taxonomic catagories are included where known, and common names appear in parentheses where applicable. Cretaceous and Eocene age microfossils were eroded from strata on the Colorado Plateau and redeposited by the ancestral Colorado River during Plio-Pleistocene time within the Salton Trough.

CRETACEOUS FLORAL ASSEMBLAGE
(Plio-Pleistocene)

Proteacidites spp. † (pollen)
Aquilapollenites spp. † (pollen)
Mancicorpus sp. † (pollen)
Tricolpites interangulus † (pollen)
Corollina sp. † (pollen)
Appendicisporites sp. † (pollen)
Cicatricosisporites sp. † (pollen)
Camarazonosporites sp. † (pollen)
Dinogymnium sp. † (pollen)
Palaeohystrichophora infusoriodes † (pollen)

Order indeterminate
  Family indeterminate
    Genera and species indeterminate (dynoflagellates and spores)

EOCENE FLORAL ASSEMBLAGE
(Plio-Pleistocene)

Pistillipollenites sp. cf. *P. mcgregorii* † (pollen)

FISH CREEK GYPSUM FLORAL ASSEMBLAGE
(late Miocene)

Hystrichokolpoma rigaudae (marine dinoflagellate)
Operculodinium centrocarpum (marine dinoflagellate)
Tectatodinium sp. (marine dinoflagellate)
Spiniferites ramosus (marine dinoflagellate)
Selenophix sp. (marine dinoflagellate)
Genera and species indeterminate (marine dinoflagellates)

Class Gymnospermae
Order Coniferales
  Family Cupressaceae
    Genus and species indeterminate (cypress, cedar or juniper)
  Family Taxodiacea
    Genus and species indeterminate (redwoods or sequoia)
  Family Pinaceae
    Genus and species indeterminate (pine or spruce)
  Family Ephedraceae
    Ephedripites sp. † (joint-fir)

Class Angiospermae
Subclass Dicotyledones
Order Fagales
  Family Juglandaceae
    Juglanspollenites sp. † (walnut)
Order Malvales
  Family Malvaceae
    Genus and species indeterminate (mallow)

Order Asterales
  Family Asteraceae
    Genus and species indeterminate (sunflower)
Order Myrtales
  Family Onagraceae
    Genus and species indeterminate (evening-primrose)

CARRIZO LOCAL FLORA
(Plio-Pleistocene)

Class Gymnospermae
Order Coniferales
  Family Cupressaceae
    Pineoxylon sp. (cedar or juniper)

Class Angiospermae
Subclass Dicotyledones
Order Laurales
  Family Lauraceae
    Persea coalingensis †(avocado)
    Umbellularia salicifolia †(bay laurel)
Order Malpighales
  Family Salicaceae
    Populus sp. (cottonwood)
    Populus sp. cf. *P. alexanderi* †(Alexander's cottonwood)
    Salix sp. (willow)
    Salix gooddingii (Gooding's willow)
Order Lamiales
  Family Oleaceae
    Fraxinus caudata (ash)
Order Fagales
  Family Juglandaceae
    Juglans pseudomorpha †(walnut)
Order Sapindales
  Family Hippocastanaceae
    Aesculus sp. (buckeye)

Subclass Monocotyledones
Order Arecales
  Family Arecaceae
    Washingtonia sp. (fan-palm)
    Genus and species indeterminate (palm)

FRESH WATER FLORAL ASSEMBLAGE
(Plio-Pleistocene)

Pediastrum spp. (algal phytoplankton)
Scenedesmus spp. (algal phytoplankton)
Chara haitensis (macrophytic algae)
Lamprothamnium sp. (macrophytic algae)
Nitellopisis sp. (macrophytic algae)

TABLE 2
Systematic List of Fossil Invertebrates from Anza-Borrego Desert State Park and the Salton Trough Region, San Diego, Imperial, and Riverside Counties.

Explanation: ★ = type specimens; † = extinct taxon; sp. = single species; spp. = two or more species present; ? = tentative assignment. Taxonomic names have been revised to conform with current usage. Higher taxonomic categories are included where known, and common names are provided where known. Subgenera are not listed. Authors and dates for species are included when known. Note that new names (nov.) of Stump (1972) and Powell (1986, 1988) remain informal. Microfossils that comprise the Cretaceous age assemblage were eroded from the Colorado Plateau and redeposited within the Salton Trough during Plio-Pleistocene time.

PALEOZOIC MOLLUSCS
(? Mississippian)

Phylum Mollusca (marine clams and snails)
Order indeterminate
  Family indeterminate
    Genus and species indeterminate

CRETACEOUS MICROFOSSILS
(Plio-Pleistocene)

Subkingdom Protozoa
PHYLUM SARCOMASTIGOPHORA
Subphylum Sarcodina (foraminifers)
Class Granuloreticulosea
Order Foraminifera
  Family Globotruncanidae
    Globotruncana globigerinoides †
  Family Planomalinidae
    Globigerinelloides aspera †
  Family Heterohelicidae
    Heterohelix globulosa †
    H. reussi †
    H. striata †
  Family Turrilinidae
    Neobulimina canadensis †
  Family Uvigerinidae
    Pseudouvigerina cretacea †

FISH CREEK GYPSUM ASSEMBLAGE
(late Miocene)

Division Haptophyta
Class Coccolithophyceae (calcareous nannoplankton)
Order Isochrysidales
  Family Gephrocapsaceae
    Dictyococcittes scrippsae (?)
    D. minutes
    Reticulofenestra pseudoumbilica
    Crenalithus doronicoides (?)
Order Discoasterales
  Family Sphenolithaceae
    Spenolithus abies
    S. moriformis
  Family Braarudosphaeracea
    Braarudosphaera bigelowii
Order Eiffellithales
  Family Helicosphaeracea
    Helicosphaera kamptnen

Order Coccolithales
  Family Coccolithaceae
    Calcidiscus macintyrei
    Coccolithus pelagicus

IMPERIAL GROUP: WESTERN SALTON TROUGH, COYOTE MOUNTAINS, AND VALLECITO CREEK/FISH CREEK BASIN ASSEMBLAGES
(late Miocene)

Subkingdom Protozoa
PHYLUM SARCOMASTIGOPHORA
Subphylum Sarcodina (foraminifers)
Class Granuloreticulosea
Order Foraminifera
  Family Amphisteginidae
    Amphistegina gibbosa
    A. lessoni
  Family Miliolidae
    Quinqueloculina sp.
  Family Buliminidae
    Bolivina interjuncta
    B. subaenariensis mexicana
    B. vaughani
    Reussella pacifica
    Uvigerina peregrina
    Trifarina bella
    T. angulosa
  Family Cassidulinidae
    Cassidulina delicata
    C. laevigata
    C. subglobssa
    C. tortuosa
  Family Anomalinidae
    Cibicides fletcheri
    Hanzawaia nitidula
    H. basiloba
    Planulina ornata
  Family Discorbidae
    Epistominella subperuviana
  Family Globigerinidae
    "*Globigerina pachyderma*"
    G. quinqueloba
    Globigerinita uvula
    Sphaeroidinella subdhiscens
  Family Nonionidae
    Elphidium gunteri
    Elphidium sp.
    Nonion basispinata

N. miocenica stella
N. glabratalia californica
Family Planorbulinidae
Planorbulina acervalis
Family Rotaliidae
Ammonia beccarii
Ammonia sp. cf. *A. parkinsonia*
Siphonina pulchra
Family Textulariidae
Textularia schencki
Textularia spp.

PHYLUM CNIDARIA

Class Anthozoa
Subclass Zoantharia (corals)
Order Scleractinia
Family Meandrinidae (brain and flower corals)
Meandrina bowersi (Vaughan, 1917)★ †
Dichocoenia merriami (Vaughan, 1900)★ †
D. eminens Weisbord, 1974†
Eusmilia carrizensis Vaughan, 1917★ †
Family Siderastreidae (starlet corals)
Siderastrea mendenhalli Vaughan, 1917★ †
Family Rhizangiidae (= Astrangiidae) (cup corals)
Astrangia haimei Verrill, 1866 †
Family Faviidae (star corals)
Manicina sp.
Plesiastrea californica (Vaughan, 1917) †
Solenastrea fairbanksi (Vaughan, 1900)★ †
Family Poritidae (finger corals)
Porites carrizensis Vaughan, 1917★ †

PHYLUM MOLLUSCA

Class Gastropoda (marine snails)
Order Patellogastropoda
Family Lottiidae (limpets)
"Patella" sp. †
Order Vetigastropoda
Family Fissurellidae (keyhole limpets)
Diodora alta (C.B. Adams, 1852)
D. cayenesis (Lamarck, 1822)
Diodora sp.
Fissurella sp.
Hemitoma emarginata (Blainville, 1825)
Rimula cancellata Schremp, 19981★ †
Family Haliotidae (abalones)
Haliotis pourtalesii Dall, 1881
Haliotis sp.
Family Turbinidae (turban snails)
Mirachelus imperialis Schremp, 1981★ †
Turbo magnificus Jonas, 1844
T. mounti Schremp, 1981★ †
Turbo sp.
Homalopoma maiquetiana (Weisbord, 1962) †
Family Liotiidae
Arrene stephensoni Schremp, 1981 ★ †
Family Trochidae (pearly top snails)
Calliostoma bonita Strong, Hanna, and Hertlein, 1933
C. olssoni Schremp, 1981★ †
Tegula mariana Dall, 1919

Order Neritopsina
Family Nertidae (nerites)
Nerita scabricosta Lamarck, 1822
N. funiculata Menke, 1851
Theodoxus luteofasciatus Miller, 1879
Order Neotaenioglossa
Family Cerithiidae (ceriths)
Cerithium incisum Sowerby, 1855
Liocerithium judithae Keen, 1971
Family Potamididae (horn snails)
Cerithidea mazatlanica Carpenter, 1857
Family Modulidae (button snails)
Modulus catenulatus (Philippi, 1849)
Family Turritellidae (tower snails)
Turritella gonostoma Valenciennes, 1832
T. imperialis Hanna, 1926★ †
Vermicularia pellucida (Broderip and Sowerby, 1829)
Family Littorinidae (periwinkles)
Littorina varia Sowerby, 1832
Family Strombidae (conchs)
Strombus galeatus Swainson, 1823
S. granulatus Swainson, 1822
S. gracilior Sowerby, 1825
S. obliteratus Hanna, 1926 †
Family Hipponicidae (horse-hoof limpets)
Cheilea cepacea (Broderip, 1834)
Hipponix panamensis C.B. Adams, 1852
Family Calyptraeidae (cup and saucer snails)
Crepidula onyx Sowerby, 1824
Crucibulum scutellatum (Wood, 1828)
C. spinosum (Sowerby, 1824)
Family Cypraeidae (cowries)
Macrocypraea cervinetta (Kiener, 1843)
Muracypraea sp.
Zonaria sp.
Family Naticidae (moon snails)
Natica chemnitzii Pfeiffer, 1840
N. unifasciata Lamarck, 1822
Polinices bifasciatus (Griffith and Pidgeon, 1834)
P. uber (Valenciennes, 1832)
Family Cassididae (helmet snails)
Cassis subtuberosus Hanna, 1926★ †
Family Tonnidae (tuns)
Malea ringens (Swainson, 1822)
Family Ficidae (fig snails)
Ficus ventricosa (Sowerby, 1825)
Family Personidae (distorsio snails)
Distorsio constricta (Broderip, 1833)
Family Epitoniidae (wentletraps)
Epitonium efferum Bramkamp in Durham, 1937
Order Neogastropoda
Family Muricidae (murex snails)
Hexaplex brassica (Lamarck, 1822)
Eupleura muriciformis (Broderip, 1833)
Family Turbinellidae (vase snails)
Vasum pufferi Emerson, 1964★ †
Family Buccinidae (whelks)
Solenosteira anomala (Reeve, 1847)
Family Melongenidae (crown conchs)
Melongena patula (Broderip and Sowerby, 1829)
Family Fascolariidae (tulip snails and horse conchs)
Fusinus dupetitthouarsii (Kiener, 1840)

Pleuroploca princeps (Sowerby, 1825)
Latrius concentricus (Reeve, 1847)
Leucozonia cerata (Wood, 1828)
Family Nassariidae
Nassarius sp.
Family Columbellidae (dove shells)
Strombina solidula (Reeve, 1859)
Family Olividae (olive snails)
Oliva incrasatta [Lightfoot, 1786]
O. porphyria (Linnaeus, 1758),
O. spicata (Röding, 1798)
Olivella gracilis (Broderip and Sowerby, 1829)
Family Mitridae (miter snails)
Mitra crenata Broderip, 1836
Subcancilla longa (Gabb, 1873) †
S. sulcata (Swainson in Sowerby, 1825)
Family Cancellariidae (nutmeg snails)
Cancellaria cassidiformis (Sowerby, 1832)
C. coronadoensis Durham, 1950 †
C. obesa Sowerby, 1832
C. urceolata Hinds, 1843
Family Terebridae (auger snails)
Terebra dislocata (Say, 1822)
T. elata Hinds, 1844
T. robusta Hinds, 1844
Family Turridae (turrid snails)
Knefastia olivacea (Sowerby, 1833)
Polystira oxytropis (Sowerby, 1834)
Family Conidae (cone snails)
Conus arcuatus Broderip and Sowerby, 1829
C. bramkampi Hanna and Strong, 1949 ★ †
C. californicus Reeve, 1844
C. durhami Hanna and Strong, 1949 ★ †
C. gladiator Broderip, 1833
C. fergusoni Sowerby, 1873
C. patricius Hinds, 1843
C. regularis Sowerby, 1833
C. ximenes Gray, 1839
Order Heterostropha
Family Architectonicidae (sundials)
Architectonica nobilis discus Grant & Gale, 1931 †
Order Cephalaspidea
Family Bullidae (bubble snails)
Bulla punctulata A. Adams in Sowerby, 1850
B. striata Bruguière, 1792
Order Basommatophora
Family Melampidae
Melampus sp.

Class Scaphopoda (tusk shells)

Order Dentalioida
Family Dentaliidae
Dentalium sp.

Class Bivalvia (marine clams)

Order Nuculoida
Family Nuculidae (nut clams)
Nucula sp.
Family Nuculanidae
Nuculana acuta (Conrad, 1831)
N. santarosaensis Perrilliat, 1976 †
Order Arcoida
Family Arcidae (ark clams)
Arca mutabilis (Sowerby, 1833)

A. pacifica (Sowerby, 1833)
Barbatia reeveana (Orbigny, 1846)
Anadara carrizoensis Reinhart, 1943 ★ †
A. concinna (Sowerby, 1833)
A. formosa (Sowerby, 1833)
A. multicostata (Sowerby, 1833)
A. reinharti (Lowe, 1935)
Family Glycymerididae (bittersweet clams)
Glycymeris bicolor (Reeve, 1843)
G. delessertii (Reeve, 1843)
G. gigantea (Reeve, 1843)
G. maculata (Broderip, 1832)
G. multicostata (Sowerby, 1833)
Order Mytiloida
Family Mytilidae (mussels)
Lithophaga sp. aff. *L. plumula* (Hanley, 1844)
Order Pterioida
Family Pinnidae (penshells)
Pinna latrania Hanna, 1926 ★ †
P. mendenhalli Hanna, 1926 ★ †
Atrina stephensi Hanna, 1926 ★ †
Order Limoidea
Family Limidae (file shells)
Ctenoides floridana (Olsson and Harbison, 1953)
Limaria sp. †
Order Ostreoida
Family Ostreidae (true oysters)
Myrakeena angelica (Rochebrune, 1895)
"Dendostrea" angermani (Hertlein and Jordan, 1927)
Dendostrea? vespertina (Conrad, 1854) ★†
Crassostrea columbiensis (Hanley, 1846)
Saccostrea palmula (Carpenter, 1857)
Undulostrea megodon (Hanley, 1846)
Family Gryphaeidae (oysters)
Pycnodonte heermanni (Conrad, 1855) ★ †
Family Pectinidae (scallops)
Antipecten? praevalidys (Jordan and Hertlein, 1926)
Argopecten circularis bramkampi (Durham, 1950) ★ †
A. deserti (Conrads, 1855) ★ †
A. mendenhalli (Arnold, 1906) †
A. sverdrupi (Durham, 1950) †
A. ventricosus (Sowerby, 1842)
Chlamys corteziana Durham, 1950 †
C. lowei (Hertlein, 1935)
C. mediacostata (Hanna, 1926) ★ †
Cyclopecten pernomus (Hertlein, 1935)
Euvola keepi (Arnold, 1906) †
Flabellipecten carrizoensis (Arnold, 1906) ★ †
Flabellipecten sp.
Leptopecten palmeri (Dall, 1897)
L. velero (Hertlein, 1935)
Lyropecten tiburonensis Smith, 1991 †
Family Plicatulidae (kittenpaws)
Plicatula inezana Durnam, 1950
P. penicillata Carpenter, 1857
Plicatula sp.
Family Spondylidae (thorny osters)
Spondylus bostrychites Guppy, 1867 †
S. calcifer Carpenter, 1857
S. princeps Broderip, 1833
Family Anomiidae (jingles)
Anomia subcostata Conrad, 1855 ★ †
Placunanomia hannibali Jordan and Hertlein, 1926 †

Order Veneroida
  Family Lucinidae (lucine clams)
    Calucina quincula Olsson, 1961 ????
    Codakia distinguenda (Tryon, 1872)
    Miltha xantusi (Dall, 1905)
    Pegophysema edentuloides (Verrill, 1870)
    Divalinga eburnea (Reeve, 1850)
    Parvilucina mazatlanica Carpenter, 1857
  Family Crassatellidae (crasstellas)
    Crassinella mexicana Pilsbry and Lowe, 1932
    Eucrassatella digueti Lamy, 1917
    E. subgibbosa (Hanna, 1926) ★ †
  Family Chamidae (jewelbox clams)
    Chama frondosa Broderip, 1835
    Arcinella arcinella (Linnaeus, 1767)
    A. californica (Dall, 1903)
  Family Carditidae (carditas)
    Cardites crassicostata (Sowerby, 1825)
    C. megastropha (Gray, 1825)
    Carditamera laticostata Sowerby, 1833
  Family Corbiculidae (marshclams)
    Polymesoda notabilis (Deshayes, 1855)
  Family Cardiidae (cockles or heart clams)
    Trigoniocardia sp. aff. *T. guanacastensis* (Hertlein and Strong, 1947)
    Laevicardium sp.
    Lophocardium gurabicum (Maury, 1917)
  Family Veneridae (Venus clams)
    Chione hannai Parker, 1949
    Chione sp.
    Cyclinella cyclica (Guppy, 187x)
    Dosinia dunkeri (Philippi, 1844)
    D. ponderosa (Gray, 1838)
    Gouldia californica Dall, 1917
    Irus ellipticus (Sowerby, 1834)
    Megapitaria sp.
  Periglypta multicostata (Sowerby, 1835)
    Pitar sp. ? *P. catharius* (Dall, 1902)
    Globivenus isocardia (Verrill, 1870)
  Family Tellinidae (tellens)
    Tellina ulloana Hertlein, 1968
    T. pristiphora Dall, 1900
    T. ochracea Carpenter, 1864
    Florimetis dombei (Hanley, 1844)
    Macoma siliqua (C.B. Aams, 1852)
  Family Donacidae (bean or wedge clams)
    Donax sp. cf. *D. gracilis* Hanley, 1845
  Family Semelidae (semele clams)
    Semele bicolor (C.B. Adams, 1852)
    S. sayi Toula, 1909
  Family Solecurtidae (tagelus clams)
    Tagelus californianus (Conrad, 1837)
    T. violascens (Carpenter, 1857)
    Solecurus gatunensis Toula, 1909
Order Myoida
  Family Myidae (softshell clams)
    Crpytomya sp.
  Family Corbulidae (corbulas)
    Corbula mexicana Perrilliat, 1984 †
    Corbula sp. †
  Family Haitellidae (geoducks)
    Panopea abrupta (Conrad, 1849)

Family Pholadidae (rock piddocks)
    Cyrtopleura costata (Linnaeus, 1758)
Order Pholadomyoida
  Family Thraciidae (thracia clams)
    Cyathodonta undulate Conrad, 1849

PHYLUM ECHINODERMATA

Class Asteroidea (sea stars)
Order Paxillosida
  Family Astropectinidae (sand stars)
    Astropecten armatus Gray, 1840
Subclass Ophiuroidea
  Family, Genus and species indeterminate (brittle stars)

Class Echinoidea (sea urchins and sand dollars)
Order Cidaroida
  Family Cidaridae (club-spined urchins)
    Cidaris sp.
    Eucidaris thouarsii (Valenciennes, 1846)
Order Diadematoida
  Family Diadematidae (ea urchins)
    Centrostephanus sp.
Order Arbacioida
  Family Arbaciidae (regular sea urchins)
    Arbacia incisa (Agassiz, 1863)
Order Temnopleuroida
  Family Toxopneustidae (white urchins)
    Lytechinus sp. cf. *L. anamesus* Clark, 1912
    Toxopneustes roseus (Agassiz, 1863)
    Tripneustes californicus (Kew, 1914) †
Order Echinoida
  Family Strongylocentrotidae (sea urchins)
    Strongylocentrotus purpuratus (Stimpson, 1857)
Order Clypeasteroida
  Family Clypeasteridae (Sea biscuits)
    Clypeaster bowseri Weaver, 1908 ★ †
    C. carrizoensis Kew, 1914 ★ †
    C. deserti Kew, 1915 ★ †
    Clypeaster sp. cf. *C. rotundus* (Agassiz, 1863)
  Family Echinarachniidae (sand dollar)
    Vaquerosella sp. †
  Family Mellitidae (Key-hole sand dollar)
    Encope arcensis Durham, 1950 †
    E. sverdrupi Durham, 1950 †
    E. tenuis Kew, 1914 ★ †
  Family Echinoneidae (regular heart urchins)
    Echinoneus burgeri Grant and Hertlein, 1938 †
Order Spatangoida
  Family Schizasteridae (puffball heart urchins)
    Agassizia scrobiculata Valenciennes, 1846 †
    Agassizia sp.
    Schizaster morlini Grant and Hertlein, 1956 †
  Family Brissidae (heart urchins)
    Brissus obesus Verrill, 1867
    Metalia spatagus Linnaeus, 1758
    Meoma sp.
  Family Loveniidae (Porcupine heart urchins)
    Lovenia hemphilli Israelsky, 1923 †

PHYLUM BRYOZOA

Class Gymnolaemata (bryozoans)
Order Cheilostomata
    Family Membraniporidae
        Conopeum commensale Kirkpatrick and Metzelaar, 1922

PHYLUM BRACHIOPODA

Class Inarticulata
Order Artremata
    Family Lingulidae
        Glottidia? sp.

PHYLUM ARTHROPODA
Subphylum Crustacea

Class Maxillopoda
Order Sessilia
    Family Balanidae
        Megabalanus tintinnabulum (Linnaeus, 1758) (acorn barnacle)

Class Malacostraca
Order Decapoda
    Family Goneplacidae
        Speocarcinus berglundi Tucker, Feldman, and Powell, 1994 (crab)

Class Ostracoda (marine water fleas, mussel shrimps)
Order Podocopida
    Family Trachyleberididae
        Puriana sp.
        Hermanites sp.
    Family Loxoconchidae
        Loxocorniculum sp.
        Loxoconcha sp.
    Family Xestoleberideidae
        Xestoleberis sp.
    Family Cytheruridae
        Cytheura sp.
        Anterocythere sp.
    Family Hemicytheridae
        ?Aurila sp
        ?Ambostracon sp.
        Caudites sp.
    Family Cytheridae
        Perissocytheridea sp.
    Family Microcytheridae
        Microcytherua sp.

TYPANITES MARINE ICHNOFAUNA (borings and tracks)
(late Miocene)

Entobia (clionid sponge boring)
cf. *Gastrochaenolites torpedo* (clam boring)
Maeandro polydora (polychaete worm boring)
Typanites (polychaete worm or barnacle boring)
Echinoid (boring)

FRESH WATER LACUSTRINE & TERRESTRIAL ASSEMBLAGES
(Plio-Pleistocene)

PHYLUM MOLLUSCA
Class Bivalvia (fresh water clams)
Order Unionoida
    Family Unionidae (freshwater mussels)
        Anodonta californiensis Lea, 1852

Order Veneroida
    Family Sphaeriidae (pea clams)
        Pisidium compressum Prime, 1852
    Family Mactridae (rangia clams)
        Rangia lecontei (Conrad, 1853) ★ †

Class Gastropoda (fresh water & land snails)
Order Neotaenioglossa
    Family Hydrobiidae (various freshwater snails)
        Amnicola sp.
        Pyrgulopsis longinqua (Gould, 1855)
        Fluminicola sp.
        Hydrobia sp.
        Tyronia sp.
Order Basommatophora
    Family Physidae (physa snails)
        Physa humerosa (Gould, 1855)
        P. virgata (Gould, 1855)
    Family Lymnaeidae (lymnaea snails)
        Fossaria techella Haldeman, 1867
    Family Planorbidae (rams-horn snails)
        Gyraulus parvus (Say, 1817)
        Planorbella tenuis (Dunker, 1850)
Order Stylommatophora
    Family Zonitidae (zonite land snails)
        Zonitoides arboreus (Say, 1816)
    Family Limacidae (slugs)
        Deroceras sp.

PHYLUM ARTHROPODA
Subphylum Crustacea

Class Ostracoda
Order Podocopia (fresh water fleas)
    Family Candonidae
        Candona patzcuaro var. *C. p. mexico*
        C. sigmoides
        ? Lineocypris sp.
    Family Limnocytheridae
        Elkocythereis bramlettei
        Limnocythere bradburyi
        L. camera
        L. herricki
        L. inopinata
        L. itasca
        L. pseudoreticulata
        L. staplini
        L. verrucosa
    Family Ilyocyprididae
        Ilyocypris sp.
Order Decapoda
    Suborder Pleocyemata
        Family, genus, and species indeterminate (crayfish)

Class Insecta (insects)
Order Coleoptera
    Family Bostrichidae
        cf. *Lyctus* sp. (powder-post beetle)

TABLE 3
Systematic List of Fossil Vertebrates from Anza-Borrego Desert State Park Region.

Explanation: ★ = type specimens; †= extinct taxon; cf. = compares favorably with; nr. = near; sp. = single species; spp. = two or more species present; ? = tentative assignment. Taxonomic names have been revised to conform with current usage, and common names are provided in parentheses where applicable.

SANTA ROSA METAMORPHICS ASSEMBLAGE
(middle Ordovician)

"Class" Euconodonta ("agnathan fish")
Paltodus sp. ? *P. spurius*
Paltodus sp. ? *P. bassleri*
"Scolopodus" sp. ? *"S."* quadraplicatus
? *Ulrichodina* sp.

IMPERIAL GROUP MARINE ASSEMBLAGE
(late Miocene)

Class Chrondricthyes (sharks, skates, and rays)
Order Galeomorpha
Family Cetorhinidae
Cetorhinus sp. (basking sharks)
Family Carcharinidae
Carcharhinus sp. (requiem sharks)
Family Lamnidae
Carcharocles megalodon † (great white shark)
Isurus sp. (mako shark)
Family Odontaspididae
Odontaspis sp. (sand shark)
Order Myliobatiformes
Family Myliobatidae
Myliobatis sp. (bat rays)

Class Actinopterygii (bony fish)
Order Clupeiformes
Family Clupeidae
Genus and species indeterminate (herrings)
Order Tetraodontiformes
Family Tetraodontidae
Arothron sp. (puffer fish)
Family Balistidae
Genus and species indeterminate (triggerfish)
Order Perciformes
Family Labridae
Semicossyphus sp. (sheepshead)
Family Sphyraenidae
Sphyraena sp. (barracudas)

Class Reptilia (reptiles)
Order Testudines
Family Cheloniidae
Genus and species indeterminate (sea turtles)

Class Mammalia (mammals)
Order Carnivora
Family Odobenidae
Valenictus imperialensis ★ † (Imperial walrus)
Order Sirenia

Family Dugongidae
Genus and species indeterminate (dugong)
Order Cetacea
Family Balaenopteridae
Genus and species indeterminate (baleen whale)
Family Cetotheriidae
Genus and species indeterminate (whale-bone whale)
Family Delphinidae and/or Phocaenidae
Genus and species indeterminate (dolphins and porpoises)

FISH CREEK, VALLECITO CREEK, AND BORREGO AQUATIC AND TERRESTRIAL ASSEMBLAGES
(Plio-Pleistocene)

Class Actinopterygii (bony fish)
Order Salmoniformes
Family Salmonidae
? Oncorhynchus sp. (Pacific salmons and trouts)
Order Cypriniformes
Family Catostomidae
Genus and species indeterminate (suckers)
Xyrauchen sp. cf. *X. texanus* (razorback sucker)
Family Cyprinidae
Gila sp. (chub)
Ptychochelius lucius (pike-minnow)

Class Amphibia (amphibians)
Order Anura
Family Bufonidae
Bufo sp. (toads)

Class Reptilia (reptiles)
Order Testudines
Family Kinosternidae
Kinosternon sp. cf. *K. sonoriense* (Arizona mud turtle)
Family Testudinidae
Hesperotestudo sp. (giant tortoises)
Xerobates agassizii (desert tortoise)
Family Emydidae
Clemmys marmorata (western pond turtle)
Trachemys scripta (common slider)
Order Squamata
Suborder Lacertilia
Family Iguanidae
Dipsosaurus dorsalis (desert iguana)
Gambelia corona ★ † (crowned leopard lizard)
Phrynosoma sp. (horned lizard)
Phrynosoma anzaense ★ † (Anza horned lizard)
Pumilia novaceki ★ † (Novacek's small iguana)
cf. *Sceloporus* sp. A (spiny lizard)
cf. *Sceloporus* sp. B (spiny lizard)
Uta sp. ? *U. stansburiana* (side-blotched lizard)

Family Teiidae
    Ameiva sp. or *Cnemidophorus* sp. (ground lizard or whiptail)
Family Scincidae
    Eumeces sp. (skinks)
Family Xantusiidae
    Xantusia downsi ★ † (Downs' night lizard)
Family Anguidae
    cf. *Gerrhonotus* sp. (alligator lizards)
Family Helodermatidae
    Heloderma sp. (Gila monsters)
Suborder Serpentes
  Family Colubridae
    Hypsiglena sp. (night snakes)
    Lampropeltis getulus (common king snake)
    Masticophis flagellum (coachwhip snake)
    Thamnophis sp. (garter snakes)
  Family Crotalidae
    Crotalus sp. (rattlesnakes)

Class Aves (birds)

Order Gaviformes
  Family Gaviidae
    Gavia sp. (loons)
Order Podicipediformes
  Family Podicipedidae
    Aechmophorus occidentalis (western grebe)
    Podiceps sp. † (grebes)
    Podiceps nigricollis (eared grebe)
Order Pelecaniformes
  Family Pelecanidae
    Pelecanus sp. (pelican)
Order Ciconiiformes
  Family Teratornithidae
    Aiolornis incredibilis † (incredible wind god)
  Family Vulturidae (Cathartidae)
    Gymnogyps sp. (condors)
Order Phoenicopteriformes
  Family Phoenicopteridae
    Phoenicopterus sp. † (flamingos)
Order Anseriformes
  Family Anatidae
    Anas sp. ? *A. acuta* (pintail duck)
    A. clypeata (shoveller duck)
    Anser sp. (large goose)
    Branta sp. cf. *B. canadensis* (Canada goose)
    Brantadorna downsi ★ † (Downs' gadwall duck)
    Bucephala fossilis ★ † (fossil goldeneye)
    Chen rossii (Ross' goose)
    Cygnus sp. ? *C. paloregonus* † (fossil Oregon swan)
    Melanitta sp. ? *M. persipicillata* (surf scoter)
    Oxyura bessomi ★ † (Bessom's stiff-tailed duck)
Order Accipitriformes
  Family Accipitridae
    Aquila sp. ? *A. chrysaëtos* (golden eagle)
    Buteo sp. ? *B. lineatus* (red-shouldered hawk)
    Neophrontops vallecitoensis ★ † (Vallecito neophron)

Family Falconidae
    ? *Falco* sp. (falcon)
Order Galliformes
  Family Phasianid
    Meleagris anza ★ † (Anza turkey)
  Subfamily Odontophorinae
    Callipepla sp. ? *C. californica* (California quail)
    Callipepla gambelii (Gambel's quail)
Order Gruiformes
  Family Gruidae
    Grus canadensis (sandhill crane)
  Family Rallidae
    Fulica americana (American coot)
    F. hesterna ★ † (yesterday's coot)
    Rallus sp. (rails)
    Rallus limicola (Virginia rail)
Order Charadriiformes
  Family Charadriidae
    Charadrius vociferus (killdeer)
Order Strigiformes
  Family Strigidae
    Asio sp. †(eared owls)
Order Piciformes
  Family Picidae
    Genus and species indeterminate (woodpeckers or wryneck)
Order Passeriformes
  Family indeterminate
    Genera and species indeterminate (song birds)
  Family Corvidae
    Corvus sp. (crows)

Class Mammalia (mammals)

Order Insectivora
  Family Soricidae
    Notiosorex jacksoni †(Jackson's desert shrew)
    Sorex sp. (red-toothed shrews)
  Family Talpidae
    Scapanus malatinus ★ † (mole)
Order Chiroptera
  Family Vespertilionidae
    Anzanycteris anzensis ★ † (Anza-Borrego bat)
    Myotis sp. (mouse-eared bats)
Order Xenarthra
  Family Megalonychidae
    Megalonyx leptostomus † (narrow-mouthed ground sloth)
    M. jeffersonii † (Jefferson's ground sloth)
    M. wheatleyi † (Wheatley's ground sloth)
  Family Megatheriidae
    Nothrotheriops sp. cf. *N. shastensis* † (Shasta ground sloth)
  Family Mylodontidae
    Paramylodon sp. ? *P. harlani* † (Harlan's ground sloth)
Order Lagomorpha
  Family Leporidae
  Subfamily Archaeolaginae
    Hypolagus edensis †(Eden rabbit)
    H. vetus †(ancient rabbit)
    Pewelagus dawsonae ★ † (Dawson's rabbit)
  Subfamily Leporinae
    ? *Nekrolagus* sp. † (rabbits)
    ? *Sylvilagus* sp. (cottontail rabbits)
    Sylvilagus hibbardi ★ † (Hibbard's cottontail)

Order Rodentia
  Family Erethizontidae
    Coendou stirtoni ★ † (Stirton's coendou)
  Family Sciuridae
    Eutamias sp. (chipmunks)
    Spermophilus (*Otospermophilus*) sp. (ground squirrels)
  Family Geomyidae
    Geomys anzensis ★ † (Anza pocket gopher)
    G. garbanii ★ † (Garbani's pocket gopher)
    Thomomys sp. (western pocket gopher)
  Family Heteromyidae
    Dipodomys sp. (A) † (kangaroo rat)
    Dipodomys sp. (B) † (kangaroo rat)
    Dipodomys sp. cf. *D. minor* † (small kangaroo rat)
    D. compactus (gulf coast kangaroo rat)
    D. hibbardi † (Hibbard's kangaroo rat)
    Microdipodops sp. † (kangaroo mice)
    Perognathus nr. *P. hispidus* (hispid pocket mouse)
    Prodipodomys sp. † (extinct kangaroo rats)
  Family Castoridae
    Castor sp. (beaver)
  Family Cricetidae
    Subfamily Sigmodontinae
      Baiomys sp. (pygmy mice)
      Calomys (*Bensonomys*) sp. † (extinct mice)
      Neotoma (*Paraneotoma*) sp. (woodrat)
      Neotoma (*Parahodomys*) sp. (A) † (woodrat)
      Neotoma (*Parahodomys*) sp. (B) † (woodrat)
      Onychomys sp. (grasshopper mice)
      Peromyscus sp. (white-footed mice)
      Reithrodontomys sp. (harvest mice)
      Repomys sp. † (pygmy woodrats)
      Sigmodon curtisi † (Curtis cotton rat)
      S. lindsayi ★ † (Lindsay's cotton rat)
      S. medius and/or *S. minor* † (intermediate and/or small cotton rat)
    Subfamily Arvicolinae
      Lasiopodomys sp. cf. *L. deceitensis* † (Cape Deceit vole)
      Microtus sp. ? *M. californicus* (California meadow vole)
      Mimomys (*Ophiomys*) sp. cf. *M.(O.) parvus* † (small Snake River vole)
      M. (*Cosomys*) sp. † (Coso vole)
      ? *Ondatra idahoensis* † (Idaho muskrat)
      Pitymys meadensis † (Mead's vole)
      Pliopotamys minor † (pygmy muskrat)
      Synaptomys (*Mictomys*) *anzaensis* ★ † (Anza bog lemming)
Order Carnivora
  Family Canidae
    Borophagus diversidens † (bone-eating dog)
    Canis armbrusteri † (Armbruster's wolf)
    C. priscolatrans (= *C.edwardii*, Edward's dog) † (wolf coyote)
    C. lepophagus † (Johnston's wolf)
    C. latrans (coyote)
    Urocyon sp. ? *U. progressus* † (extinct gray fox)
    ? *Vulpes* sp. (fox)
  Family Ursidae
    Arctodus simus † (short-faced bear)
    Tremarctos floridanus † (Florida cave bear)
    ? *Ursus* sp. cf. *U. americanus* (American black bear)
  Family Procyonidae
    Bassaricus casei † (Case's ringtail)
    Procyon sp. cf. *P. rexroadensis* † (Rexroad raccoon)
    Nasua sp. (coati)

Family Mustelidae
  cf. *Satherium piscinarium* (fish-eating river otter)
  Mustela sp. cf. *M. frenata* (long-tailed weasel)
  Spilogale sp. cf. *S. putorius* (spotted skunk)
  Taxidea taxus (badger)
  ? *Gulo* sp. (wolverine)
  Subfamily Grisoninae
    Trigonictis macrodon † (large-toothed grison)
Family Felidae
  Subfamily Felinae
    Miracinonyx inexpectatus (cheetah-like cat)
    Panthera onca † (jaguar)
    Felis rexroadensis † (Rexroad cat)
    F. (*Lynx*) *rufus* (bobcat)
    Felis sp. (cat)
  Subfamily Machairodontinae
    Smilodon gracilis † (gracile sabertooth cat)
Order Proboscidea
  Family Gomphotheriidae
    Gomphotherium sp. † (gomphothere)
    Stegomastodon sp. † (stegomastodont)
  Family Elephantidae
    Mammuthus meridionalis † (southern mammoth)
    M. columbi † (Columbian mammoth)
Order Perissodactyla
  Family Equidae
    cf. *Dinohippus* sp. † (Pliocene horse)
    E. enormis ★ † (giant zebra)
    E. scotti † (Scott's horse)
    E. (*Plesippus*) sp. cf. *E.* (*P.*) *simplicidens* † (American zebra)
    Equus (*Equus*) sp. A † (medium-sized horse)
    Hippidion sp. † (South American equid)
  Family Tapiridae
    Tapirus merriami † (Merriam's tapir)
Order Artiodactyla
  Family Tayassuidae
    Platygonus spp. † (peccaries)
  Family Camelidae
    Tribe Camelini
      Gigantocamelus spatula † (giant camel)
    Tribe Lamini
      Blancocamelus meadi † (Mead's camel)
      Camelops hesternus or *C. huerfanensis* † (yesterday's camel or Huerfano camel)
      Camelops minidokae † (Minidoka camel)
      Hemiauchenia blancoensis † (Blanco llama) or
      H. macrocephala † (large-headed llama)
      H. vera † (vera llama)
      Hemiauchenia sp. † (gracile llama)
      Paleolama sp. † (ancient llama)
  Family Cervidae
    Cervus sp. (elks)
    Odocoileus sp. cf. *O. hemionus* (mule deer)
    cf. *Navahoceros* sp. † (mountain deer)
  Family Antilocapridae
    Antilocapra sp. (pronghorn)
    Capromeryx sp. † (diminutive pronghorn)
    Stockoceros sp. † (Stock's pronghorn)
    Tetrameryx sp. † (four-horned pronghorn)
  Family Bovidae
    ? *Euceratherium* sp. † (shrub-oxen)

**VERTEBRATE ICHNITES (FOOT PRINTS/TRACKS)
FISH CREEK CANYON AND SAN FELIPE ICHNOFAUNAS**

**Division Vertebratichnia
Class Avipedia (bird tracks)**

Order Gruiformipeda
  Morphofamily Gruipedidae
    Gruipeda diabloensis ★ † (Diablo least sandpiper track)

Class Mammalipedia (mammal tracks)
Order Carnivoripedida
  Morphofamily Mustelipedidae
    Mustelidichnum vallecitoensis ★ † (Vallecito river otter track)
  Morphofamily Canipedidae
    Chelipus therates ★ † (claw-footed dog track)
    Chelipus sp. † (coyote-like track)
  Morphofamily Felipedidae
    Pumaeichnum milleri ★ † (Miller's lynx-sized cat track)

    P. stouti ★ † (Stout's cheetah track)
    Pumaeichnum sp. † (bobcat-sized felid track)
Order Proboscidipedida
  Morphofamily Gomphotheriipedidae
    Stegomastodonichnum garbanii ★ †
      (Garbani's gomphothere track)
    Mammuthichnum sp. † (mammoth track)
Order Perissodactipedida
  Morphofamily Hippipedidae
    Hippipeda downsi ★ † (Downs' horse track)
Order Artiodactipedida
  Morphofamily Tayassuipedidae
    Tayassuichnum sp. (peccary track)
  Morphofamily Pecoripedidae
    Lamaichnum borregoensis ★ † (Borrego small llama track)
    Megalamaichnum albus ★ † (White's large llama track)
    Camelopichnum sp. † (camel track)
  Morphofamily Cervipedidae
    Odocoilichnum sp. (deer track)

TABLE 4
Stratigraphic Names Correlation Table – Vallecito Creek-Fish Creek Basin.

Hanna (1926)	Woodring (1931)	Dibblee (1954)	Woodard (1963)	Winker and Kidwell (1996)	Remeika (1998)	Cassiliano (2002)	STRATIGRAPHIC NAMES USED IN THIS VOLUME
							terrace gravels and alluvium
				Hueso Formation		Hueso Formation	Hueso Formation
			Vallecito member / Huesos member				
			Tapiado member	Tapiado Formation		Tapiado Claystone	Tapiado Claystone
			lower / Diablo member	Diablo Formation	Palm Spring Formation	Arroyo Diablo Formation	Arroyo Diablo Formation (Olla Formation)
	Palm Spring Formation	Palm Spring Formation	upper / Camels Head member	Olla Formation		Olla Fm.	
				Camels Head member		Unit F / Unit D [Camels Hd]	Camels Head member (Olla Formation)
			upper	Yuha member	Yuha Formation	Unit C [Deguynos]	Yuha member
			Deguynos member	Deguynos Formation			Deguynos Formation
			lower	Mud Hills member	Coyote Mountain Clays	Unit B [Mud Hills]	Mud Hills member
				Wind Caves member		Unit A [Lycium]	sandy turbidites
Yuha Reefs			Lycium member	Upper Megabrec. / Lycium member	Latrania / Fish Creek sturzstrom Formation	Unit E [Terms in brackets are Cassiliano 1999]	upper megabreccia
Coyote Mountain Clays	siltstone member	Imperial Formation	upper fanglomerate / marine arenite member	Latrania Formation		Imperial Formation	sandy turbidites / Latrania Formation
Latrania Sands	basal member		Fish Creek Gypsum	Fish Creek Gypsum	Fish Creek Gypsum		Fish Creek Gypsum
		Fish Creek Gypsum	lower fanglomerate	Lower Megabrec.	Split Mtn. sturzstrom	Split Mountain Formation	lower megabreccia
			upper member	Elephant Trees Formation	Elephant Trees member		Elephant Trees Conglom.
		Alverson Andesite	lower member	Alverson Formation			Alverson volcanics
			Anza Fm.	Red Rock Formation	Split Mountain Formation / Jacumba Basalt	Anza Formation	Red Rock Formation

Group assignments (left-hand side of chart): Palm Spring Group; Canebrake Conglomerate; Imperial Group; Split Mountain Group.

Igneous and metamorphic basement rocks of the Peninsular Ranges

(Table by L.E. Lindsay)

NOTES:

1- The column on the right hand side of the chart, based on Winker and Kidwell (1996), is the general standard for this volume.

2- The convention herein for capitalization of stratigraphic units is to follow that published by the author.

3- Units on the left hand side of selected boxes (e.g., Olla Fm.) imply a proximal, basin-margin provenance (Winker- L suite, locally derived) with respect to right-hand units (e.g., Arroyo Diablo Fm.) which imply a distal, basin-center, or extra-regional provenance (Winker-C suite, direction of Colorado River).

TABLES 5.A-C
Biostratigraphic Ranges of Mammals in the Vallecito Creek-Fish Creek Basin.

The ranges are plotted against several metrics: collecting units, thickness of the sedimentary rocks containing mammal fossils, magnetostratigraphic pattern according to Opdyke et al. (1977) and Johnson et al. (1983), lithostastigraphic unit, NALMA, epoch, and period. The numerical dates are based on Cande and Kent (1995).

TABLE 5.A

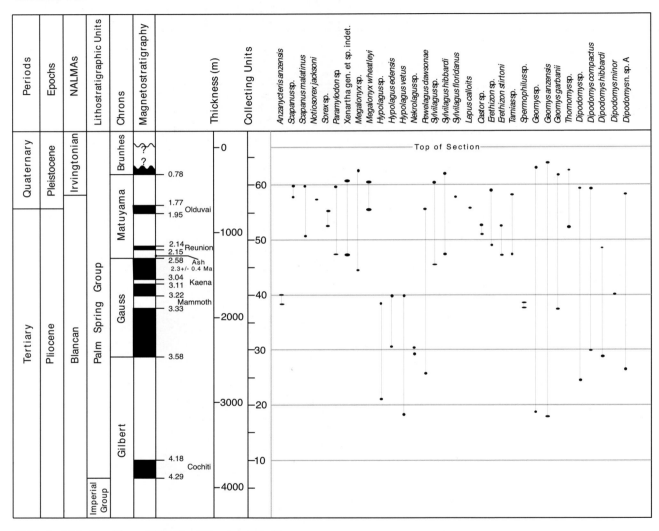

(Table by M.L. Cassiliano, 1999. See *Literature Cited*.)

TABLE 5.B

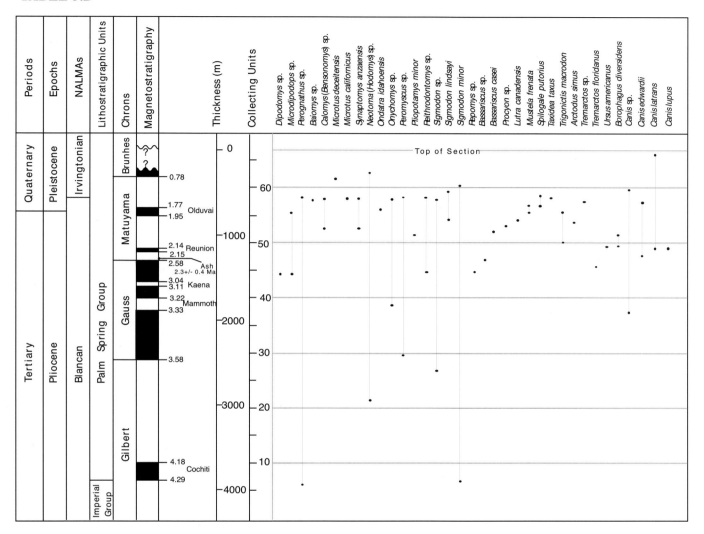

TABLE 5.C

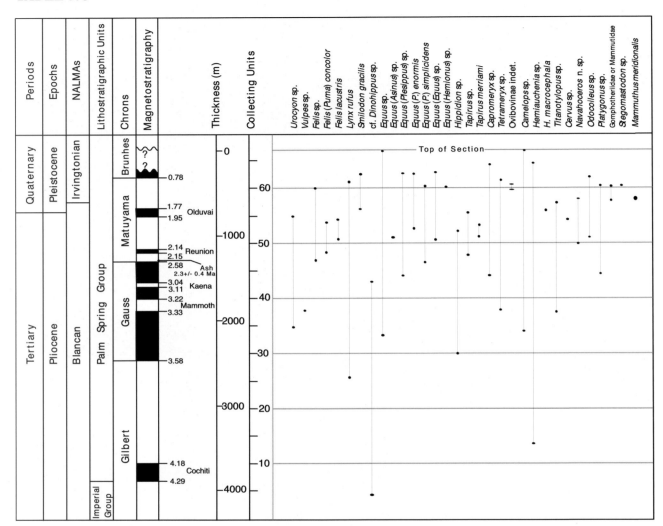

Glossary

alluvium or alluvial – Sedimentary deposits from streams and rivers.

anterior – The direction toward the front or nose of an animal.

arboreal – Adapted to tree living, describes certain animals.

assemblage – The fossils collected from a limited geographic area or sequence of strata.

biochronology or biochronologic – The relative age dating of geological events by paleontologic methods or biostratigraphy.

biostratigraphy or biostratigraphic – Stratigraphy based on the paleontologic aspects of sedimentary rocks, recognition of strata by the fossils they contain.

Blancan – The latest Miocene through Pliocene North American Mammal Age that spans the time from about 4.9 to 1.4-1.7 million years ago, follows the Hemphillian Mammal Age and precedes the Irvingtonian Mammal Age, equivalent to the Chapadmalalan Age of South America.

body fossils – The fossil remains of the tissues or structures of plants and animals, including both hard structures such as bones, shells or seeds, and soft tissues such as skin, leaves and cells.

browse – leafy young shoots and twigs of shrubs and trees, relatively near to the ground and accessible to foraging herbivores.

calcite – A mineral, one of the two crystalline forms of calcium carbonate, the other being aragonite.

carrion – The remains of recently dead animals that provide food for scavengers.

cellulose – Principal chemical constituent of a plant cell wall.

cementum – A tough mineral-like substance that is deposited on the outer enamel surfaces and inhibits tooth wear

Cenozoic – The latest geological Era in the Phanerozoic Eon, includes the Tertiary and Quaternary Periods, also called the age of mammals, spans approximately 65 million to 10 thousand years ago.

chronostratigraphic unit – A set of rock strata formed during a specific interval of geologic time.

Clarendonian – The late Miocene North American Mammal Age that spans the time from about 12.6 to 8.9 million years ago, follows the Barstovian Mammal Age and precedes the Hemphillian Mammal Age.

conglomerate – A coarse-grained sedimentary rock composed of clasts ranging in size from granules and pebbles through boulders.

conspecific – Two or more organisms that are placed within the same species.

coprolite – The fossilized excrement or dung of animals that may contain remnants of consumed food or prey items.

copse – A thicket or patch of trees, usually on a grassy plain.

coquina – A sedimentary deposit or rock primarily consisting of fragmented shells.

Cretaceous – The latest geological Period of the Mesozoic Era and the last period in the age of dinosaurs, spans approximately 145 to 65 million years ago.

cursorial – Adapted for running or walking, describing the limbs of certain animals.

deciduous – Falling off or shed at maturity or at specific times, such as plant leaves or petals, or animal teeth or hair.

deltaic – Pertaining to the features, ecological conditions, sedimentary deposits or organisms that inhabit a river delta.

dentine – The hard, dense mineral-containing tissue forming the body of a tooth.

dermal ossicles – small bones embedded as armor in the skin of a mylodont sloth, which are usually scattered following death of the animal.

ecomorph or ecomorphic – A specialization of a vertebrate body part to a particular (often extreme) type of environment that is shared in two or more unrelated taxa. Independent and parallel ecological adaptations in two separate taxa.

ecotone – The boundary or transitional area between two adjacent habitats, ecological zones or communities.

endemic – A taxon that is known only from one locality or assemblage, indigenous or native.

Eocene – The geological Epoch of the Tertiary Period that spans from about 55 to 33.5 million years ago, precedes the Oligocene.

epifauna, epifaunal – Aquatic organisms that live in the water on the water/sediment interface.

equids – Members of the genus *Equus*, horses, zebras and asses.

fault – Fracture or break in the earth's crust or a large section of bedrock along a fault line or fault plane . (See also strike-slip faulting.)

fluvial (fluviatile) – Any features or deposits of rivers and streams, the habitats of or organisms that inhabit streams or rivers.

fossil – Any remains, trace, or imprint of an organism, plant or animal that has been preserved in the earth's crust from some past geologic time, usually older than 5,000 years.

fossorial – Adapted for digging or burrowing, describes the behavior of certain animals.

Hemphillian – The late Miocene North American Mammal Age that spans the time from about 8.9 to 4.9 million years ago, precedes the Blancan Mammal Age, equivalent to the Huayquerian Age of South America.

holotype – The single specimen chosen as the representative type of a new species or subspecies in its original description.

ichnite or ichnofossils – A trace fossil, any markings left in sediment by an organism, including tracks, burrows and borings.

infauna, infaunal – Aquatic organisms that live within the sediment near or at the sediment/water interface.

intraspecific – Characters of features shared between two or more different species.

intertidal – Describing the narrow coastal area, habitats, and organisms between the normal limits of high and low tide.

Irvingtonian – The early Pleistocene North American Mammal Age that spans the time from about 1.4 to 0.5-0.2 million years ago, follows the Blancan Mammal Age and precedes the Rancholabrean Mammal Age.

incisor/s – A nipping tooth; any of the front teeth between the canines in either jaw.

in situ – In the natural or original position or place, for example a fossil in matrix.

interfinger, interdigitate – The gradation between of two types of sedimentary deposits through a series of interpenetrating wedge-shaped layers.

intertidal – Describing the narrow coastal area, habitats, and organisms between the normal limits of high and low tide.

invertebrate – An animal without a backbone, for example clams, snails or crabs and shrimp.

ipsilateral – On the same side of the body; the right hand is ipsilateral to the right foot. Also a pacing gait. The opposite term is contralateral.

junior synonym – When two named species are found to be equivalent, priority goes to the name and original description that was published first. The earlier name is called the "senior synonym" and the latter the "junior synonym." Only the senior synonym should be used in reference to the species.

lacustrine – Pertaining to the features, ecological conditions, sedimentary deposits or organisms that inhabit lakes.

lectotype – In taxonomy, a specimen which is designated to serve as the type specimen for a taxon subsequent to the publication of its original description.

littoral – Pertaining to the depth zone between high water and low water, or to the organisms of that environment; intertidal.

local fauna – A group of similar fossil assemblages from a restricted area and geologic time.

macroscopically – Visible to the naked eye without instrumentation.

magnetostratigraphy – The age dating of strata based on their original preserved magnetic polarity.

matrix – The material containing or surrounding fossils or clasts in a sedimentary rock.

megafauna – Referring to the larger animals in a paleofauna.

mesic – Referring to an environment that is neither extremely wet (hydric) nor extremely dry (xeric).

Mesozoic – The geological Era of the Phanerozoic Eon that includes the Triassic, Jurassic and Cretaceous Periods, also called the age of dinosaurs, spans approximately 251 to 65 million to years ago.

Miocene – The geological Epoch of the Tertiary Period that spans from about 23.5 to 5.3 million years ago, precedes the Pliocene.

monospecific – Pertaining to or containing a single species.

monotypic – A Family that includes only one genus or a genus with only one species.

multivariate biometric analysis – The analysis of biological data through statistical and quantitative means using a number of independent mathematical or statistical variables. In paleontology, this usually involves statistical computer analyses of numerous linear measurements of the same skeletal features on many fossils and recent specimens.

Neogene – An interval of time consisting of the Miocene and Pliocene of the Tertiary period.

niche (ecological niche) – The habitat in which an organism lives, the periods of time during which it occurs and is active there, and the resources it obtains there.

nocturnal – An animal that is active primarily at night.

nomenclature – A naming system used in a specific branch of learning or activity, such as the Linnaean system for classifying plants and animals with a double name that indicates genus and species.

North American Land Mammal Age – (see Barstovian, Clarendonian, Hemphillian, Blancan, Irvingtonian and Rancholabrean).

Oligocene – The geological Epoch of the Tertiary Period that spans from about 33.5 to 23.5 million years ago, precedes the Miocene.

omnivore (omnivorous) – An animal that consumes a broad variety of plant and animal foods.

ontogenetic – Relating to the life cycle or biological development of an individual organism.

Order – A level of organization in the classification of organisms that consists of a group of families.

paleoecology or paleoecological – The study of ancient ecologic conditions and systems, or the relationship between ancient organisms and their environment.

paleoenvironment/s – The entire set of physical, geologic and climatic conditions that prevailed in a region during some ancient period, surrounding and affecting the development of living organisms.

paleoichnology – The study of trace fossils such as footprints, tracks and/or burrows.

paleofauna – The animals that lived in a local area during a specific geologic time, the paleontologic equivalent of a modern fauna.

paleomagnetism – The study of natural remanent magnetization of rocks in order to determine the intensity and direction of the earth's magnetic field in the geologic past.

paleontology – The study of preserved organisms or their traces from past geologic times, the study of fossils.

Paleozoic – The earliest geological Era within the Phanerozoic Eon, spans from about 544 to 251 million years ago.

phalanx/phalanges – The individual bones of the toes are phalanges; the toes themselves are digits.

photosynthesis or photosynthetic – The chemical process by which green plants synthesize organic compounds from carbon dioxide and water in the presence of sunlight.

phylum/phyla – One of the broad principal divisions of plant or animal kingdoms.

Pleistocene – The geological Epoch that spans from about 1.8 million to 10 thousand years ago, follows the Pliocene and precedes the Holocene or Recent Epochs. Also referred to as the Ice Ages.

plesimorph/plesimorphies – Referring to a character state inherited from a common ancestor and shared by different groups of organisms. Primitive. The opposite of apomorphic.

Pliocene – The geological Epoch of the Tertiary Era that spans from about 5.3 to 1.8 million years ago, follows the Miocene Epoch and precedes the Pleistocene Epoch.

Plio-Pleistocene – That part of the Tertiary that includes the Pliocene and Pleistocene epochs.

polyphyletic – A taxonomic group that is derived from more than one ancestral type.

postcrania or postcranial – Situated below or behind the cranium, or head; bones of the body.

posterior or posteriorly – The direction towards the back or tail of an animal.

proboscis – An animals nose, usually large and flexible, or the trunk in elephants.

proximal – Near to, as opposed to distal, or far from; anatomically, proximal means toward the center of the body.

Quaternary – The latest geological Period, includes the Pleistocene and Holocene Epochs, approximately 1.8 million to 10 thousand years ago.

radioisotopic – Relating to unstable radioactive forms of certain elements; as they decay at a characteristic fixed rate, they emit radiation, a property useful in determining the age of materials.

Rancholabrean – The late Pleistocene North American Mammal Age that spans the time from about 0.3 million to 10 thousand years ago, follows the Irvingtonian Age.

raptor – A bird of prey.

riparian – Areas, habitats and/or organisms located adjacent to a body of water, usually streams or rivers.

rostrum/rostra – A forward projection or extension of the snout.

ruminant – A cud-chewing herbivorous mammal having a stomach with four chambers.

savanna – A grassland with scattered patches of trees (see copse).

sexual dimorphism – A difference in size or physical characteristics (other than sex organs) between the females and males of a species.

sediment – Sediment is material derived from pre-existing rock or from biological sources, or precipitated by chemical processes and deposited at, or near, the surface of the earth.

sister groups – Taxa that share a common ancestor and are more closely related to each other than to any other taxa that do not share a common ancestor. They are related by recognizable shared characteristics.

specimen – Any collected fossil or related material/s including petrologic samples, microfossil matrix samples, botanic, invertebrate and vertebrate body and trace fossils.

step over faulting – The faulting process in which separate fault blocks are downthrown systematically in one direction, forming a staircase.

strike (on strike) – The compass direction of a horizontal line on a geological structural surface, such as a bedding or fault plane. For example, a strike ridge is an elongated hill developed along the strike of a bed that is more resistant than its adjacent strata or is developed along a strike-slip fault.

strike-slip faulting – Horizontal displacement along a fault in a right-lateral (dextral) or left-lateral (sinistral) direction. The sense (right or left) is relative to an object on the opposite side of the fault from the observer (i.e., an object on the opposite side of the fault from the observer would appear to move to the right in a right-lateral fault.)

stratigraphy (stratigraphic) – The study of sedimentary rock strata, or layers, including features such as composition, fossil content, thickness, geographic distribution, and depositional environment.

sublittoral – In a marine environment, the seashore zone lying immediately below the intertidal (littoral) zone, and extending to a depth of about 200 meters or to the edge of the continental shelf.

synonymy – The placing of two previously described species together under a single species name.

systematics – The classification of organisms into a hiarchy starting with genus and species based on physical characterisatics and genetics.

taphonomy, taphonomic – The study of the conditions and setting/s of an animal's death, burial and fossilization.

taxon – The systematic classification or name for a type of organism or form, such as in a Family, Genus, or species. Plural is taxa.

taxonomy – The science of classification; in biology, a system of arranging animals and plants into natural, related groups, based on some factor common to each, such as structure, biochemistry, embryology, etc.

tectonic/s – The study of the origin and history of deformational features and movements in the Earth's crust.

Tertiary – The earlier and longest of two Periods of the Cenozoic, approximately 65 million to 1.8 million years ago.

trophic diversity – An animal with a wide variety of food preferences.

Tribe/s – Next systematic division under Family and above genus.

ungulate – A mammal having hoofs.

vertebrate – An animal with a backbone.

xerophytic – Plants that prefer dry or arid conditions, and are adapted to an environment deficient in moisture.

xeric – Arid or dry desert conditions.

zooxanthella/ae – Single-celled photosynthetic animals (dinoflagellates and cryptomonads) with flagellae that live in symbiosis with certain corals and giant clams (Tridacnidae). Zooxanthellae provide the nutrients that corals need to build their calcium carbonate skeletons.

Literature Cited

The following bibliography contains references cited by the authors, and relevant works from important graduate student papers, technical reports and published literature that were used to compile Tables 1-3. For other bibliographic sources on regional geologic and paleontologic topics, including student research papers and government agency reports, see the *Selected Bibliography of the Western Salton Trough Detachment and Related Subjects, ABDSP, California* (Jefferson and Remeika, 1995).

Abbreviations:

AAPG, American Association of Petroleum Geologists

ABDSP, Anza-Borrego Desert State Park

AMNH, American Museum of Natural History

CIWP, Carnegie Institution of Washington Publication

GSA, Geological Society of America

JVP, Journal of Vertebrate Paleontology

MNH, Museum of Natural History

MST, Master of Science Thesis

NHMLA (formerly LACM), Natural History Museum of Los Angeles County

PPP, Palaeogeography, Palaeoclimatology, Palaeoecology

SBCMAQ, San Bernardino County Museum Association Quarterly

SDSU, San Diego State University

UCLA, University of California, Los Angeles

Abbott, P. and D. Seymour (editors), 1996. Sturzstroms and Detachment Faults, ABDSP, California, South Coast Geological Society Annual Field Trip Guide Book Number 24.

Abbott, W.O. 1969. Salton Basin, a model for Pacific Rim diastrophism. GSA Special Paper 121:1.

Agenbroad, L.D. 2003. New localities, chronology and comparisons for *Mammuthus exilis*. In Reumer, J.W.F. 9:1-16.

Agenbroad, L.D., D.P. Hallman, and C.N. Jass 1998. Description and comparitive morphometrics of a Stegomastodon mirificus from the Verde Formation, Yavapi County, Arizona. JVP 18(3):23A.

Agenbroad, L.D., and J.I. Mead. 1986. Large carnivores from Hot Springs Mammoth Site, South Dakota. National Geographic Research 2(4):508-516.

Akersten, W.A. 1985. Canine function in Smilodon (Mammalia; Felidae; Machairodontinae). NHMLA Contributions in Science 356:1-22.

Akersten, W.A., T. Foppe, and G.T. Jefferson 1988. New source of dietary data for extinct herbivores. Quaternary Research 30(1):92-97.

Albright, L.B., III 1999a. Biostratigraphy and vertebrate paleontology of the San Timoteo Badlands, southern California. University of California Publications Geological Sciences 144:1-121.

Albright, L.B., III 1999b. Magnetostratigraphy and biochronology of the San Timoteo Badlands, southern California. GSA Bulletin 111(9):1256-1293.

Albright, L.B., III, and M.O. Woodburne 1993. Refined chronologic resolution of the San Timoteo Badlands, Riverside County, California. In Reynolds, R.E. and J. Reynolds 1993, 93(1):104-105.

Anderson, A.E., and O.C. Wallmo 1984. *Odocoileus hemionus*. Minlian Species 219:1-9.

Anderson, E. 1984. Review of the small carnivores of North America during the last 3.5 million years. In Contributions in Quaternary Vertebrate Paleontology, Carnegie MNH Special Publication 8:257-266.

— 1996. A preliminary report on the Carnivora of Porcupine Cave, Park County, Colorado. In Stewart, K.M. and K.L. Seymour 1996, pp. 259-282.

Anonymous 1990. Paleontologist George Miller leaves legacy of desert finds. Borrego Sun, Borrego Springs, California 4 January p. 5.

Arnold, R.E. 1904. The faunal relations of the Carrizo Creek beds of California. Science 19:1-503.

— 1906. The Tertiary and Quaternary pectens of California. USGS Professional Paper 47:84-85.

— 1909. Environment of the Tertiary faunas of the Pacific Coast of the United States. Journal of Geology 17:509-533.

Arnold, T.S. 1998. Muracypraea Woodring, 1957 (Gastropoda: Cyraeidae) in the upper Miocene and lower Pliocene Latrania Formation (Imperial Group) of Imperial County, southern California. The Festivus 30:89-93.

Atwater, T. 1970. Implications of plate tectonics for the Cenozoic evolution of western North America. GSA Bulletin 81:3513-3536.

Atwater, T., and J. Stock 1998. Pacific-North America plate tectonics on the Neogene southwestern United States – an update. International Geology Review 40:375-402.

Axelrod, D.I. 1937. A Pliocene flora from the Mount Eden beds, southern California. CIWP 476:125-183.

— 1939. A Miocene flora from the western border of the Mojave Desert. CIWP 516:1-129.

— 1940. The Pliocene Esmeralda Flora of west-central Nevada. Washington Academy of Science Journal 30:163-174.

— 1944a. The Mulholland Flora. CIWP 553:103-145.

— 1944b. The Sonoma Flora. CIWP 553:167-206.

— 1950a. The evolution of desert vegetation in western North America. CIWP 590:215-306.

— 1950b. The Piru Gorge Flora of southern California. CIWP 590:119-158.

— 1950c. The Anaverde Flora of southern California. CIWP 590:119-158.

— 1950d. Further studies of the Mount Eden flora, southern California. CIWP 590:73-117.

— 1958. Evolution of the Madro-Tertiary Geoflora. Botanical Review 24:433-509.

— 1966. The Pleistocene Soboba flora of southern California. University of California Publications in the Geological Sciences 60:1-79.

— 1979. Age and origin of the Sonoran Desert vegetation. Occasional Papers of the California Academy of Sciences 132:1-74.

Axelrod, D.I., and T.A. Deméré 1984. A Pliocene flora from Chula Vista, San Diego County, California. Transactions of the San Diego Society of Natural History 20(15):277-300.

Axen, G.J. 1995. Extensional segmentation of the main Gulf escarpment, Mexico and United States. Geology 23:515-518.

Axen, G.J., and J.M. Fletcher 1998. Late Miocene-Pleistocene extensional faulting, northern Gulf of California, Mexico and Salton Trough, California. International Geology Review 40:217-244.

Axen, G.J., Kairouz M., Alexander N., Janecke, S.U., and Dorsey, R.J. 2004. Structural expression of low-angle normal faults developed in wrench settings. GSA Abstracts with Programs 36(5):548-549.

Azzaroli, A. 1995. A synopsis of the Quaternary species of *Equus* in North America. Bolletino della Socirtà Paleontologica Italiana 34(2):205-221.

— 1998. The genus *Equus* in North America. Palaeontographica Italica 85:1-60.

Azzaroli, A., and Voorhies, M.R. 1993. The genus *Equus* in North America. The Blancan species. Paleontographica Italiana 34:175-198.

Barbour, E.H., and C.B. Schultz 1934. A new giant camel, *Titanotylopus nebraskensis*, gen. sp. nov. Bulletin University Nebraska State Museum 1:291-294.

Barnosky, A.D., and C.J. Bell 2004. Age and correlation of key fossil sites in Porcupine Cave. In Biodiversity Response to Climate Change in the Middle Pleistocene. University of California Press, Berkeley pp. 64-73.

Barnosky, A.D., and D.L. Rasmussen 1988. Middle Pleistocene arvicoline rodents and environmental change at 2900 meters elevation, Porcupine Cave, South Park, Colorado. Annals of Carnegie MNH 57:267-292.

Bartholomew, M.J. 1968. Geology of the northern portion of Seventeen Palms and Font's Point quadrangles, Imperial and San Diego Counties, California. MST, UCLA 60 pp.

— 1970. San Jacinto Fault Zone in the northern Imperial Valley. GSA Bulletin 81:3161-3166.

Baskin, J.A. 1982. Tertiary Procyoninae (Mammalia: Carnivora) of North America. JVP 2:71-93.

— 1998. Mustelidae. In Janis, C.M., K.M. Scott and L.L. Jacobs 1998, pp. 152-173.

Beale, D.M., and A.D. Smith 1970. Forage use, water consumption, and productivity of pronghorn antelope in western Utah. Journal of Wildlife Management 34:570-582.

Becker, J.J., and J.A. White 1981. Late Cenozoic geomyids (Mammalia: Rodentia) from the Anza-Borrego Desert, southern California. JVP 1:211-218.

Bell, C.J. 2000. Biochronology of North American microtine rodents. In Quaternary Geochronology: Methods and Applications, edited by S. Noller, J.M. Sowers, and W.R. Lettis, AGU Reference Shelf 4:379-406.

Bell, C.J., and A.D. Barnosky 2000. The microtine rodents from the Pit Locality in Porcupine Cave, Park County, Colorado. Annals of the Carnegie Museum 69(2):93-134.

Bell, C.J., E.L. Lundelius Jr., A.D. Barnosky, R.W. Graham, E.H. Lindsay, D.R. Ruez Jr., H.A. Semken Jr., S.D. Webb, and J.R. Zakrzewski 2004. The Blancan, Irvingtonian, and Rancholabrean mammal ages. In Late Cretaceous and Cenozoic Mammals of North America. Columbia University Press, New York pp.232-314.

Bell, C.J., and J.I. Mead 1993. Fossil lizards from the Elsinore Fault Zone, Riverside County, California. Abstracts of Proceedings Desert Symposium, SBCMAQ 40(2):20-21.

— 1995. A fossil *Sigmodon* (Mammalia: Rodentia) from the San Francisco Bay area, Solano Co., California, with comments on additional fossil material from Kern County, California. PaleoBios 16(4):9-12.

Bennett, D.K. 1980. Stripes do not a zebra make, Part I - a cladistic analysis of *Equus*. Systematic Zoology 29:272-287.

Berggren, W.A., D.V. Kent, C.C. Swisher, III, and M-P. Aubry 1995. A revised Cenozoic geochronology and chronostratigraphy. Society for Sedimentary Geology Special Publication 54:129-212.

Berta, A. 1981. The Plio-Pleistocene hyaena *Chasmaporthetes ossifragus* from Florida. JVP 1(3-4):341-356.

— 1987. The sabercat *Smilodon gracilis* from Florida and a discussion of its relationships (Mammalia, Felidae, Smilodontini). Bulletin of the Florida State MNH, Biological Sciences 31(1):1-63.

— 1995. Fossil carnivores from the Leisey Shell Pits, Hillsborough County, Florida. In Paleontology and geology of the Leisey Shell Pits, early Pleistocene of Florida, Part II, Bulletin of the Florida State MNH 37(14):463-499.

— 1998. Hyaenidae. In Janis, C.M., K.M. Scott and L.L. Jacobs 1998, Chapter 14, pp. 243-246.

Berta, A. and H. Galiano 1983. *Megantereon hesperus* from the late Hemphillian of Florida with remarks on the phylogenetic relationships of machairodonts (Mammalia, Felidae, Machairodontinae). JVP 57(5):892-899.

Blake, W.P. 1853. (untitled personal diary). Unpublished Manuscript on File, Arizona Historical Society, Southern Arizona Division 26, book 12.

— 1854. Ancient lake in the Colorado Desert. American Journal of Science, Second Series 27:435-438.

— 1914. The Cahuilla Basin and desert of the Colorado. In The Salton Sea, A Study of the Geography, the Geology, the Floristics, and the Ecology of a Desert Basin, CIWP 193:1-12.

Bond, S.I. 1977. An annotated list of the mammals of San Diego County. The San Diego Society of Natural History Transactions 18(14):229-245.

Bowers, S. 1901. Reconnaissance of the Colorado Desert Mining District. California State Mining Bureau, Special Publication 11:1-19.

Bradley, R.S. (editor) 1985. Quaternary paleoclimateology, methods of paleoclimatic reconstruction. Allen and Unwin, Boston, Massachusetts 472 pp.

Brattstrom, B.H. 1961. Some new fossil tortoises from western North America with remarks on the zoogeography and paleoecology of toroises. Journal of Paleontology 35(3):543-560.

Brogenski, C.B. 2001. Terrestrial evidence for Plio-Pleistocene climate change in oxygen and carbon isotope values of southern California fossil horse teeth. MST, Department of Geology, UC Davis 91 pp.

Brogenski, C.B., and H.J. Spero 2000. Terrestrial evidence for Plio-Pleistocene boundary climate change in oxygen and carbon isotopic values of southern California fossil horse teeth. GSA Annual Meeting, Abstract 52172.

Brown, N.D., M.D. Fuller, and R.H. Sibson 1991. Paleomagnetism of the Ocotillo Badlands, southern California. Earth and Planetary Science Letters 102:277-288.

Buchheim, P.B., V.L. Leggitt, R.E. Biaggi, and M.A. Loewen 1999. Paleoenvironments of avian trackway sites within the "Olla member," Palm Springs Formation, ABDSP, California. In Fossil Footprints, The Desert Research Symposium Proceedings, SBCMAQ 46(2):47-53.

Campbell, K.E., E. Scott Jr., and K.B. Springer 1999. A new genus for the incredible teratorn (Aves: Teratornithidae). In Avian Paleontology at the Close of the 20th Century, Smithsonian Contributions to Paleobiology 89:169-175.

Cande, S.C., and D.V. Kent 1995. Revised calibration of the geomagnetic polarity timescale for the late Cretaceous and Cenozoic. Journal of Geophysical Research 100(B4):6093-6095.

Carroll, R.L. 1988. Vertebrate paleontology and evolution. W.H. Freeman and Company.

Cassiliano, M.L. 1994. Palecology and taphonomy of vertebrate faunas from the Anza-Borrego Desert of California. Doctoral Dissertation, Department of Geosciences, University of Arizona Tucson 421 pp.

— 1997. Taphonomy of mammalian fossils across the Blancan-Irvingtonian boundary. PPP 129(1997)81-108.

— 1998a. Stable community structure of a terrestrial fauna in southern California across the Blancan-Irvingtonian boundry. JVP 18(3):32A.

— 1998b. Stratigraphic patterns and depositional environments in the Huesos Member (new lithostratigraphic unit) of the Palm Spring Formation of southern California. Contributions to Geology, University of Wyoming 32(2):133-157.

— 1999. Biostratigraphy of Blancan and Irvingtonian mammals in the Fish Creek-Vallecito Creek section, southern California, and a review of the Blancan-Irvingtonian boundary. JVP 19:169-186.

— 2002. Revision of stratigraphic nomenclature of the Plio-Pleistocene Palm Spring Group (new rank), Anza-Borrego Desert, southern California. Proceedings of the San Diego Society of Natural History 38:1-30.

— (in review). Corrected biostratigraphy of the Fish Creek-Vallecito Creek section, Blancan and Irvingtonian of southern California. JVP

Chaney, R.W. 1944. The Troutdale Flora. In Pliocene Floras of California and Oregon, edited by R.W. Chaney, CIWP 533:323-351.

Collins, G.E. 1981. Fossil camel tracks, Anza-Borrego Desert. Imperial Valley College Museum Society Newsletter 9(10):3-4.

Coltrain, J.B., J.M. Harris, T.E. Cerling, J.R. Ehleringer, M.-D. Dearing, J. Ward, and J. Allen 2004. Rancho La Brea stable isotope biogeochemistry. PPP 205:199-219.

Condit, C. 1944. The Table Mountain Flora. In Pliocene Floras of California and Oregon, edited by R.W. Chaney, CIWP 533:57-90.

Conrad, T.A. 1854. Descriptions of new fossil shells of the United States. Journal of the Academy of Natural Sciences Philadelphia Series 2, 2(4):299-300.

— 1855. Report on the fossil shells collected by W.P. Blake, United States Topographical Engineers, 1852. In W.P. Blake, Preliminary Geological Report, U.S. Pacific Railroad Exploration, U.S. 33rd Congress, 1st session, House Executive Document 129:5-21, appendix.

Cope, E.D. 1892. A contribution to the vertebrate paleontology of Texas. American Philosophical Society Proceedings 30(137):123-131.

Corner, R.G., and M.R. Vorhies 1998. Titanotylopus (= Gigantocamelus): a gracile giant camelid of late Blancan-early Irvingtonian age in North America. JVP 18(3):36A.

Cosma, T.N., A.J. Smith, and D.F. Palmer 2001. Late Tertiary climate variations inferred from ostracode data, Anza-Borrego Desert. GSA Abstracts 33(6):A196.

Cox. B.F., J.C. Matti, T. King, and D.M. Morton 2002. Neogene strata of southern Santa Rosa Mountains, California. GSA Abstracts with Programs 34(6):124.

Crouch, R.W., and C.W. Poag 1979. Amphistegina gibbosa d'Orbigny from the California borderlands, the Caribbean connection. Journal of Foraminiferal Research 9(2):85-105.

Croxen III, F.W., and D.R. Sussman 2000. Vertebrate paleontology of the middle Pleistocene Colorado River delta, northwestern Sonora, Mexico - GPS and GIS applications. JVP 20(3):37A.

Croxen III, F.W., D.R. Sussman, and C.A. Shaw 2003. Vertebrate paleontology of the middle Pleistocene Colorado River Delta, northwestern Sonora, Mexico – new discoveries 2000-2002. In Plio-Pleistocene environments of Arizona and adjacent regions. Abstract of the Southwest Paleontological Symposium.

Cunningham, G.D. 1984a. The Plio-Pleistocene Dipodomyinae and geology of the Palm Spring Formation, Anza-Borrego Desert, California. MST, Department of Geology, Idaho State University, Pocatello 115 pp.

— 1984b. Geology and paleoecology of the Palm Spring Formation at the Plio-Pleistocene boundary. GSA Special Paper 15:219.

Curtain, C.C., and H.H. Fudenberg 1973. Evolution of the immunoglobulin antigens in the Ruminantia. Biochemical Genetics 8:301-308.

Dalquest, W.W. 1975. Vertebrate fossils from the Blanco local fauna of Texas. Occasional Papers, The Museum of Texas Tech University 30:1-52.

— 1978. Early Blancan mammals of the Beck Ranch local fauna of Texas. Journal of Mammalogy 59:269-298.

— 1988. Astrohippus and the origin of Blancan and Pleistocene horses. Occasional Papers, The Museum of Texas Tech University 116:1-23.

Dalquest, W.W., and G.E. Schultz 1992. Ice age mammals of northwestern Texas. Midwestern State University Press, Wichita Falls, Texas.

Dagg, A.I. 1974. The locomotion of the camel (Camelus dromedarius). Journal of Zoology 174:67-78.

Davis, W.B., and D.J. Schmidly 1994. The mammals of Texas. Texas Parks and Wildlife Press, Austin 338 pp.

Dean, M.A. 1988. Genesis, mineralogy, and stratigraphy of the Neogene Fish Creek Gypsum, southwestern Salton Trough, California. MST, SDSU, California 150 pp.

— 1990. The Neogene Fish Creek Gypsum; a forerunner to the incursion of the Gulf of California into the western Salton Trough. GSA Abstracts with Programs 22(3):18.

— 1996. Neogene Fish Creek gypsum and associated stratigraphy and paleontology, southwestern Salton Trough, California. In Abbott, P. and D. Seymour 1996 24:123-148.

Deméré, T.A. 1993. Fossil mammals from the Imperial Formation (upper Miocene/lower Pliocene, Coyote Mountains, Imperial County, California. In Reynolds, R.E. and J. Reynolds 1993, 93(1):82-85.

DeMets, C. 1995. A reappraisal of seafloor spreading lineations in the Gulf of California. Geophysical Research Letters 22:3545-3548.

Dibblee, T. 1954. Geology of the Imperial Valley region. In Geology of Southern California, edited by R.H. Jahns, California Division of Mines Bulletin 170:22-23.

— 1984. Stratigraphy and tectonics of the San Felipe Hills, Borrego Badlands, Superstition Hills, and vicinity. In The Imperial Basin, Tectonics, Sedimentation and Thermal Aspects, Society of Economic Paleontologists and Mineralogists 40:31-44.

— 1996a. Stratigraphy and tectonics of the San Felipe Hills, Borrego Badlands, Superstition Hills and vicinity. In Abbott, P. and D. Seymour 1996 24:45-58.

— 1996b. Stratigraphy and tectonics of the Vallecito-Fish Creek Mountains, Vallecito Badlands, Coyote Mountains, and Yuha Desert, southwestern Imperial Basin. In Abbott, P. and D. Seymour 1996 24:59-80.

Dickerson, R.E. 1918. Mollusca of the Carrizo Creek beds and their Caribbean affinities. GSA Special Paper 29:148.

Dickinson, W.R. 1996. Kinematics of transrotational tectonism in the California transverse Ranges and its contribution to cumulative slip along the San Andreas transform fault system. GSA Special Paper 305:1-46.

Dockum, M.S., and R.H. Miller 1982. Ordovician conodonts from the greenschist facies carbonates, western Imperial County, California. GSA Abstracts with Programs 14:160-161.

Dorf, E. 1930. Pliocene floras of southern California. CIWP 412:1-112.

Dorsey, R.J. 2002. Stratigraphic record of Pleistocene initiation and slip on the Coyote Creek Fault, southern California. In Tectonic Evolution of Southern and Baja California, Sonora and Environs, GSA Special Paper 365:251-269.

Dorsey, R.J., and J.J. Roering (in review). Quaternary landscape evolution in the San Jacinto fault zone, Peninsular Ranges of southern California. Submitted to Geomorphology, September 2004.

Dorsey, R.J., S.U. Janecke 2002. Late Miocene to Pleistocene West Salton Trough Detachment Fault system and basin evolution, southern California. GSA Abstracts with Programs 34(6):248.

Dorsey, R.J., S.U. Janecke, S.M. Kirby, G.J. Axen, and A.N. Steely 2004. Pliocene lacustrine trangression in the western Salton trough, southern California. GSA Abstracts with Programs 36(5):317-318.

Downs, T. 1954-1960. (field notes). Unpublished Manuscript on File Colorado Desert District Stout Research Center, Paleontology Archives, and NHMLA, Vertebrate Paleontology Section (unpaginated).

— 1957. Late Cenozoic vertebrates from the Imperial Valley region, California. GSA Bulletin 68(12,2):1822.

— 1967. Airlift for fossils. NHMLA Alliance Quarterly 6(1):20-25.

Downs, T., and G.J. Miller 1994. Late Cenozoic equids from Anza-Borrego Desert, California. NHMLA Contributions in Science 440:1-90.

Downs, T., and J.A. White 1965a. Late Cenozoic vertebrates of the Anza-Borrego Desert area, southern California. American Association for the Advancement of Science Meeting, Abstract Section E, 1965:10-11.

— 1965b. Pleistocene vertebrates of the Colorado Desert, California. International Association for Quaternary Research, VII International Congress, General Session Program p. 107.

— 1968. A vertebrate faunal succession in superposed sediments from late Pliocene to middle Pleistocene in California. In Tertiary/Quaternary Boundary, International Geological Congress 23, Prague 10:41-47.

Downs, T., and G.D. Woodard 1961a. Stratigraphic succession of the western Colorado Desert, San Diego and Imperial Counties, California. GSA Special Paper 68:73-74.

— 1961b. Middle Pleistocene extension of the Gulf of California into the Imperial Valley. GSA Abstracts with Programs 68(12):21.

Dronyk, M.P. 1977. Stratigraphy, structure and seismic refraction survey of a portion of the San Felipe Hills, Imperial Valley, California. MST, Department of Geological Sciences, University of California Riverside 141 pp.

Durham, J.W. 1944. Lower California-Colorado Desert area. In Correlation of the marine Cenozoic formations of western North America, Bulletin of the GSA 55:574-575.

— 1950. 1940 E. W. Scripps cruise to the Gulf of California. GSA Memoir 43:1-216.

— 1954. The marine Cenozoic of southern California. California Division of Mines Bulletin 170:23-31.

Durham, J.W., and E.C. Allison 1961. Stratigraphic position of the Fish Creek gypsum at Split Mountain Gorge, Imperial County, California. GSA Special Paper 68:32.

Eisenmann, V., M.T. Alberdi, C. de Guili, and U. Staesche 1988. Volume I: methodology. In Studying Fossil Horses, Leiden: E.J. Brill 71 pp.

Eisenmann, V., and M. Baylae 2000. Extant and fossil Equus (Mammalia, Perrisodactyla) skulls. Zoologica Scripta 29(2):89-100.

Elders, W.A., R.W. Rex, T. Meidav, P.T. Robinson, and S. Biehler 1972. Crustal Spreading in southern California. Science 178:15-24.

Elders, W.A., and J.H. Sass 1988. The Salton Sea scientific drilling project. Journal of Geophysical Research 93:12,953-12,968.

Emerson, W.K. 1964. Results of the Puritan-American Museum of Natural History Expedition to western Mexico. 20 The Recent mollusks. American Museum Novitates 2202:1-23.

Emslie, S. D. 1995. The fossil record of Arctodus pristinus (Ursidae: Tremarctinae) in Florida. In Paleontology and geology of the Leisey Shell Pits, early Pleistocene of Florida,Part II, Bulletin of the Florida State MNH 37(15):501-514.

Emslie, S. D. and N. J. Czaplewski. 1985. A new record of giant short-faced bear, Arctodus simus, from western North America with a re-evaluation of its paleobiology. NHMLA, Contributions in Science 371:1-12.

Ewer, R.F. 1973. The Carnivores. Cornell University Press, Ithaca.

Fairbanks, H.W. 1893. Geology of San Diego County, also portions of Orange and San Bernardino Counties, California. Annual Report of the California State Mineral Bureau 11:88-90. (also listed as California State Mineralogist Eleventh Report pp. 88-90).

Feragen, E.S. 1986. Geology of the southeastern San Felipe Hills, Imperial Valley, California. MST, SDSU, California 144 pp.

Feranec, R.S. 2003. Stable isotopes, hypsodonty, and the paleodiet of Hemiauchenia (Mammalia: Camelidae). Paleobiology 29:230-242.

Ferretti, M.P. 1999. Tooth enamel structure in the hyaenid Chasmaporthetes lunensis lunensis from the late Pliocene of Italy. JVP 19(4):767-770.

Fisher, D.C. 1996. Extinction of Proboscideans in North America. In The Proboscidea. Oxford University Press pp. 296-315.

Fleming, R.F. 1993a. Palynological data from the Imperial and Palm Spring Formations, ABDSP, California. USGS Open-File Report 93-678:1-29.

— 1993b. Cretaceous pollen and Pliocene climate. The American Association of Stratigraphic Palynologists, Program and Abstracts, Annual Meeting 26:24.

— 1994a. Cretaceous pollen in Pliocene rocks. Geology 22:787-790.

— 1994b. Palynological records from Pliocene sediments in the California region. In Pliocene Terrestrial Environments and Data/Model Comparisons, USGS Open-File Report 94.

Fleming, R. F., and P. Remeika 1997. Pliocene climate of the Colorado Plateau and age of the Grand Canyon. In Partners in Paleontology, U.S. Department of the Interior Natural Resources Report 97/01:73.

Forsten, A. 1986. Equus lambei Hay, the Yukon wild horse, not ass. Journal of Mammalogy 67:422-423.

— 1988. Middle Pleistocene replacement of stenonid horses by caballoid horses. Palaeography, Palaeoclimatology, Palaeoecology 65:23-33.

Forsten, A., and V. Esienmann 1995. Equus (Pleisippus) simplicidens Cope, not Dolichohippus. Mammalia 59(1):85-89.

Foster, A.B. 1979. Environmental variation in a fossil scleractinian coral. Lethaia 12:245-264.

— 1987. Pliocene reef-corals north of the Gulf of California and their biogeography. GSA Abstracts with Programs 19(7):666-667.

Frazier, M.K. 1981. A Revision of the fossil Erethizontidae of North America. Bulletin of the Florida State Museum Biological Sciences 27(1):1-76.

Frick, C. 1921. Extinct vertebrate faunas of the badlands of Bautista Creek and San Timoteo Canyon, southern California. University of California Publications, Department of Geological Sciences Bulletin 12:277-409.

— 1937. Horned ruminants of North America. Bulletin of the AMNH69:1-669.

Friedmann, S.J., and D.W. Burbank 1995. Rift basins and supradetachment basins. Basin Research 7:109-127.

Frost, E.G., M.J. Fattahipour, and K.L. Robinson 1996b. Neogene detachment and strike-slip faulting in the Salton Trough region. In Field Conference Guide 1996, Pacific Section AAPG, Guide Book 73, Society of Economic Paleontologists and Mineralogists, Pacific Section 80:263-276.

Frost, E.G., S.C. Suitt, and M.J. Fattahipour 1996a. Emerging perspectives of the Salton Trough region with an emphasis on extentional faulting. In Abbott, P. and D. Seymour 1996 24:81-122.

Fruis, G.S., and W.M. Kohler 1984. Crustal structure and tectonics of the Imperial Valley region, California. In The Imperial Basin – Tectonics, Sedimentation, and Thermal Aspects, Society of Economic Mineralogists and Paleontologists, Pacific Section 40:1-13.

Fruis, G.S., W.D. Mooney, J.J. Healey, G.A. McMechan, and W.J. Lutter 1982. Crustal structure of the Imperial Valley region. USGS Professional Paper 1254:25-49.

Galiano, H., and D. Frailey 1977. Chasmaporthetes kani, new species from China, with remarks on phylogenetic relationships of genera within the Hyaenidae (Mammalia, Carnivora). American Museum Novitates 2632:1-16.

Gastil, R.G., J. Neuhaus, M. Cassidy, T.J. Smith, J.C. Ingle, Jr., and D. Krummenacher 1999. Geology and paleontology of southwestern Isla Tiburon, Sonora, Mexico. Revista Mexicana de Ciencias Geologicas 16(1):1-34.

Gaudin, T.J. 1999. The morphology of xenarthrous vertebrae (Mammalia: Xenarthra). Feldiana Geology n.s. 41:1-38.

Gauthier-Pilters, H., and A.I. Dagg 1981. The camel. University of Chicago Press.

Gazin, C.L. 1936. A study of fossil horse remains from the Upper Pliocene of Idaho. Proceedings of the U.S. National Museum 83(2985):281-320.

Geiger, D.L., and L.T. Groves 1999. Review of fossil abalone (Gastropoda: Vetigastropoda: Haloitidae) with comparison to Recent species. JP 73(5):872-885.

Gensler, P.A. 2001. The first fossil *Heloderma* from the mid-Pleistocene (late Irvingtonian) Coyote Badlands, ABDSP, southern California. In The Changing Face of the East Mojave Desert p. 71.

— 2002. Late Irvingtonian (mid-Pleistocene) vertebrate fauna from Ash Wash, Coyote Badlands, ABDSP, California. MST, Quaternary Studies Program, Northern Arizona University, Flagstaff, Arizona 104 pp.

Gidley, J.W. 1901. Tooth characters and revision of the North American species of the genus *Equus*. Bulletin of the AMNH14:91-142.

Gidley, J.W., and C.L. Gazin 1938. The Pleistocene vertebrate fauna from Cumberland Cave Maryland. U.S. National Museum Bulletin 171:1-99.

Gillette, D.D., and D.B. Madsen 1992. The short-faced bear *Arctodus simus* from the late Quaternary in the Wasatch Mountains of central Utah. JVP 12(1):107-112.

Girty, G.H., and A. Armitage 1989. Composition of Holocene Colorado River sand. Journal of Sedimentary Petrology 59:597-604.

Gjerde, M.W. 1982. Petrology and geochemistry of the Alverson Formation, Imperial County, California. MST, SDSU, California 85 pp.

Gompper, M.E. 1995. *Nasua narica*. Mammalian Species 487:1-10.

Gompper, M.E., and D.M. Decker 1998. *Nasua nasua*. Mammalian Species 580:1-9.

Gonyea, W.D. 1976. Behavioral implications of saber-toothed felid morphology. Paleobiology 2(4):332-342.

— 1978. Functional implications of felid forelimb anatomy. Acta Anatomica 102:111-121.

Grant, U.S., IV, and H.R. Gale 1931. Pliocene and Pleistocene mollusca of California and adjacent regions. San Diego Society of Natural History Memoirs 1:1-1036.

Grant, U.S., IV, and L.G. Hertlein 1956. *Schizaster morlini*, a new species of echinoid from the Pliocene Imperial County, California. Southern California Academy of Sciences Bulletin 55:107-110.

Groves, C.P., and D.P. Willoughby 1981. Studies on the taxonomy and phylogeny of the genus *Equus* 1. subgeneric classification of the recent species. Mammalia 45(3):321-354.

Guilday, J.E., and D.C. Irving 1967. Extinct Florida spectacled bear *Tremarctos floridanus* (Gidley) from central Tennessee. The National Speleological Society Bulletin 29(4):149-162.

Guthrie, L.L. 1990. An internally standardized study of Cenozoic sand and sandstone compositions, Salton basin, southern California. MST, SDSU, California 180 pp.

Hanna, G.D. 1926. Paleontology of Coyote Mountain, Imperial County, California. Proceeding of the California Academy of Sciences, 4th Series 14:427-503.

Hanna, G.D., and A.M. Strong 1949. West American mollusks of the genus *Conus*. Proceedings of the California Academy of Sciences, 4th Series 26:247-322.

Hall, E.R. 1981. Mammals of North America; 2nd edition. John Wiley and Sons.

Hansen, R.M. 1978. Shasta ground sloth food habits, Rampart Cave, Arizona. Paleobiology 4(3):302-319.

Haq, B.U., J. Hardenbol, and P.R. Vail 1987. Chronology of fluctuating sea levels since the Triassic. Science 235:1156-1167.

Harrison, J.A. 1979. Revision of the Camelinae (Atriodactyla, Tylopoda) and description of the new genus *Alforjas*. The University of Kansas Paleontological Contributions 57:1-20.

— 1985. Giant camels from the Cenozoic of North America. Smithsonian Contributions to Paleobiology 57:1-29.Haynes, G. 1991. Mammoths, mastodons, and elephants. Cambridge University Press.

Heald, F. and C. Shaw 1991. Sabertooth cats. In Great Cats. Majestic Creatures of the Wild, edited by J. Seidensticker and S. Lumpkin, Rodale Press, New York pp. 26-27.

Heitmann, E.A. 2002. Characteristics and structural fabric developed at the termination of a major wrench fault. MST, SDSU, California 77 pp.

Heller, E. 1912. New genera and races of African ungulates. Smithsonian Miscellaneous Collections, Publication 2148, 60(8):3.

Hermanson, J.W. and B.J. MacFadden 1992. Evolutionary and functional morphology of the shoulder region and stay-apparatus in fossil and extant horses (Equidae) JVP 12(3): 377-386

Hertlein, L.G., and E.K. Jordan 1927. Paleontology of the Miocene of lower California. Proceedings of the California Academy of Sciences, 4th Series 16(19):607-611.

Hibbard, C.W. 1955. Pleistocene vertebrates from the upper Becerra (Becerra Superior) Formation, Valley of Tequixquiac, Mexico. Contributions of the Museum of Paleontology, University of Michigan 12:47-96.

Hibbard, C.W., C.E. Ray, D.E. Savage, D.W. Taylor, and J.E. Guilday 1965. Quaternary mammals of North America. In The Quaternary of the United States. Princeton University Press pp. 509-525.

Hoetker, G.M., and K.W. Gobalet 1999. Fossil razorback sucker (Pisces: Catostomidae, *Xyrauchen texanus*) from southeastern California. Copeia (3):755-759.

Hoffmeister, D.F. 1986. Mammals of Arizona. The University of Arizona Press, Tucson.

Hoffstetter, R. 1950. La structure des incisives inférieures chez les Equidés modernes: importance dans la classification des Zèbres-Couggas. Bulletin, Paris MNH 22:684-692.

Honey, J.G., J.A. Harrison, D.R. Prothero, and M.S. Stevens 1988. Camelidae. In Janis, C.M., K.M. Scott and L.L. Jacobs 1998, pp. 439-462.

Hoover, R.A. 1965. Areal geology and physical stratigraphy of a portion of the southern Santa Rosa Mountains, San Diego County, California. MST, Department of Geological Sciences, University of California Riverside 81 pp.

Howard, H. 1947. A preliminary survey of trends in avian evolution from Pleistocene to Recent time. Condor 49:10-13.

— 1963. Fossil birds from the Anza-Borrego Desert. Los Angeles County Museum Contributions in Science 73:1-33.

— 1972a. Type specimens of avian fossils in the collections of the NHMLA. NHMLA Contributions in Science 228:1-27.

— 1972b. The incredible teratorn again. Condor 74(3):341-344.Hudnut, K.W., L. Seeber, and J. Pacheco 1989. Cross-fault triggering in the November 1987 Superstition Hills earthquakes sequence, southern California. Geophysical Research Letters 16:199-202.

Hudnut, K.W., and K.E. Sieh 1989. Behavior of the Superstition Hills Fault during the past 330 years. Seismological Society of America Bulletin 79:403-329.

Hulbert, R.C. Jr. 1995. *Equus* from the Leisey Shell Pit 1A and other Irvingtonian localities from Florida. Bulletin of the Florida MNH 37 Part II (17):553-602.

Hunt, R.M., Jr. 1998. Ursidae. In Janis, C.M., K.M. Scott and L.L. Jacobs 1998, Chapter 10, pp. 174-195.

Hutchison, J.H. 1987. Moles of the *Scapanus latimanus* group (Talpidae, Insectivora) from the Pliocene and Pleistocene of California. NHMLA Contributions in Science 386:1-15.

Ingle, J.C. Jr. 1974. Paleobathymetric history of Neogene marine sediments, northern Gulf of California. In Geology of Peninsular California, AAPG, Society of Economic Paleontologists and Mineralogists, Society of Exploration Geophysicists Fieldtrip Guidebook pp. 121-138.

Ingles, L.G. 1965. Mammals of the Pacific states. Stanford University Press, California.

Jameson, E.W., Jr., and H.J. Peeters 1988. California Mammals. University of California Press, Berkeley 403 pp.

Janecke, S.U., S.M. Kirby, and R. J. Dorsey 2003. New strand of the San Jacinto fault zone SW of the Salton Sea and a possible contractional step-over in the San Felipe Hills. GSA Abstracts with Programs 35(4):26.

Janecke, S.U., S.M. Kirby, V.E.Langenheim, B. Housen, R.J. Dorsey, R.E. Crippen, and R.G. Blom 2004. Kinematics and evolution of the San Jacinto fault zone in the Salton Trough. GSA Abstracts with Programs 36(5):317.

Janis, C.M., J.A. Baskin, A. Berta, J.J. Flynn, G.F. Gunnell, R.M. Hunt Jr., L.D. Martin and K. Munthe 1998. Carnivorous mammals. In Janis, C.M., K.M. Scott and L.L. Jacobs 1998, Chapter 4, pp. 73-90.

Janis, C.M., J.A. Effinger, J.A. Harrison, J.G. Honey, D.G. Kron, B. Lander, E. Manning, D.R. Prothero, M.S. Stevens, R.K. Stucky, S.D. Webb, and D.B. Wright 1998. Artiodactyla. In Janis, C.M., K.M. Scott and L.L. Jacobs 1998, pp. 337-357.

Janis, C.M. and E. Manning 1998. Antilocapridae. In Janis, C.M., K.M. Scott and L.L. Jacobs 1998, pp. 491-507.

Janis, C.M., K.M. Scott and L.L. Jacobs 1998. Evolution of Tertiary Mammals of North America, Volume I: Terrestrial Carnivores, Ungulates and Ungulate-Like Mammals, Cambridge University Press.

Janis, C.M., J.M. Theodor, and B. Boisvert 2002. Locomotor evolution in camels revisited. JVP 22(1):110-121.

Janis, C.M. and P.B. Wilhelm 1993. Were there mammalian pursuit predators in the Tertiary? Dances with wolf avatars. Journal of Mammalian Evolution 1:103-126.

Jefferson, G.T. 1989. Late Cenozoic tapirs (Mammalia: Perissodactyla) of western North America. NHMLA Contributions in Science 406:1-21.

— 1995. An additional avian specimen referrable to Teratornis incrediblis from the early Irvingtonian, Vallecito-Fish Creek Basin, ABDSP, California. In Remeika, P., and A. Sturz 1995 1:94-96.

— 1999. A late Miocene terrestrial vertebrate assemblage from ABDSP. In The 1999 Desert Research Symposium, SBCMAQ 46(3):109-111.

— 2001. Paleontological resources of ABDSP, a management perspective. In The Human Journey and Ancient Life in California's Deserts, Desert Managers Group Millennium Conference, Abstracts with Program p. 8.

— 2003. Stratigraphy and paleontology of the middle to late Pleistocene Manix Formation. In Paleoenvironments and Paleohydrology of the Mojave and Southern Great Basin Deserts, GSA Special Paper 368:43-60.

— 2004. Paleontological resources of ABDSP, a management perspective. In The Human Journey and Ancient Life in California's Deserts, Maturango Museum Publications 15:39-64.

Jefferson, G.T., and G.E. McDaniel Jr. 2002. A new gomphothere from the mid-Pliocene ancestral Colorado River deltaic deposits in ABDSP, California. In Between the Basins, California State University Desert Studies Consortium, Desert Symposium Abstracts 2002 p. 77.

Jefferson, G.T., and D.G. Peterson, Jr. 1998. Hydrothermal origin of the Fish Creek Gypsum, Imperial County, southern California. In Lindsay, L. and W.G. Hample, 1988 pp. 40-51.

Jefferson, G.T., and P. Remeika 1995a. Selected bibliography of the western Salton Trough detachment and related subjects, ABDSP, California. In Remeika, P., and A. Sturz 1995:1-88.

— 1995b. The mid-Pleistocene stratigraphic co-occurrence of Mammuthus columbi and Mammuthus imperator in the Ocotillo Formation, Borrego Badlands, ABDSP, California. Revised from Current Research in the Pleistocene 11:89-92, In Remeika, P., and A. Sturz 1995 1:104-108.

Jefferson, G.T., and A. Tejada-Flores 1995. Late Blancan Acinonyx (Carnivora, Felidae) from the Vallecito Creek Local Fauna of ABDSP, California. SBCMAQ 42(2):33.

Johnson, N.M., C.B. Officer, N.D. Opdyke, G.D. Woodard, P.K. Zeitler, and E.H. Lindsay 1983. Rates of Cenozoic tectonism in the Vallecito-Fish Creek basin, western Imperial Valley, California. Geology 11:664-667.

Johnson, N.M., N.D. Opdyke, and E. Lindsay 1975. Magnetic polarity stratigraphy of Pliocene-Pleistocene terrestrial deposits and vertebrate fauna, San Pedro Valley, Arizona. GSA Bulletin 86:5-11.

Johnson, W.E., and S.J. O'Brien, 1997. Phylogenetic reconstruction of the Felidae using 16S rRNA and NADH-5 Mitochondrial Genes, Journal of Molecular Evolution, 44 (Suppl 1): S98-S116

Jolly, D.W. 2000. Fossil turtle and tortoises of Anza Borrego Desert State Park, California. MST, Quaternary Studies, Northern Arizona University 197 pp.

Keen, A.M. 1971. Seashells of tropical west America. Second Edition. Stanford University Press, California 1064 pp.

Keen, M., and H. Bentson 1944. Check list of California Tertiary marine mollusca. GSA Special Paper 56:21-22.

Kerr, D.R. 1982a. Early Neogene continental sedimentation, western Salton Trough, California. MST, SDSU, California 138 pp.

— 1982b. Early Neogene continental sedimentation, western Salton Trough, California. GSA Abstracts with Programs 14:177.

— 1982c. Early Neogene continental sedimentation in the Vallecito and Fish Creek Mountains, western Salton Trough, California. Sedimentary Geology 38:217-246.

Kerr, D.R., and P.L. Abbott 1996. Miocene suberial sturzstrom deposits, Split Mountain, ABDSP. In Abbott, P. and D. Seymour 1996 24:149-163.

Kerr, D.R., and S.M. Kidwell 1991. Late Cenozoic sedimentation and tectonics, western Salton Trough, California. In Geological Excursions in Southern California and Mexico, SDSU, California pp. 397-416.

Kerr, D.R., S. Pappajohn, and G.L. Peterson 1979. Neogene stratigraphic section at Split Mountain, eastern San Diego County, California. In Tectonics of the Juncture Between the San Andreas Fault System and the Salton Trough, Southeastern California, GSA Annual Meeting Guidebook pp. 111-124.

Kew, W.S.W. 1914. Tertiary echnoids of the Carrizo Creek region in the Colorado Desert. California University Publications, Bulletin of the Department of Geology 8:39-60.

— 1920. Cretaceous and Cenozoic echinoidea of the Pacific coast of North America. University of California Publication of the Geology Department 12(2):32-137.

Kidwell, S.M. 1988. Taphonomic comparison of passive and active continental margins. PPP 63:201-223.

— 1996. Neogene record of rifting and breakup in the Salton Trough, SE California. AAPG Annual Meeting Abstracts pp. 499-522.

Kidwell, S.M., C.D. Winker, and E. D. Gyllenhaal 1988. Transgressive stratigraphy of a marine rift basin. GSA Abstracts with Programs 20:A380.

Kirby, S.M., 2005. Quaternary tectonic evolution of the San Felipe Hills, California, MST, Utah State University.

Kirby, S.M., S.U. Janecke, R.J. Dorsey, B.A. Housen, and K. McDougall 2004a. A 1.07 Ma change from persistent lakes to intermittent flooding and desiccation in the San Felipe Hills, Salton Trough, southern California. GSA Abstracts with Programs 36(5):318.

Kirby, S.M., S.U. Janecke, R.J. Dorsey, and E.B. Layman 2004b. Reorganization or initiation of the San Jacinto fault zone at 1 Ma. GSA Abstracts with Programs 36(4):37.

Kitchen, D.W. 1974. Social behavior and ecology of the pronghorn. Wildlife Monograph 38:1-96.

Kurtén, B. 1965. The Pleistocene Felidae of Florida. Bulletin of the Florida State Museum, Biological Sciences 9(6):215-273.

— 1966. Pleistocene bears of North America, Part 1. Acta Zoologica Fennica 115:1-120.

— 1967. Pleistocene bears of North America, Part II. Acta Zoologica Fennica 117:1-60.

— 1973. Pleistocene jaguars in North America. Commentationes Biologicae 62:1-23.

— 1974. A history of coyote-like dogs (Canidae, Mammalia). Acta Zoologica Fennica 140:1-38.

Kurtén, B., and E. Anderson 1980. Pleistocene mammals of North America. Columbia University Press, New York.

Kurtén, B. and L. Werdelin 1988. A review of the genus *Chasmaporthetes* Hay, 1921 (Carnivora, Hyaenidae). JVP 8(1):46-66.

Lambert, W.D. 1996. The biogeography of the gomphotherid proboscideans of North America. In The Proboscidea, edited by J. Shoshani and P. Tassy, Oxford University Press pp. 143-148.

Larivière, S., and L.R. Walton 1998. *Lontra canadensis*. Mammalian Species 587:1-8.

Laudermilk, J.D., and P.A. Munz 1934. Plants in the dung of Nothrotheriops from Gypsum Cave, Nevada. CIWP 453:31-37.

— 1938. Plants in the dung of *Nothrotheriops* from Rampart and Muav Caves, Arizona. CIWP 487:271-281.

Leyhausen, P. 1979. Cat behavior. Garland Press, New York.

Lindsay, E.H. 1984. Late Cenozoic mammals from northwestern Mexico. JVP 4:208-215.

— 1984. The Plio-Pleistocene boundary in terrestrial deposits of southwestern U.S.A. Proceedings of the 27th International Geological Congress 3:35-48.

Lindsay, E.H., N.M. Johnson, and N.D. Opdyke 1975. Preliminary correlation of North American land mammal ages and geomagnetic chronology. In Studies on Cenozoic Paleontology and Stratigraphy, University of Michigan Papers in Paleontology 12:111-119.

Lindsay, E.H., N.M. Johnson, N.D. Opdyke, and R.F. Butler 1987. Mammalian chronology and the magnetic time scale. In Woodburne, M.O. 1987 pp. 269-290.

Lindsay, E.H., N.D. Opdyke, and N.M. Johnson 1984. Blancan-Hemphillian land mammal ages and late Cenozoic mammal dispersal events. Annual Review of Earth and Planetary Sciences 12:445-448.

Lindsay, E.H., and J.S. White 1993. Biostratigraphy and magnetostratigraphy in the southern Anza-Borrego area. In Reynolds, R.E. and J. Reynolds 1993, 93-1:86-87.

Lindsay, L. and W.G. Hample (editors), 1998. Geology and Hydrothermal Resources of the Imperial and Mexicali Valleys, San Diego Association of Geologists Annual Field Trip Guidebook.

Lister, A.M. 2001. "Graded" evolution and molar scaling in the evolution of the mammoth. In The World of Elephants, Proceedings of the 1st International Conference, Rome pp. 648-651.

Lister, A.M., and A.V. Sher 2001. The origins and evolution of the wolly mammoth. Science 294:1094-1097.

Lonsdale, P. 1989. Geology and tectonic history of the Gulf of California. In The Eastern Pacific Ocean and Hawaii, GSA, Boulder, Colorado pp. 499-522.

Lough, C.F. 1993. Structural evolution of the Vallecito Mountains, Colorado Desert and Salton Trough geology. San Diego Association of Geologists pp. 91-109.

— 1998. Detachment faulting around Borrego Valley. In Lindsay, L. and W.G. Hample, 1988 pp. 40-51.

Lough, C.F., and A.L. Stinson 1991. Structural evolution of the Vallecito Mountains, southwest California. GSA Abstracts with Programs 23(5):246.

Lundelius, E.L., Jr., T. Downs, E.H. Lindsay, H.A. Semken, R.J. Zakrzewski, C.S. Churcher, C.R. Harrington, G.E. Schultz, and S.D. Webb 1987. The North American Quaternary sequence. In Woodburne, M.O. 1987 pp. 211-235.

Lutz, A.T., 2005. Tectonic controls on Pleistocene basin evolution in the central San Jacinto fault zone, southern California. MST, University of Oregon.

Lutz, A.T., and R.J. Dorsey 2003. Stratigraphy of the Pleistocene Ocotillo Conglomerate, Borrego Badlands, southern California. GSA Abstracts with Programs 35(6):248.

Lutz, A.T., R.J. Dorsey, and B.A. Housen 2004. 0.5-0.6 Ma onset of uplift and transpressive deformation in the Borrego Badlands, southern California. Transactions of the American Geophysical Union 85:F1760.

Lyell, C. 1833. Principles of geology, First Edition, Volume 3, John Murray.

MacFadden, B.J. 1992. Fossil horses. Cambridge University Press 369 pp.

— 1997. Pleistocene horses from Tarija, Boliva, and validity of the genus † *Onohippidium* (Mammalia, Equidae). JVP 17(1):199-218.

MacFadden, B.J., and M.F. Skinner 1979. Diversification and biogeography of the one-toed horses *Onohippidium* and *Hippidion*. Postilla 175:7.

Maglio, V.J. 1973. Origin and evolution of the Elephantidae. Transactions of the American Philosophical Society, New Series 63:1-199.

Marshall, L.G. 1981. The Great American Interchange – an invasion induced crisis for South American mammals. In Biotic Crises in Ecological Evolutionary Time, Academic Press, Inc. Burlington, MA pp. 133-229.

Marshall, L.G., S.D. Webb, J.J. Sepkoski, and D.M. Raup 1982. Mammalian evolution and the great American interchange. Science 215(4538):1351-1357.

Martin, L.D. 1998. Felidae. In Janis, C.M., K.M. Scott and L.L. Jacobs 1998, pp. 236-242.

Martin, P.S., B.E. Sables, and D. Shutler Jr. 1961. Rampart Cave coprolite and ecology of the Shasta ground sloth. American Journal of Science 259:102-127.

Martin, R.A. 1974. Fossil mammals from the Coleman IIA fauna, Sumter County. In Pleistocene Mammals of Florida, edited by S.D. Webb, The University of Florida, Gainesville, Chapter 3, pp. 35-99.

— 1979. Fossil history of the rodent genus *Sigmodon*. University of Chicago Evolutionary Monographs 2:1-36.

— 1986. Energy, ecology, and cotton rat evolution. Paleobioloby 12:370-382.

— 1993. Late Pliocene and Pleistocene cotton rats in the southwestern United States. In Reynolds, R.E. and J. Reynolds 1993, 93(1):88-89.

Martin, R.A., and R.H. Prince 1989. A new species of early Pleistocene cotton rat from the Anza-Borrego Desert of southern California. Southern California Academy of Sciences Bulletin 88(2):80-78.

Martin, R.A., and S.D. Webb 1974. Late Pleistocene mammals from the Devil's Den fauna, Levy County. In Pleistocene Mammals of Florida, edited by S.D. Webb, The University of Florida, Gainesville pp. 114-145.

Matthew, W.D. 1924. A new link in the ancestry of the horse. AMNH Novitates 131:2 p.

— 1930. The phylogeny of dogs. Journal of Mammalogy 11:117-138.

Matti, J.C., B.F. Cox, D.M. Morton, R.V. Sharp, and T. King 2002. Fault-bounded Neogene sedimentary deposits in the Santa Rosa Mountains, southern California. GSA Abstracts with Programs 34(6):124.

May, S.R., and C.A. Repenning 1982. New evidence for the age of the Mount Eden fauna, southern California. JVP 2(1):109-113.

Mayer, J.J. and R.M. Wetzel 1986. *Catagonus wagneri*. Mammalian Species 259:1-5.

— 1987. Tayassu pecari. Mammalian Species 293:1-7.

McDaniel, G., Jr., and G.T. Jefferson 1997. A nearly complete skeleton of *Mammuthus meridionalis* from the Borrego Badlands, ABDSP, California. In Memories, Minerals, Fossils, and Dust, SBCMAQ 44(2):35-36.

— 1999a. Distribution of proboscideans in ABDSP. In The 1999 Desert Research Symposium, SBCMAQ 46(3):61-62.

— 1999b. *Mammuthus meridionalis* (Proboscidea: Elephantidae) from the Borrego Badlands of ABDSP, California. In The Second International Mammoth Conference, Official Conference Papers, Rotterdam pp. 41-42.

— 2000. The hyoid bone of *Mammuthus meridionalis* from the mid-Pleistocene of ABDSP. In Empty Basins, Vanished Lakes, SBCMAQ 47(2):80-81.

— 2001. A late Pleistocene proboscidean site in Death Valley Lake Tecopa beds near Shoshone, California. In The Changing Face of the Mojave Desert, California State University Desert Studies Consortium and Western Center for Archaeology and Paleontology, 2001 Desert Symposium Abstracts p. 72.

— 2002. Mammoths in our midst. Abstracts with Programs, 3rd International Mammoth Conference, Dawson City, Canada pp. 85-87.

— 2003. *Mammuthus meridionalis* (Proboscidea: Elephantidae) from the Borrego Badlands of ABDSP, California. In Advances in Mammoth Research, Proceedings of The 2nd International Mammoth Conference, Rotterdam, DEINSEA 9:239-252.

— 2005. Mammoths in our midst. In The 3rd International Mammoth Conference, Dawson, Yukon, Canada. Quaternary International 142/143:124-129.

McDaniel, G.E., Jr., G.T. Jefferson, and H.G. McDonald 2001. A large *Paramylodon harlani* osteoderm layer from the Irvingtonian of ABDSP, California. JVP 21(3):79A.

McDonald, H.G. 1993. Harlan's Ground Sloth, *Glossotherium harlani*, from Pauba Valley, Riverside County, California. In Reynolds, R.E. and J. Reynolds 1993, 93-1:101-103.

— 1995. Gravigrade xenarthrans from the early Pleistocene Leisey Shell Pit 1A, Hillsborough County, Florida. Bulletin of the Florida MNH 37(2):345-374.

— 1996. Population structure of the late Pliocene (Blancan) zebra *Equus simplicidens* (Perrisodactyla: Mammalia) from the Hagerman Horse Quarry, Idaho. In Stewart, K.M. and K.L. Seymour 1996, pp. 134-155.

— 1996. Biogeography and paleoecology of ground sloths in California, Arizona, and Nevada. The 1996 Desert Research Symposium, Abstracts of Papers Submitted to the Meetings, SBCMAQ 43(1):61-65.

— 2002. *Platygonus compressus* from Franklin County, Idaho and a review of the genus in Idaho. In And Whereas . . . Papers on the Vertebrate Paleontology of Idaho, Volume 2, Idaho MNH Occasional Paper 37:141-149.

McDonald, H.G., and D.C. Anderson 1983. A well-preserved ground sloth (*Megalonyx*) cranium from Turin, Monoma County, Iowa. Proceedings of the Iowa Academy of Science 90(4):134-140.

McDonald, H.G., C.R. Harrington, and G. De Iuliss 2000. The ground sloth *Megalonyx* from Pleistocene deposits of the Old Crow Basin, Yukon Territory, Canada. Arctic 53(3):231-220.

McDonald, H.G., G.T. Jefferson, and C. Force 1996. Pleistocene distribution of the ground sloth *Nothrotheriops shastense* (Xenarthra, Megalonychidae). SBCMAQ 43(1,2):151-152.

McDonald, H.G., W.E. Miller, and T.H. Morris 2001. Taphonomy and significance of Jefferson's ground sloth (Xenarthra: Megalonychida from Utah. Western North American Naturalist 61(1):64-77.

McDougall, K.A., R.Z. Poore, and J.C. Matti 1999. Age and paleoenvironment of the Imperial Formation near San Gorgonio Pass, southern California. Journal of Foraminiferal Research 29:4-25.

Mech, L.D. 1974. *Canis lupus*. Mammalian Species 37:1-6.

Mendenhall, W.C. 1910. Notes on the geology of Carrizo Mountain and vicinity, San Diego County, California. Journal of Geology 18:336-355.

Merriam, J. C. 1912. The fauna of Rancho La Brea. Part II. Canidae. Memoirs of the University of California 1(2):217-272.

— 1913. Preliminary report on the horses of Rancho La Brea. University of California Publications, Bulletin of the Department of Geology 7(21):397-418.

Merriam, J.C. and C. Stock 1925. Relationships and structure of the short-faced bear, *Arctotherium*, from the Pleistocene of California. CIWP 347:1-35.

Merriam, R.H., and O.L. Bandy 1965. Source of upper Cenozoic sediments in the Colorado River delta region. Journal of Sedimentary Petrology 35:384-399.

Miller, J.M.G., and B.E. John 1999, Sedimentation patterns support seismogenic low-angle normal faulting, southeastern California and western Arizona. GSA Bulletin 111:1350–1370.

Miller, R.H., and M.S. Dockum 1983. Ordovician conodonts from meta-morphosed carbonates of the Salton Trough, California. Geology 11:410-412.

Miller, W.E. 1971. Pleistocene vertebrates of the Los Angeles basin and vicinity (exclusive of Rancho La Brea). Bulletin of the Los Angeles County Museum of Natural History, Science 10: 1-124.

— 1980. The late Pliocene Las Tunas local fauna from southernmost Baja California, Mexico. JP 54:762-805.

Mitchell, E.D. Jr. 1961. A new walrus from the Imperial Pliocene of southern California. Los Angeles County Museum Contributions in Science 44:1-28.

Moore, D.M. 1978. Post-glacial vegetation in the South American territory of the giant ground sloth, *Mylodon*. Botanical Journal of the Linnean Society 77(3):177-202.

Morley, E.R., Jr. 1963. Geology of the Borrego Mountain quadrangle and the western portion of the Shell Reef quadrangle, San Diego County, California. Master of Arts Thesis, UCLA 138 pp.

Morton, D.M., and J.C. Matti 1993. Extension and contraction within an evolving divergent strike-slip fault complex. In The San Andreas Fault System, GSA Memoir 178:217-230.

Mount, J.D. 1974. Molluscan evidence for the age of the Imperial Formation, southern California. Southern California Academy of Sciences Abstracts with Programs p. 9.

— 1988. Molluscan fauna of the Imperial Formation. In Geologic Field Guide to the Western Salton Basin Area, edited by S.M. Testa and K.E. Green, American Institute of Professional Geologists pp. 23-24.

Muffler, L.P.J., and B.R. Doe 1968. Composition and mean age of detritus of the Colorado River delta in the Salton Trough, southeastern California. Journal of Sedimentary Petrology 38:384-339.

Muir, J. 1911. My first summer in the Sierra. Houghton Mifflin Company, Boston, Massachuetts 272 pp.

Munthe, K. 1989. The skeleton of the Borophaginae (Carnivora, Canidae). University of California Publications in Geological Sciences 133:1-115.

– 1998 Canidae. In Janis, C.M., K.M. Scott and L.L. Jacobs 1998, pp. 124-143.

Murphy, M.A. 1977. On time-stratigraphic units. JP 51:213-219.

— 1986. The Imperial Formation at Painted Hill, near Whitewater, California. In Geology Around the Margins of the Eastern San Bernardino Mountains, Inland Geological Society Publications, Redlands, California 1:63-70.

Murray, L.K., and G.T. Jefferson 1996. A review of Procyonids (Carnivora, Procyonidae) from late Blancan and early Irvingtonian Vallecito Creek Local Fauna of ABDSP, California. SBCMAQ 43(2):153.

Naples, V.L. 1989. The feeding mechanism in the Pleistocene ground sloth *Glossotherium*. NHMLA Contributions in Science 415:1-23.

Nations, J.D., and D. Gauna 1998. Stratigraphic, sedimentologic, and paleobotanical investigations of terrace gravels, U.S. Army Yuma Proving Ground. Department of Defence Legacy Resource Management Program 1.1-9.3.

Nelson, M.E. and J.H. Madsen Jr. 1983. A giant short-faced bear (*Arctodus simus*) from the Pleistocene of northern Utah. Transactions of the Kansas Academy of Science 86(1):1-9.

Norell, M.A. 1989. Late Cenozoic lizards of the Anza-Borrego Desert, California. NHMLA Contributions in Science 414:1-31.

Nowak, R.M. 1979. North American Quaternary Canis. Monograph of the MNH, University of Kansas, Lawrence 6:1-154.

— 1991. Walker's mammals of the World; 5th edition. The Johns Hopkins Press, Baltimore, Maryland 1629 pp.

— 2002. The original status of wolves in eastern North America. Southeastern Naturalist 1:95-130.

Nyborg, T.G., and V.I. Santucci 1999. The Death Valley National Park paleontology survey. National Park Service, Geological Resources Technical Report NPS/NRGRDTR-99/01:1-66.

O'Gara, B.W. 1978. *Antilocapra americana*. Mammalian Species 90:1-7.

O'Gara, B.W., and G. Matson 1975. Growth and casting of horns by pronghorns and exfoliation of horns by bovids. Journal of Mammalogy 56:829-846.

Olsen, S.L. 1974. The Pleistocene rails of North America. Condor 76(2):169-175.

Opdyke, N.D., E.H. Lindsay, N.M. Johnson, and T. Downs 1974. The magnetic polarity stratigraphy of the mammal bearing sedimentary sequence at Anza-Borrego State Park, California. GSA Abstracts with Programs 6:901.

— 1977. The paleomagnetism and magnetic polarity stratigraphy of the mammal-bearing section of Anza Borrego State Park, California. Quaternary Research 7:316-329.

Orcutt, C.R. 1890. The Colorado Desert. California Journal of Mines and Geology 10:899-919.

— 1901. The Colorado Desert. West American Scientist 21(1):2-14.

Oskin, M.E., and J.M. Stock 2003. Marine incursion syncronous with plate-boundary localization in the Gulf of California. Geology 31:23-26.

Oskin, M.E., J.M. Stock, and A. Martin-Barajas 2001. Rapid localization of Pacific-North American plate motion in the Gulf of California. Geology 29:459-462.

Pajak, A.F., III 1993. The second record of *Tapirus* from the Temecula Valley, southern California, and biostratigraphic implications. SBCMAQ 40(2):30.

— 1997. The Irvingtonian General Kearney Local Fauna, Riverside County, California and comparisons to Pleistocene faunas of southern California. JVP 17(3):68A.

Pajak, A.F.,III, E. Scott, and C.J. Bell 1996. A review of the biostratigraphy of Pliocene and Pleistocene sediments in the Elsinore Fault Zone, Riverside County, California. PaleoBios 17(2-4):28-49.

Pappajohn, S. 1980. Description of Neogene marine section at Split Mountain, easternmost San Diego County, California. MST, SDSU, California 77 pp.

Parks, J., B. Stout, G.J. Miller, P. Remeika, and V.E. Waters 1989. A progress report on half-million year old marks on mammoth bones from the Anza-Borrego Desert Irvingtonian. SBCMAQ 36(2):63.

Pasini, G., and M.L. Colalongo 1997. The Pliocene-Pleistocene boundary-stratotype at Vrica, Italy. In The Pleistocene Boundary and the Beginning of the Quaternary, Cambridge University Press pp. 15-45.

Peterson, O.A. 1914. A mounted skeleton of *Platigonus leptorhinus* in the Carnegie Museum. Annals of the Carnegie Museum 9:114-117.

Petersen, M.D., L. Seeber, L.R. Sykes, L.J. Nabelek, J.G. Armbruster, J. Paceco, and K. Hudnut 1991. Seismicity and fault interaction, southern San Jacinto fault zone, southern California. Tectonics 10:1187-1230.

Pettinga, J.R. 1991. Structural styles and basin margin evolution adjacent to the San Jacinto Fault Zone, southern California. GSA Abstracts with Programs 23:(5)257.

Poglayen-Neuwall, I., and D.E. Toweill 1988. *Bassariscus astutus*. Mammalian Species 327:1-8.

Powell, C.L., II 1984a. Bivalve molluscan paleoecology on northern exposures of the marine Neogene Imperial Formation in Riverside County, California. Western Society of Malacologists Annual Report 17:1-32.

— 1984b. Review of the marine Neogene Imperial Formation, southern California. Society of Economic Paleontologists and Mineralogists, Pacific Section p. 29.

— 1987a. Correlation between sea level events and deposition of marine sediments in the proto-Gulf of California during the Neogene. GSA Special Paper 19(7):809.

— 1987b. Paleogeography of the Imperial Formation of southern California and its molluscan fauna. Western Society of Malacologists Annual Report 20:11-18.

— 1995. Preliminary report on the Echinodermata of the Miocene and Pliocene Imperial Formation in southern California. In Remeika, P., and A. Sturz 1995 1:55-63.

Presley, S.J. 2000. *Eira barbara*. Mammalian Species 636:1-6.

Proctor, R.J. 1968. Geology of the Desert Hot Springs-upper Coachella Valley area, California. California Division of Mines and Geology, Special Report 94:1-50.

Prothero, D.R. 1998. The chronological, climatic, and paleogeographic background to North American mammalian evolution. In Janis, C.M. K.M. Scott, and L.L. Jacobs, pp. 9-36.

Quinn, H.A., and T.M. Cronin 1984. Micro-paleontology and depositional environments of the Imperial and Palm Spring Formations, Imperial Valley, California. In The Imperial Basin, Tectonics, Sedimentation and Thermal Aspects, Society of Economic Paleontologists and Mineralogists 40:71-85.

Rabinowitz, A.R. 1986. Jaguar predation on domestic livestock in Belize. Wildlife Society Bulletin 14:170-174.

Randall, K. 2001. Stratigraphic correlation and vertebrate paleontology of the Pleistocene Ocotillo Conglomerate and Bautista beds in northern ABDSP, California. In The Changing Face of the East Mojave Desert, Abstracts from the 2001 Desert Symposium p. 74.

Randall, K., and G.T. Jefferson 2002. A preliminary examination of the Plio-Pleistocene camelids from ABDSP, California. JVP 22(3):98A.

Remeika, P. 1991. Formational status of the Diablo Redbeds. Symposium on the Scientific Value of the Desert, Anza-Borrego Foundation, Borrego Springs, California, Abstracts p. 12.

— 1992. Preliminary report on the stratigraphy and vertebrate fauna of the middle Pleistocene Ocotillo formation, Borrego Badlands, ABDSP. SBCMAQ 39(2):25-26.

— 1994. Lower Pliocene angiosperm hardwoods from the Vallecito-Fish Creek Basin, ABDSP, California. SBCMAQ 41(3):26-27.

— 1995. Basin tectonics, stratigraphy, and depositional environments of the western Salton Trough detachment. In Remeika, P., and A. Sturz 1995 1:3-54.

— 1997a. The Neogene Vallecito-Fish Creek Basin. In Geology and Paleontology of the Anza-Borrego Region, California, National Association of Geology Teachers, Far Western Section, Spring Conference Field Guide I:1-32.

— 1997b. Roadside basin-margin outcrop geology of the Borrego Badlands. In Geology and Paleontology of the Anza-Borrego Region, Califonia, National Association of Geology Teachers, Spring Conference Field Guide, Field Trip IV:1-20.

— 1998a. Interdisciplinary age control of the western Borrego Badlands, ABDSP, California. National Park Service, Abstracts for the Fifth Fossil Conference, Rapid City, South Dakota S18.

— 1998b. Marine invertebrate paleontology and stratigraphy of the Vallecito-Fish Creek Basin. In Lindsay, L. and W.G Hample, 1988 pp. 59-92.

— 1999. Identification, stratigraphy, and age of Neogene vertebrate footprints from the Vallecito-Fish Creek Basin, ABDSP, California. In Fossil Footprints, SBCMAQ 46(2):37-46.

— 2001. The Fish Creek Canyon ichnofauna. In Proceedings of the 6th Fossil Resource Conference, National Parks Service, Geologic Resources Division Technical Report 01/01:55-75.

Remeika, P., and S. Beske-Dehl 1996. Magnetostratigraphy of the western Borrego Badlands, ABDSP, California. In Geology of Neogene Faulting and Catastrophic Events in the Split Mountain Area, ABDSP, California, South Coast Geological Society Field Trip Guidebook Number 24:209-220.

Remeika, P., I.W. Fischbein, and S.A. Fischbein 1986. Lower Pliocene petrified wood from the Palm Spring Formation, ABDSP, California. In Geology of the Imperial Valley, California, South Coast Geological Society, Annual Field Trip Guidebook, Santa Ana 14:65-83.

— 1988. Lower Pliocene petrified wood from the Palm Spring Formation, ABDSP, California. Review of Palaeobotany and Palynology 56:183-198.

Remeika, P., and R.F. Fleming 1994. Pliocene climate of the Colorado Plateau and age of the Grand Canyon. In Partners in Paleontology Proceedings of the Fourth Conference on Fossil Resources, U.S. Department of the interior Natural Resources Report 97/01:17.

— 1995. Cretaceous palynoflora and Neogene angiosperm woods from ABDSP, California. In Remeika, P. and A. Sturz 1995 1:64-81.

Remeika, P., and G.T. Jefferson 1993. The Borrego Local Fauna. In Reynolds, R.E. and J. Reynolds 1993, 93(1):90-93.

Remeika, P., G.T. Jefferson, and L.K. Murray 1995. Fossil vertebrate faunal list for the Vallecito-Fish Creek and Borrego-San Felipe Basins, ABDSP and vicinity, California. In Remeika, P., and A. Sturz 1995 1:82-93.

Remeika, P., and J.C. Liddicoat (in preparation). Magnetostratigraphy and tephrochronology of the western half of the Borrego Badlands, ABDSP, California.

Remeika, P., and L. Lindsay 1993. Geology of Anza-Borrego. Sunbelt Publications, San Diego.

Remeika, P., and J.R. Pettinga 1991. Stratigraphic revision and depositional environments of the middle to late Pleistocene Ocotillo Conglomerate, Borrego Badlands, ABDSP, California. Symposium on the Scientific Value of the Desert, Anza-Borrego Foundation, Borrego Springs, California, Abstracts p. 13.

Remeika, P., and A. Sturz (editors) 1995. Paleontology and geology of the western Salton Trough Detachment, ABDSP, California. Field Trip Guidebook and volume for the 1995 San Diego Association of Geologist's field trip to ABDSP, Volume 1.

Repenning, C.A. 1987. Biochronology of the microtine rodents of the United States. In Woodburne, M.O. 1987 pp. 236-268.

— 1992. Allophaiomys and the age of the Olyor Suite, Krestovka Sections, Yakutia. USGS Bulletin 2037:1-98.

Repenning, C.A., T.R. Weasma, and G.R. Scott 1995. The early Pleistocene (latest Blancan-earliest Irvingtonian) Froman Ferry fauna and history of Glens Ferry Formation, southwestern Idaho. USGS Bulletin 2105:1-86.

Reumer, J.W.F and J. De vos (editors) 2003, Advances in Mammoth Research, Proceedings of the Second International Mammoth Official Conference, Rotterdam, The Neatherlands, DEINSEA.

Reynolds, R.E., L.P. Fay, and R.L. Reynolds 1990. California Oaks Road. Mojave Desert Quaternary Research Symposium, Abstracts of Proceedings, SBCMAQ 37(2):35.

Reynolds, R.E., and W.A. Reeder 1986. Age and fossil assemblages of the San Timoteo Formation, Riverside County, California. In Geology Around the Margins of the Eastern San Bernardino Mountains, Inland Geological Society Publications, California 1:51-56.

— 1992. The San Timoteo Formation, Riverside County, California. In Inland Southern California, SBCMAQ 38(3&4):44-48.

Reynolds, R.E., and P. Remeika 1993. Ashes, faults and basins: the 1993 Mojave Desert Quaternary Research Center field trip. In Reynolds, R.E. and J. Reynolds 1993, 93(1):3-33.

R.E. Reynolds and J. Reynolds (editors) 1993, Ashes, Faults and Basins, SBCM Association Special Publication.

Reynolds, R.E., and R.L. Reynolds 1990a. New late Blancan faunal assemblage from Murrieta, Riverside County, California. SBCMAQ 37(2):34.

— 1990b. Irvingtonian? faunas from the Pauba Formation, Temecula, Riverside County, California. SBCMAQ 37(2):37.

— 1993. Rodents and rabbits from the Temecula Arkose. In R.E. Reynolds and J. Reynolds 1993, 93-1:98-100.

Richards, R.L., C.S. Churcher and W.D. Turnbull 1996. Distribution and size variation in North American short-faced bears, Arctodus simus. In Stewart, K.M. and K.L. Seymour 1996, pp. 191-246.

Rightmer, D.A., and P.L. Abbott 1996. The Pliocene Fish Creek sturzstrom, ABDSP, southern California. In Abbott, P. and D. Seymour 1996 24:165-184.

Rockwell, T., C. Loughman, and P. Merifield 1990. Late Quaternary rate of slip along the San Jacinto fault zone near Anza, southern California. Journal of Geophysical Research B, 95(6):8593-8605.

Root, R.B. 1967. The niche exploitation pattern of the blue-gray gnatcatcher. Ecological Monographs 37:317-350.

Ruisaard, C.I. 1979. Stratigraphy of the Miocene Alverson Formation, Imperial County, California. MST, SDSU, California 125 pp.

Rymer, M.J. 1991. The Bishop ash bed in the Mecca Hills. In Geological excursions in southern California and Mexico, SDSU, California pp. 388-396.

Ryter, D.W. 2002. Late Pleistocene kinematics of the central San Jacinto fault zone, southern California. Doctoral Dissertation, University of Oregon, Eugene 137 pp.

Salles, L.O. 1992. Felid phylogenetics. American Museum Novitates 3047:1-67.

Sanders, C.O. 1993. Interaction of the San Jacinto and San Andreas fault zones, southern California. Science 260:973-976.

Sanders, C.O., and H. Magistrale 1997. Segmentation of the northern San Jacinto fault zone, southern California. Journal of Geophysical Research 102:27,453-27,467.

Sarna-Wojcicki, A.M., and M.S. Pringle Jr. 1992. Laser-fusion 40 Ar/39 Ar ages of the Tuff of Taylor Canyon and Bishop Tuff, E. California - W. Nevada. EOS, Transactions of the American Geophysical Union p. 241.

Saunders, J.J. 1996. North American Mammutidae. In The Proboscidea, edited by J. Shoshani and P. Tassy, Oxford University Press pp. 271-279.

— 1970. The distribution and taxonomy of Mammuthus in Arizona. MST, Department of Geolochronology, University of Arizona 148 pp.

Savage, D.E. 1951. Late Cenozoic vertebrates of the San Francisco Bay region. University of California Publications of the Department of Geological Science 28:215-314.

— 1962. Cenozoic geochronology of the fossil mammals of the Western Hemishpere. Revista del Museo Argentino de Ciencias Naturales "Bernardino Rivadavia," Ciencias Zoológicas 8:53-67.

Savage, D.E., and D.E. Russell 1983. Mammalian Paleofaunas of the World. Addison-Wesley Publishing, Boston.

Schaeffer, G.C. 1857. Description of the structure of fossil wood from the Colorado Desert. In 1855 Report of the Explorations and Survey for a Railroad Route from the Mississippi to the Pacific, U.S. Congress, 2nd Session, Senate Executive Document 91, 5(2):338-339.

Scheuing, D.F., and L. Seeber 1991. Magnetostratigraphy of Neogene sediments in the San Jacinto fault zone, southern California, and paleomagnetic evidence for block rotation. AAPG Bulletin 75(3):666-667.

Scheuing, D.F., L. Seeber, and M. Van Fossen 1990. Structural and sedimentologic effects on detrital paleomagnetic directions and their bearing on Neogene block rotation in the San Jacinto fault zone, southern California. Transactions of the American Geophysical Union 71:1632.

Schremp, L.A. 1981. Archaeogastropoda from the Pliocene Imperial Formation of California. JP 55(5):1123-1136.

Schultejann, P.A. 1984. The Yaqui Ridge antiform and detachment fault. Tectonics 3:677-691.

Scott, E. 1998. *Equus scotti* from southern California. JVP, Abstracts 18(3):76A.

— 2004. Pliocene and Pleistocene horses from Porcupine Cave. Biodiversity Response to Environmental Change in the Middle Pleistocene. Berkeley: University of California Press, p. 264-279.

Scrivner, P.J., and D.J. Bottjer 1986. Neogene avian and mammalian tracks from Death valley National Monument, California. PPP 57:285-331.

Seymour, K. 1989. *Panthera onca*. Mammalian Species 340:1-9.

— 1993. Size change in North American Quaternary jaguars. In Morphological Change in Quaternary Mammals of North America, Cambridge University Press, pp. 343-372.

Shaller, P.J., and A.J. Shaller 1996. Review of proposed mechanisms for sturzstroms (long-runout landslides). In Abbott, P. and D. Seymour 1996 24:185-202.

Sharp, R.V. 1967. San Jacinto fault zone in the Peninsular Ranges of southern California. GSA Bulletin 78:705-730.

— 1982. Tectonic setting of the Imperial Valley region. In The Imperial Valley, California Earthquake of October 15, 1979, USGS Professional Paper 1254:5-14.

Shaw, C.A. 1981. The middle Pleistocene El Golfo Local Fauna from northwestern Sonora, Mexico. MST, Department of Biology, California State University, Long Beach 141pp.

Shaw, C.A. and F.W. Croxen III 2000. A new species of antilocaprid (Mammalia) from the El Golfo local fauna, Sonora Mexico. JVP 20(3):69A.

Sheffield, S.R., and H.H. Thomas 1997. *Mustela frenata*. Mammalian Species 570:1-9.

Shoshani, J., and P. Tassy 1996. The Proboscidea. Oxford University Press.

Shoshani, J., R.M. West, N. Court, R.J.G. Savage, and R.M. Harris 1996. The earliest proboscideans. In The Proboscidea, Oxford University Press pp. 57-75.

Skinner, M.F. 1972. Order Perrisodactyla. In Early Pleistocene Pre-glacial and Glacial Rocks and Faunas of North-central Nebraska, Bulletin of the AMNH148:1-148.

Smith, W.P. 1991. *Odocoileus virginianus*. Mammalian Species 388:1-13.

Solounias, N. 1988. Evidence from horn morphology on the phylogenetic relationships of the pronghorn (*Antilocapra americana*). Journal of Mammalogy 69:140-143.

Steely, A.N., S.U. Janecke, and R.J. Dorsey 2004b. Evidence for syn-depositional folding of Imperial-age synrift deposits above the west Salton detachment fault, Borrego Mountain area, southern California. GSA Abstracts with Programs 36(5):317.

Steely, A.N., S.U. Janecke, R.J. Dorsey, and G.J. Axen 2004a. Evidence for Late Miocene-Quaternary low-angle oblique strike-slip faulting on the West Salton detachment fault, southern California. GSA Abstracts with Programs 36(5):317.

Stefen, C. 1999. Enamel microstructure of Recent and fossil Canidae (Carnivora: Mammalia). JVP 19(3):576-587.

Stehli, F.G. and S.D. Webb 1985. The Great American Biotic Interchange. Topics in Geobiology, Vol 4, Plenum Press, New York.

Steno, N. 1669. De solido intra solidum naturaliter contento dissertationis prodromus. Ex Typographia Sub Signo Stellae,Florence, Italy.

Stewart, J.D., and M. Roeder 1993. Razorback sucker (*Xyrauchen*) fossils from the Anza-Borrego Desert and the ancestral Colorado River. In Reynolds, R.E. and J. Reynolds 1993, 93(1):94-96.

Stewart, K.M. and K.L. Seymour (editors) 1996, Paleoecology and Paleoenvironments of Late Cenozoic Mammals: Tributes to the Career of CS Churcher, University of Toronto Press.

Stinson, A.L. 1990. Structural deformation within the Pinyon Mountains, San Diego County, California. MST, SDSU, California 133 pp.

Stinson, A.L., and R.G. Gastil 1996. Mid- to Late-Tertiary detachment faulting in the Pinyon Mountains, San Diego County, California. In Abbott, P. and D. Seymour 1996 24:221-224.

Stock, C., and J.M. Harris 1992. Rancho La Brea, a record of Pleistocene life in California, 7th Ed. NHMLA Science Series 37:1-113.

Stock, J.M., and K.V. Hodges 1989. Pre-Pliocene extension around the Gulf of California and transfer of Baja California to the Pacific plate. Tectonics 8:99-115.

Stout, B.W., G.J. Miller, and P. Remeika 1987. Neogene mega-vertebrate ichnites from the Vallecito Basin, ABDSP, California., Anza-Borrego Foundation, Borrego Springs, California p. 5.

Stout, B.W., and P. Remeika 1991. Status report on three major camelid tracksites in the lower Pliocene delta sequence, Vallecito-Fish Creek Basin, ABDSP. Anza-Borrego Foundation, Borrego Springs, California p. 9.

Stump, T.E. 1972. Stratigraphy and paleontology of the Imperial Formation in the western Colorado Desert. Master of Science Thesis, SDSU 132 pp.

Stump, T.E., and J.D. Stump 1972. Age, stratigraphy, paleoecology and Caribbean affinities of the Imperial fauna of the Gulf of California depression. GSA Special Paper 4(3):243.

Swift, C.C., T.R. Haglund, R. Mario, and R.N. Fisher 1993 The status and distribution of freshwater fishes of southern California. Bulletin of the Southern California Academy of Sciences 92(3):101-167.

Sykes, Godfrey, 1937. The Colorado Delta, American Geographical Society Special Publication 19: New York.

Tarbet, L.A. 1951. Imperial Valley. In Possible Future Petroleum Provinces of North America, AAPG Bulletin 35(2):260-263.

Tarbet, L.A., and W.H. Holman 1944. Stratigraphy and micropaleontology of the west side of Imperial Valley. AAPG Bulletin 28:1781-1782.

Tassy, P. 1996. The earliest gomphotheres. In The Proboscidea, edited by J. Shoshani and P. Tassy, Oxford University Press pp. 57-75.

Taylor, W.D. 1966. Summary of North American Blancan nonmarine mollusks. Malacologica 4:1-172.

Tedford, R.H., and M.E. Hunter 1984. Miocene marine-nonmarine correlations, Alantic and Gulf coastal plains, North America. PPP 47(1/2):129-151.

Tedford, R.H., and J. Martin 2001. *Plionarctos*, a tremarctine bear (Ursidae: Carnivora) from western North America. JVP 21(2):311-321.

Tedford, R.H., M.F. Skinner, R.W. Fields, J.M. Rensberger, D.P. Whistler, T. Galusha, B.E. Taylor, J.R. MacDonald, and S.D. Webb 1987. Faunal succession and biochronology of the Arikareen through Hemphillian interval (late Oligocene through earliest Pliocene epochs) in North America. In Woodburne, M.O. 1987 pp. 153-210.

Tedford, R.H., X. Wang, and B.E. Taylor 2001. History of the Caninae (Canidae). JVP, 21(Supplement to 3):107.

Testa, S.M. 1996. Early geological observations of the Colorado Desert area by William Phipps Blake, 1853 and 1905. In Abbott, P. and D. Seymour 1996 Number 24:1-43.

Thomas, H.W., and L.G. Barnes 1993. Discoveries of fossil whales in the Imperial Formation, Riverside County, California. In Reynolds, R.E. and J. Reynolds 1993, 93(1):34-36.

Thompson, R.S. 1991. Pliocene environments and climates in the western United States. In Pliocene Climates, Quaternary Science Reviews 10:115-132.

Thompson, R.S., T.R. Van Devender, P.S. Martin, T. Foppe, and A. Long 1980. Shasta ground sloth (Nothrotheriops shastense Hoffsetter) at Shelter Cave, New Mexico. Quaternary Research 14:360-376.

Thwaites, R.G. (editor) 1904. Original journals of the Lewis and Clark expedition 1804-1806; volume 1. Dodd, Mead, and Company.

Trajano, E. and H. Ferrarezzi 1994. A fossil bear from northeastern Brazil, with a phylogenetic analysis of the South American extict Tremarctinae (Ursidae). JVP 14(4):552-561.

Troxell, E.L. 1915. The vertebrate fossils of Rock Creek, Texas. American Journal of Science 39:613-638.

Tucker, A.B., R.M. Feldman, and C.L. Powell II 1994. *Speocarcinus berglundi* n. sp. (Decapoda: Brachyura), a new crab from the Imperial Formation (late Miocene-late Pliocene) of southern California. JP 68(4):800-807.

Urey, H.C. 1947. The thermodynamic properties of isotopic substances. Journal of the Chemical Society 152:190-219.

— 1948. Oxygen isotopes in nature and in the laboratory. Science 108:489-496.

Van Valkenburgh, B. 1988. Trophic diversity in past and present guilds of large predatory mammals. Paleobiology 14(2):155-173.

— 1991. Iterative evolution of hypercarnivory in canids (Mammalia: Carnivora). Paleobiology 17(4):340-362.

Van Valkenburgh, B., F. Grady and B. Kurtén 1990. The Plio-Pleistocene cheetah-like cat *Miracinonyx inexpectatus* of North America. JVP 10(4):434-454.

Van Valkenburgh, B., and R.E. Molnar 2002. Dinosaurian and mammalian predators compared. Paleobiology 28(2):527-543.

Van Valkenburgh, B. and T. Sacco 2002. Sexual dimorphism, social behavior and intrasexual competition in large Pleistocene carnivorans. JVP 22(1):164-169.

Vaughan, F.E. 1918. Evidence in the San Gorgonio Pass, Riverside County, California, of a late Pliocene extension of the Gulf of California. GSA Bulletin 29(1):164-165.

Vaughan, T.W. 1904. A California Tertiary coral reef and its bearing on American Recent coral faunas. Science (new series) 19:503.

— 1917a. The reef-coral fauna of Carrizo Creek, Imperial County, California and its significance. USGS Professional Paper 98T:355-387.

— 1917b. Significance of reef coral fauna at Carrizo Creek, Imperial Co., Calif. Washington Academy of Science Journal 7:194.

Verts, B.J., L.N. Carraway, and A. Kinlaw 2001. *Spilogale gracilis*. Mammalian Species 674:1-10.

Voorhies, M.R. and R.G. Corner 1982. Ice age superpredators. University of Nebraska State Museum, Museum Notes 70:1-4.

— 1986. *Megatylopus* (?) *cochrani* (Mammalia: Camelidae). JVP 6(1):65-75.

Vrba, E.S., G.H. Denton, T.C. Partridge, and L.H. Burckle (editors) 1995. Paleoclimate and evolution with emphasis on human orgins. Yale University Press.

Wagner, H., B.O. Riney, and D.R. Prothero 2000. A new terrestrial assemblage of middle Blancan age from the San Diego Formation, California. JVP 20(3):76A.

Wagoner, J.L. 1978. The stratigraphy and sedimentology of the Pleistocene Brawley and Borrego Formations in the San Felipe Hills area, Imperial Valley, California. GSA Abstracts with Programs 12:152.

— 1980. Grain-size distribution analysis of non-marine sandstone, Imperial Valley, California. GSA Abstracts with Programs 12:158.

Wang, X., R.H. Tedford, and B.E. Taylor 1999. Phylogenetic systematics of the Borophaginae (Carnivora: Canidae). Bulletin of the AMNH243:1-391.

Walker, E.P. 1968. Mammals of the World, 2nd ed. The Johns Hopkins Press.

Watkins, R. 1990a. Pliocene channel deposits of oyster shells in the Salton Trough region, California. PPP 79:249-262.

— 1990b. Paleoecology of a Pliocene rocky shoreline, Salton Trough region, California. Palaios 5:167-175.

— 1992. Sedimentology and paleoecology of Pliocene shallow marine conglomerates, Salton Trough region, California. PPP 95: 319-333.

Webb, S.D. 1972. Locomotor evolution in camels. Forma et Functio 5:99-112.

— 1973. Pliocene pronghorns of Florida. Journal of Mammalogy 54:203-221.

— 1974. Pleistocene llamas of Florida, with a brief review of the Lamini. In Pleistocene Mammals of Florida, The University of Florida, Gainesville, pp. 170-213.

— 1976. Mammalian faunal dynamics of the Great American interchange. Paleobiology 2(3):220-234.

— 1985. Late Cenozoic mammal dispersals between the Americas. In The Great American Biotic Interchange, Topics in Geobiology, Plenum Press, New York, 4:357-386.

— 1989. Osteology and relationships of *Thinobadistes segnis*, the first mylodont sloth in North America. In Advances in Neotropical Mammalogy, Sandhill Crane Press, pp.496-532.

— 1998. Chronology and ecology of vertebrates: who went when. American Quaternary Association. Program and Abstracts of the 15th Biennial Meeting p. 66-68.

— 1998. Cervidae and Bovidae. In Janis, C.M., K.M. Scott and L.L. Jacobs 1998, pp. 508-510.

Wells, D.L. 1987. Geology of the eastern San Felipe Hills, Imperial Valley, California. MST SDSU 140 pp.

Werdelin, L. 1985. Small Pleistocene felines of North America. JVP 5:194-210.

— 1989. Constraint and adaptation in the bone-cracking canid *Osteoborus* (Mammalia: Canidae). Paleobiology 15(4):387-401.

Werdelin, L. and N. Solounias 1991. The Hyaenidae: taxonomy, systematics and evolution. Fossils and Strata 30:1-104.

Wernicke, B. 1985. Uniform-sense normal simple shear of the continental lithosphere. Canadian Journal of Earth Science 22:108-125.

Wesnousky, S.G. 1986. Earthquakes, Quaternary faults, and seismic hazard in California. Journal of Geophysical Research 91:12, 587-12, 631.

Wetzel, R.M., R.E. Dubos, R.L. Martin, and P. Meyers 1975. *Catagonus*, an "extinct" peccary, alive in Paraguay. Science 189:379-381.

White, J.A. 1964. Kangaroo rats (Family Heteromyidae) of the Vallecito Creek Pleistocene of California. GSA Special Paper 82:288-289.

— 1968. A new porcupine from the middle Pleistocene of the Anza-Borrego Desert of California. MNHLA Contributions in Science 136:1-15.

— 1969. Late Cenozoic bats (Subfamily Nyctophylinae) from the Anza-Borrego Desert of California. University of Kansas MNH Miscellaneous Publications 51:275-282.

— 1970. Late Cenozoic porcupines (Mammalia, Erethizontidae) of North America. AMNH Novitates 241:1-15.

— 1984. Late Cenozoic Leporidae (Mammalia, Lagomorpha) from the Anza-Borrego Desert, southern California. Special Publication of the Carnegie MNH 9:41-57.

— 1987. The Archaeolaginae (Mammalia, Lagomorpha) of North America, excluding *Archaeolagus* and *Panolax*. JVP 7(4):425-450.

— 1991. North American Leporinae (Mammalia: Lagomorpha) from late Miocene (Clarendonian) to latest Pliocene (Blancan). JVP 11(1):67-89.

White, J.A., and T. Downs 1961. A new *Geomys* from the Vallecito Creek Pleistocene of California. NHMLA Contributions in Science 42:1-34.

— 1965. Vertebrate microfossils from the Canebrake Formation of the Imperial Valley region, California. Society of Economic Mineralogists and Paleontologists, Pacific Section, Abstracts with Program, Los Angeles p. 33.

Willoughby, D.P. 1974. The empire of *Equus*. A.S. Barnes and Company.

Winans, M.C. 1985. Revision of North American fossil species of the genus *Equus* (Mammalia: Perrisodactyla: Equidae). Doctoral

Dissertation, Department of Geological Sciences, University of Texas at Austin 265 pp.

— 1989. A quantative study of North American fossil species of the genus *Equus*. In The Evolution of the Perrisodactyls, Oxford University Press pp. 262-297.

Winker, C.D. 1987. Neogene stratigraphy of the Fish Creek - Vallecito section, southern California. Doctoral Dissertation, University of Arizona.

Winker, C.D., and S.M. Kidwell 1986a. Paleocurrent evidence for lateral displacement of the Pliocene Colorado River delta by the San Andreas Fault system, southeastern California. Geology 14:788-791.

— 1986b. Planispastic paleogeographic model for the Neogene Colorado delta and northern Gulf of California. GSA Abstracts with Programs 18:199.

— 1986c. Stratigraphic sequence of the Pliocene Colorado delta, Fish Creek-Vallecito section, western Salton Trough, southern California. GSA Abstracts with Programs 18:199.

— 1996. Stratigraphy of a marine rift basin. In Field Conference Guidebook and Volume for the AAPG Annual Convention, Pacific Section 73:295-336.

— 2002. Stratigraphic evidence for ages of different extensional styles in the Salton Trough, southern California. GSA Abstracts with Programs 34(6):83-84.

— 2003. Colorado River delta: 5MYR-old tide-dominated, big-river delta in a tectonically evolving, oblique rift basin. GSA Abstracts with Programs 35(4):28.

Wood, H.E., II, R.W. Chaney, J. Clark, E.H. Colbert, G.L. Jepsen, J.B. Reeside Jr., and C. Stock 1941. Nomenclature and correlation of the North American continental Tertiary. GSA Bulletin 52:1-48.

Woodard, G.D. 1962. Stratigraphic succession of the west Colorado Desert, San Diego and Imperial Counties, southern California. GSA Bulletin 68:26-31.

— 1963. The Cenozoic succession of the west Colorado Desert, San Diego and Imperial Counties, southern California. Doctoral Dissertation, University of California, Berkeley.

— 1974. Redefinition of Cenozoic stratigraphic column in Split Mountain Gorge, Imperial Valley, California. AAPG Bulletin 58:521-539.

Woodburne, M.O. 1968. Cranial myology and osteology of *Dicotles tajacu*, the collared peccary. Memoirs of the Southern California Academy of Sciences 7:1-48.

— 1977. Definition and characterization in mammalian chrono-stratigraphy. JP 51:220-234.

— 1987a. Cenozoic mammals of North America, Geochronology and Biostratigraphy. University of California Press.

— 1987b. Principles, classification, and recommendations. In Woodburne, M.O. 1987 pp. 9-17.

— 1987c. A prospectus of the North American Mammal Ages. In Woodburne, M.O. 1987 pp. 285-290.

— 1996. Precision and resolution in mammalian chronostratigraphy. JVP 16:531-555.

— 2004. Late cretaceous and cenozoic mammals of North America. Columbia University Press, New York.

Woodburne, M.O., Reynolds, R.E., and D.P. Whistler 1992. Inland Southern California. SBCMAQ 38(3&4) 115 pp.

Woodring, W.P. 1931. Distribution and age of the marine Tertiary deposits of the Colorado Desert. CIWP 418:1-25.

— 1931b. A Miocene *Haliotis* from southern California. JP 5:34-39.

Wright, D.B. 1993. Evolution of sexually dimorphic characters in peccaries (Mammalia, Tayassuidae). Paleobiology 19(1):52-70.

— 1998. Tayassuidae. In Janis, C.M., K.M. Scott and L.L. Jacobs 1998, pp. 389-401.

Yensen, E., and T. Talrifa 2003a. *Galictis cuja*. Mammalian Species 728:1-8.

— 2003b. *Galictis vittata*. Mammalian Species 727:1-8.

Zakrzewski, R.J. 1972. Fossil microtines from late Cenozoic deposits in the Anza-Borrego Desert, California. NHMLA Contributions in Science 221:1-12.

Index

This index lists proper names for people and places noted in the text (see also *Literature Cited*), geological formation names (see also *Glossary* and *Appendix* Table 4 "Stratigraphic Names"), and scientific names for plants and animals below the Family level (see also *Appendix* Tables 1-3 "Systematic Lists".) Geological age names or age terms are not included (see *Glossary*, *Appendix*, Table 5 "Biostratigraphic Ranges," and Timelines listed in "Major Maps and Illustrations.")

Camel Ridge, 312-313, 317-323
Camelopichnum, 322, 326
Camelops, 8, 118, 131, 133, 201, 289, 297-298, 300-301, 304-307, 324, 326
    hesternus, 118, 289, 300, 307, 326
    huerfanensis, 118, 300, 326
    minidokae, 300, 324
    sulcatus, 300
    traviswhitei, 300
Camelus, 295, 300, 304, 306-307, 309
    bactrianus, 304
    dromedarius, 304, 309
Cancellaria obesa, 50, 55
Canebrake Conglomerate, 100
Canis, 131, 180, 190, 198, 201-203, 210, 212-213, 277, 279, 282, 289, 318-319, 326, 337
    armbrusteri, 180, 190, 202, 277, 279
    dirus, 180, 202, 337
    edwardii, 202, 277, 279, 337
    latrans, 201-203, 210, 277, 318-319, 326
    lepophagus, 201-202, 210, 212, 277, 279, 319, 326
    lupus, 201, 282, 318
    nehringi, 180
    priscolatrans, 131, 201-203, 213, 289
    rufus, 147, 149, 180, 202-204, 277, 320
Capra, 280
Capromeryx, 190, 193, 214, 280-282, 284, 287
    arizonensis, 282
    furcifer, 282
    minor, 52, 85, 99, 186, 244, 247-249, 282, 284
    tauntonensis, 282
Carcharodon, 36-37
Carcharocles megalodon, 37
Caribbean, 30, 35-36, 44-45, 47-50, 52-55, 57-58, 60-66, 68-69, 343-344
    Basin, 52
    Current, 35, 45
    Sea, 35, 44, 47-50, 53-54, 57-58, 60, 64-66, 68-69, 343
Carnivore Ridge, 318-319, 324
Carrizo
    Badlands, 30-31, 37, 76
    Local Flora, 76, 80-81, 86-87
Cassiliano, Michael, 121
Cassis, 53
    cornuta, 53
    subtuberosa, 53
Castor, 193, 244-245
    canadensis, 147, 153, 193, 206, 244, 277
Catagonus, 275-277, 336-337
    wagneri, 275, 337
Central America, 35, 155, 179, 205, 241, 246-247, 268, 275, 286, 298, 300, 332, 336
Central American Seaway, 35-36, 45
Ceratomeryx, 280-282
Cervus, 278
Charadrius vociferus, 153
Chasmaporthetes, 126, 186, 190-191
    lunensis, 186
    ossifragus, 186
Chelipus therates, 314, 318-319, 324, 326
Chen rossii, 153
China, 114
Chula Vista, California, 82-83
Cita Canyon, Texas, 189
Clark Lake (Valley), 102-103, 200, 203, 295
Clemmys marmorata, 142-143
Clypeaster, 65-66
    bowersi, 48, 65-66
    carrizoensis, 57, 66

    caudatus, 66
    cotteaui, 66
    deserti, 38, 59, 66
    pallidus, 66, 240
    rotundus, 66
    subdepressus, 66
Cochise County, Arizona, 243
Codakia distinguenda, 60-61
Coendu, 130, 193, 241-242
    mexicanus, 241-242
    stirtoni, 130, 193, 241-242
Colombia, 35, 61-62
Colorado
    Desert, 2-5, 7-10, 14-22, 30, 32-33, 84-85, 143, 152, 236, 238, 274, 283, 321-322, 324, 327
    Desert District, 5, 15-22, 236, 274
    Desert District Stout Research Center, 15-22, 236, 274
    Plateau, 33, 76, 86, 100
    River, 4, 33, 37-38, 44-45, 69, 71, 75-79, 81, 83-87, 99-102, 131, 140-141, 143, 145, 153, 206, 208, 244-245, 247, 260, 312, 325, 344
Conifer Woodland Element, 81, 87
Conus
    arcuatus, 56
    bramkampi, 56
    durhami, 56
    fergusoni, 53, 56
    patricius, 56
    regularis, 56
    ximenes, 56
Cope, Edward, 262
Corvus, 152
Coso, 249
Costa Rica, 35, 45, 52
Couts, C.J., 3
Coyote
    Badlands, 30, 109, 144, 203, 245, 282
    Creek, 5, 34-35, 38, 92, 103, 180
    Creek fault, 92, 103
    Mountain, 102-103, 279, 316
    Mountains, 30, 34-38, 48-49, 51, 54, 61, 65-68, 129
    Wells, 5
Craig Scale, 222, 229
Crassinella mexicana, 61
Crocuta, 179
    crocuta, 179
Crotalus, 145, 289
Cudahy, Kansas, 250
Cupidinimus, 129
Curtis Ranch, Arizona, 243
Cyathodonta undulata, 64
Cygnus paloregonus, 153
Cyrtopleura, 38, 57, 63-64
    costata, 38, 57, 63
    crucigera, 64

Daniel, Harry, 17
Darwin, Charles, 109
Death Valley (National Park), 84, 153, 326
Deguynos Formation, 31, 37-40, 44-45, 47-49, 51-52, 54, 57-59, 64, 69, 97-99, 130, 209, 224, 260
Dendostrea vespertina, 57-58, 69
Dichocoenia, 47
    labyrinthiformis, 48
    merriami, 47, 147, 149, 268-270, 304, 335, 337
    stokesi, 48
Dickerson, Roy, 7
Dicotyles tajacu, 275
Dinohippus, 71, 73, 130-131, 256, 258-262, 267, 269, 322, 327